HOMEPLUG AV AND IEEE 1901

IEEE Press
445 Hoes Lane
Piscataway, NJ 08854

IEEE Press Editorial Board 2013
John Anderson, *Editor in Chief*

Linda Shafer	Saeid Nahavandi	George Zobrist
George W. Arnold	David Jacobson	Tariq Samad
Ekram Hossain	Mary Lanzerotti	Dmitry Goldgof
Om P. Malik		

Kenneth Moore, *Director of IEEE Book and Information Services (BIS)*

To our families, our mentors, and our colleagues.

CONTENTS

List of Figures xix

List of Tables xxiii

Preface xxvii

Acknowledgments xxix

Biographical Sketches of the Authors xxxi

1 Introduction 1

    1.1 HomePlug AV and Its Relationship to IEEE 1901, 2
    1.2 Focus of the Book, 3
    1.3 The HomePlug Powerline Alliance, 4
        1.3.1 HomePlug Specifications, 4
        1.3.2 How the HomePlug AV Specification Was Developed, 5
        1.3.3 The Regulatory Working Group, 6
            1.3.3.1 The United States and the FCC, 6
            1.3.3.2 Europe, CISPR, and CENELEC, 7
            1.3.3.3 Rest of the World, 8
    1.4 The Role of PLC in Multimedia Home Networking and Smart Energy Applications, 8
    1.5 Book Outline, 9

2 The HomePlug AV Network Architecture — 12

- 2.1 Introduction, 12
- 2.2 Protocol Layers, 12
- 2.3 Network Architecture, 14
  - 2.3.1 Station Roles, 16
  - 2.3.2 Bridging, 16
  - 2.3.3 Channel Access, 16
- 2.4 Summary, 17

3 Design Approach for Powerline Channels — 18

- 3.1 Introduction, 18
- 3.2 Channel Characteristics, 19
- 3.3 Frequency Band, 21
  - 3.3.1 Tone Mask, 22
  - 3.3.2 Amplitude Map, 22
- 3.4 Windowed OFDM, 23
- 3.5 Turbo Convolutional Code, 24
- 3.6 Channel Adaptation, 25
  - 3.6.1 Bit-Loading, 27
- 3.7 Beacon Period Synchronized to AC Line Cycle, 27
  - 3.7.1 AC Line Cycle Synchronization for TDMA Allocations, 28
- 3.8 TDMA with Persistent and Nonpersistent Schedules, 29
- 3.9 Data Plane: Two-Level Framing, Segmentation, and Reassembly, 30
- 3.10 PHY Clock Synchronization, 30
- 3.11 Summary, 31

4 Physical Layer — 32

- 4.1 Introduction, 32
  - 4.1.1 Transceiver Block Diagram, 33
- 4.2 PPDU, 34
  - 4.2.1 PPDU Formats, 35
  - 4.2.2 PPDU Structure, 36
  - 4.2.3 Symbol Timing, 36
- 4.3 Preamble, 37
- 4.4 Frame Control, 38
- 4.5 Payload, 39
  - 4.5.1 Scrambler, 40
  - 4.5.2 Turbo Convolutional Encoder, 41
    - 4.5.2.1 Constituent Encoders, 41
    - 4.5.2.2 Termination, 41
    - 4.5.2.3 Puncturing, 42
    - 4.5.2.4 Turbo Interleaving, 42

CONTENTS

- 4.5.3 Channel Interleaver, 44
- 4.5.4 ROBO Modes, 46
  - 4.5.4.1 ROBO Interleaver, 46
- 4.5.5 Mapping and Tone Maps, 49
  - 4.5.5.1 Empty Tone Filling, 50
  - 4.5.5.2 Last Symbol Padding, 50
  - 4.5.5.3 Mapping Reference, 51
  - 4.5.5.4 Mapping for BPSK, QPSK, 8 QAM, 16 QAM, 64 QAM, 256 QAM, 1024 QAM, 51
  - 4.5.5.5 Mapping for ROBO-AV, 53
- 4.5.6 Payload Symbols, 54
- 4.5.7 Windowed OFDM and Symbol Shaping, 55
- 4.6 Priority Resolution Symbol, 56
- 4.7 Transmit Power, Tone Mask, and Amplitude Map, 56
  - 4.7.1 Relative Power Levels, 56
  - 4.7.2 Tone Mask, 57
  - 4.7.3 Amplitude Map, 58
- 4.8 Summary, 60

5 MAC Protocol Data Unit (MPDU) Format   61

- 5.1 Introduction, 61
  - 5.1.1 General AV Frame Control, 63
- 5.2 Beacon, 64
  - 5.2.1 Beacon Frame Control, 65
    - 5.2.1.1 Beacon Time Stamp (BTS), 65
    - 5.2.1.2 Beacon Transmission Offset (BTO), 65
  - 5.2.2 Beacon Payload, 65
    - 5.2.2.1 Beacon Type, 65
    - 5.2.2.2 Addressing, 66
    - 5.2.2.3 Neighbor Network Coordination, 67
    - 5.2.2.4 Network Operation Mode, 67
    - 5.2.2.5 CCo Capability, 68
    - 5.2.2.6 Participation in Multiple Networks, 68
    - 5.2.2.7 CCo Handover, 68
    - 5.2.2.8 Beacon Management Information (BMI), 68
    - 5.2.2.9 Beacon Payload Check Sequence (BPCS), 77
- 5.3 Start-of-Frame (SOF), 77
  - 5.3.1 Start-of-Frame (SOF) Frame Control, 77
    - 5.3.1.1 Addressing-Related Fields, 78
    - 5.3.1.2 Queue-Related Fields, 79
    - 5.3.1.3 Bursting-Related Fields, 79
    - 5.3.1.4 Payload Demodulation-Related Fields, 80
    - 5.3.1.5 TDMA Allocation-Related Fields, 81

　　　　　　　5.3.1.6　SACK Retransmission-Related Fields, 81
　　　　　　　5.3.1.7　Encryption-Related Fields, 82
　　　　　　　5.3.1.8　Detection Status-Related Fields, 82
　　　　　　　5.3.1.9　Participation in Multiple Networks-Related Fields, 82
　　　　　　　5.3.1.10　Convergence Layer SAP Type (CLST), 83
　　　　5.3.2　SOF Payload, 83
　5.4　Selective Acknowledgment (SACK), 85
　　　　5.4.1　Addressing-Related Field, 85
　　　　5.4.2　Queue-Related Field, 86
　　　　5.4.3　Bursting-Related Field, 86
　　　　5.4.4　TDMA Allocation-Related Fields, 87
　　　　5.4.5　Detection Status-Related Field, 87
　　　　5.4.6　Version-Related Fields, 87
　　　　5.4.7　SACK Data, 87
　5.5　Request to Send (RTS)/Clear to Send (CTS), 88
　　　　5.5.1　Addressing-Related Fields, 88
　　　　5.5.2　Queue-Related Fields, 89
　　　　5.5.3　TDMA Allocation-Related Fields, 89
　　　　5.5.4　Detection Status Fields, 89
　　　　5.5.5　Immediate Grant-Related Fields, 90
　　　　5.5.6　Virtual Carrier Sense (VCS)-Related Fields, 90
　　　　5.5.7　RTS Flag, 91
　5.6　Sound, 91
　　　　5.6.1　Sound Frame Control, 91
　　　　　　　5.6.1.1　Addressing, 91
　　　　　　　5.6.1.2　Queue, 92
　　　　　　　5.6.1.3　Bursting, 92
　　　　　　　5.6.1.4　Payload Demodulation, 92
　　　　　　　5.6.1.5　TDMA Allocations, 93
　　　　　　　5.6.1.6　Detection Status-Related Field, 93
　　　　　　　5.6.1.7　Sound ACK, 93
　　　　　　　5.6.1.8　Sound Complete Flag, 93
　　　　　　　5.6.1.9　Sound Reason Code, 93
　　　　　　　5.6.1.10　Max Tone Maps, 94
　　　　5.6.2　Format of Sound MPDU Payload, 94
　5.7　Reverse Start-of-Frame (RSOF), 95
　　　　5.7.1　Reverse SOF (RSOF) Frame Control, 95
　　　　　　　5.7.1.1　Addressing-Related Field, 95
　　　　　　　5.7.1.2　Queue-Related Field, 95
　　　　　　　5.7.1.3　Bursting-Related Field, 96
　　　　　　　5.7.1.4　TDMA Allocation-Related Fields, 97
　　　　　　　5.7.1.5　Detection Status-Related Field, 97

		5.7.1.6	Version-Related Fields, 97
		5.7.1.7	Selective Acknowledgment-Related Field, 97
		5.7.1.8	Payload Demodulation-Related Fields, 97
	5.8	Summary, 98	

6 MAC Data Plane 99

- 6.1 Introduction, 99
- 6.2 MAC Frame Generation, 101
- 6.3 MAC Frame Streams, 102
  - 6.3.1 Priority of Management Streams, 103
- 6.4 Segmentation, 104
- 6.5 Long MPDU Generation, 104
- 6.6 Reassembly, 106
- 6.7 Buffer Management and Flow Control, 106
  - 6.7.1 Transmit Buffer Management, 107
  - 6.7.2 Receive Buffer Management, 109
- 6.8 Communication Between Associated but Unauthenticated STAs, 112
- 6.9 Communication Between STAs not Associated with the Same AVLN, 112
  - 6.9.1 Multinetwork Broadcast (MNBC), 113
- 6.10 Data Encryption, 114
- 6.11 MPDU Bursting, 114
- 6.12 Bidirectional Bursting, 115
  - 6.12.1 Bidirectional Bursting During CSMA, 116
  - 6.12.2 Connections and Links During Bidirectional Bursts, 118
  - 6.12.3 Encryption of RSOF Payload, 118
- 6.13 Automatic Repeat Request (ARQ), 118
  - 6.13.1 Request SACK Retransmission, 119
  - 6.13.2 Broadcast/Multicast and Partial Acknowledgment, 119
- 6.14 Summary, 120

7 Central Coordinator 121

- 7.1 Introduction, 121
- 7.2 CCo Selection, 122
  - 7.2.1 CCo Selection for a New AVLN, 122
  - 7.2.2 Auto-Selection of CCo, 122
    - 7.2.2.1 CCo Capability, 123
  - 7.2.3 User-Appointed CCo, 124
- 7.3 Backup CCo and CCo Failure Recovery, 125
  - 7.3.1 Backup CCo, 125
  - 7.3.2 CCo Failure Recovery, 125
- 7.4 Transfer/Handover of CCo Functions, 125
- 7.5 CCo Network Management Functions, 127
  - 7.5.1 Network Time Base Synchronization, 127

　　　　　　　7.5.1.1　Arrival Time Stamp for MSDU Jitter and Delay Control, 129
　　　　　　　7.5.1.2　PHY Clock Correction When Participating in More Than One Network, 129
　　　　7.5.2　Discover Process, 130
　　7.6　Summary, 132

8　Channel Access　　　　　　　　　　　　　　　　　　　　　　　133

　　8.1　Introduction, 133
　　8.2　Beacon Period and AC Line Cycle Synchronization, 135
　　　　8.2.1　Line Cycle Synchronization, 135
　　8.3　Beacon Period Structure, 135
　　　　8.3.1　Beacon Period Structure in CSMA-Only Mode, 139
　　　　8.3.2　Beacon Period Structure in Uncoordinated Mode, 141
　　　　8.3.3　Beacon Period Structure in Coordinated Mode, 142
　　8.4　CSMA Channel Access, 143
　　　　8.4.1　Carrier Sense Mechanism, 144
　　　　　　　8.4.1.1　MAC-Level Acknowledgments, 144
　　　　　　　8.4.1.2　Setting of Virtual Carrier Sense (VCS) Timer, 145
　　　　　　　8.4.1.3　RTS/CTS, 146
　　　　8.4.2　Contention Procedure, 146
　　　　　　　8.4.2.1　Priority Contention, 148
　　8.5　TDMA Channel Access, 148
　　　　8.5.1　Admission Control and Scheduling (Persistent and Nonpersistent), 148
　　8.6　Summary, 149

9　Connections and Links　　　　　　　　　　　　　　　　　　　150

　　9.1　Introduction, 150
　　9.2　Packet Classification, 151
　　9.3　Connection Specification (CSPEC), 152
　　9.4　Connections and Links, 154
　　　　9.4.1　Link Identifiers, 156
　　　　　　　9.4.1.1　Assignment of LIDs, 157
　　　　9.4.2　Connection Identifiers, 157
　　9.5　Connection Services, 157
　　　　9.5.1　Connection Setup, 159
　　　　9.5.2　Connection Monitoring, 161
　　　　9.5.3　Connection Teardown, 161
　　　　9.5.4　Connection Reconfiguration, 164
　　　　9.5.5　Global Link Reconfiguration Triggered by CCo, 167
　　　　　　　9.5.5.1　Squeeze and De-Squeeze, 167

CONTENTS

- 9.6 Bandwidth Management by CCo, 168
  - 9.6.1 Scheduler and Bandwidth Allocation, 168
  - 9.6.2 Connection Admission Control, 171
  - 9.6.3 Beacon Period Configuration, 171
- 9.7 Summary, 171

10 Security and Network Formation — 172

- 10.1 Introduction, 172
- 10.2 Power-on Network Discovery Procedure, 172
  - 10.2.1 Unassociated STA Behavior, 174
  - 10.2.2 Unassociated CCo Behavior, 175
  - 10.2.3 Behavior as an STA in an AVLN, 176
  - 10.2.4 Behavior as a CCo in an AVLN, 177
- 10.3 Forming or Joining an AVLN, 178
  - 10.3.1 AVLN Overview, 178
    - 10.3.1.1 Network Identification, 178
    - 10.3.1.2 Human-Friendly Station and AVLN Names, 178
    - 10.3.1.3 Get Full AVLN Information, 178
    - 10.3.1.4 Get Full STA Information, 178
  - 10.3.2 Association, 179
    - 10.3.2.1 TEI Assignment and Renewal, 179
  - 10.3.3 Method for Authentication, 181
  - 10.3.4 Forming a New AVLN, 181
    - 10.3.4.1 Two Unassociated STAs with Matching NIDs, 183
    - 10.3.4.2 Two Unassociated STAs Form an AVLN Using a DAK-Encrypted NMK, 183
    - 10.3.4.3 Two Unassociated STAs: One in SC-Add and One in SC-Join, 186
    - 10.3.4.4 Two Unassociated STAs: Both in SC-Join, 186
  - 10.3.5 Joining an Existing AVLN, 188
    - 10.3.5.1 Matching NIDs, 189
    - 10.3.5.2 DAK-Encrypted NMK, 189
    - 10.3.5.3 SC-Join and SC-Add, 190
  - 10.3.6 Leaving an AVLN, 192
  - 10.3.7 Removing a Station from an AVLN, 193
- 10.4 Security Overview, 193
  - 10.4.1 Encryption Keys, Pass Phrases, Nonces, and Their Uses, 194
    - 10.4.1.1 Device Access Key (DAK), 194
    - 10.4.1.2 Device Password (DPW), 194
    - 10.4.1.3 Network Membership Key (NMK), 194
    - 10.4.1.4 Network Password (NPW), 194
    - 10.4.1.5 Network Encryption Key (NEK), 194

                10.4.1.6   Temporary Encryption Key (TEK), 195
                10.4.1.7   Nonces, 195
        10.4.2 Methods for Authorization (NMK Provisioning), 195
                10.4.2.1   Security Level, 196
                10.4.2.2   Preloaded NMK, 198
                10.4.2.3   Direct Entry of the NMK, 198
                10.4.2.4   Distribution of NMK Using DAK, 199
                10.4.2.5   Distribution of NMK Using Unicast Key Exchange (UKE), 200
                10.4.2.6   Distribution of NMK Using Other Key Management Protocols, 202
                10.4.2.7   Changing the NMK, 203
        10.4.3 NEK Provisioning, 203
                10.4.3.1   Provisioning NEK for New STA, 203
                10.4.3.2   Provisioning NEK for Part or All of the AVLN, 203
        10.4.4 Encryption Key Uses and Protocol Failures, 204
        10.4.5 AES Encryption Algorithm and Mode, 207
                10.4.5.1   PHY Block-Level Encryption, 207
                10.4.5.2   Payload-Level Encryption, 207
        10.4.6 Generation of AES Encryption Keys, 208
                10.4.6.1   Generation from Passwords, 208
                10.4.6.2   Automatic Generation of AES Keys, 208
                10.4.6.3   Generation of Nonces, 208
        10.4.7 Encrypted Payload Message, 209
        10.4.8 User Interface Station (UIS), 210
  10.5 Summary, 210

11 Additional MAC Features   211

  11.1 Introduction, 211
  11.2 Channel Estimation, 211
        11.2.1 Channel Estimation Procedure, 212
        11.2.2 Initial Channel Estimation, 213
        11.2.3 Dynamic Channel Adaptation, 214
        11.2.4 Maintenance of Tone Maps, 217
        11.2.5 Tone Map Intervals, 218
        11.2.6 Priority of Channel Estimation Response, 219
        11.2.7 Channel Estimation with Respect to the AC Line Cycle, 219
  11.3 Bridging, 219
        11.3.1 Acting as an AV Bridge, 220
        11.3.2 Communicating Through an AV Bridge, 221
                11.3.2.1   Communication with a Known DA, 222
                11.3.2.2   Communicating with an Unknown DA, 222

11.4 HomePlug 1.0.1 Coexistence, 223
    11.4.1 HomePlug AV Coexistence Modes, 223
    11.4.2 Detection and Reporting of Active HomePlug 1.0.1, 224
    11.4.3 HomePlug 1.0.1/1.1 Coexistence Mode Changes, 224
    11.4.4 HomePlug 1.0.1-Compatible Frame Lengths, 225
11.5 Proxy Networking, 225
    11.5.1 Identification of Hidden Stations, 227
    11.5.2 Association of Hidden Station, 227
    11.5.3 Instantiation of Proxy Network, 229
    11.5.4 Proxy Beacons, 229
    11.5.5 Provisioning the NMK to Hidden Stations, 229
    11.5.6 Provisioning NEK for Hidden Stations (Authenticating the HSTA), 230
    11.5.7 Exchange of MMEs Through a PCo, 230
    11.5.8 Transitioning from Being a STA to Being an HSTA, 231
    11.5.9 Transitioning from Being an HSTA to Being a STA, 231
    11.5.10 Recovering from the Loss of a PCo, 232
    11.5.11 Proxy Network Shutdown, 232
11.6 Summary, 232

12 Neighbor Networks   **233**

12.1 Introduction, 233
    12.1.1 CSMA-Only Mode, 233
    12.1.2 Uncoordinated Mode, 234
    12.1.3 Coordinated Mode, 234
12.2 Transition Between Neighbor Network Operating Modes, 234
12.3 Coordinated Mode, 236
    12.3.1 Interfering Network List, 237
    12.3.2 Group of Networks, 237
    12.3.3 Determining a Compatible Schedule, 237
        12.3.3.1 Computing the INL Allocation, 238
    12.3.4 Communication Between Neighboring CCos, 239
    12.3.5 Neighbor Network Instantiation, 240
        12.3.5.1 Procedure to Establish a New Network in Coordinated Mode, 240
        12.3.5.2 Changing the Number of Beacon Slots, 242
        12.3.5.3 Setting the Value of SlotUsage Field, 244
    12.3.6 Procedure to Share Bandwidth in Coordinated Mode, 244
    12.3.7 Bandwidth Scheduling Rules, 246
    12.3.8 Procedure to Shut Down an AVLN, 246
    12.3.9 AC Line Cycle Synchronization in Coordinated Mode, 247
12.4 Passive Coordination in CSMA-Only Mode, 248
12.5 Neighbor Network Bandwidth Sharing Policy, 248
12.6 Summary, 249

13 Management Messages — 250

- 13.1 Introduction, 250
- 13.2 Management Message Format, 250
  - 13.2.1 Original Destination Address (ODA), 250
  - 13.2.2 Original Source Address (OSA), 251
  - 13.2.3 VLAN Tag, 251
  - 13.2.4 MTYPE, 251
  - 13.2.5 Management Message Version (MMV), 251
  - 13.2.6 Management Message Type (MMTYPE), 251
  - 13.2.7 Fragment Management Information, 252
  - 13.2.8 Management Message Entry Data (MME), 254
  - 13.2.9 MMEPAD, 254
- 13.3 Station–Central Coordination (CCo), 254
- 13.4 Proxy Coordinator (PCO) Messages, 260
- 13.5 Central Coordinator–Central Coordinator, 260
- 13.6 Station–Station, 262
- 13.7 Manufacturer-Specific Messages, 266
- 13.8 Vendor-Specific Messages, 267
- 13.9 Summary, 267

14 IEEE 1901 — 268

- 14.1 Introduction, 268
- 14.2 FFT, 269
  - 14.2.1 30–50 MHz Frequency Band, 269
  - 14.2.2 Additional Guard Intervals, 270
  - 14.2.3 4096 QAM, 271
  - 14.2.4 16/18 Code Rate, 271
  - 14.2.5 SNID Reuse, 271
  - 14.2.6 Repeating and Routing, 272
    - 14.2.6.1 Repeating and Routing of Unicast MSDUs, 272
    - 14.2.6.2 Repeating and Routing of Broadcast/Multicast MSDUs, 273
- 14.3 Wavelet, 274
  - 14.3.1 Baseband PHY, 274
  - 14.3.2 Bandpass PHY, 274
    - 14.3.2.1 Wavelet MAC, 274
  - 14.3.3 Transceiver Block Diagram, 275
  - 14.3.4 PPDU Format, 276
    - 14.3.4.1 Overview of the PPDU Encoding/Decoding Process, 277
    - 14.3.4.2 Modulation-Dependent Parameters, 278
  - 14.3.5 PHY Encoder, 278
    - 14.3.5.1 Generator for RCE Frame, 278
    - 14.3.5.2 Scrambler, 278

                14.3.5.3   CRC Encoder for FL, 279
                14.3.5.4   Concatenated Encoder, 279
                14.3.5.5   Convolutional Codes Defined by Low-Density Parity-Check Polynomials (Optional), 280
                14.3.5.6   FEC Type Field, 281
                14.3.5.7   Interleaver, 281
                14.3.5.8   Wavelet Process, 282
                14.3.5.9   Major Specifications, 292
                14.3.5.10 Notch and Power Control, 293
                14.3.5.11 System Clock Frequency Tolerance, 294
    14.4   Coexistence, 294
            14.4.1   Coexistence Signals, 294
            14.4.2   ISP Signaling Scheme, 295
                14.4.2.1   ISP Fields, 296
                14.4.2.2   Network Status, 298
            14.4.3   Coexistence Resources, 298
                14.4.3.1   FDM, 298
                14.4.3.2   TDM, 298
            14.4.4   ISP Resource Allocation, 299
            14.4.5   ISP Parameters, 301
            14.4.6   Management Messages, 301
    14.5   Summary, 301

15   HomePlug Green PHY   302

    15.1   Introduction, 302
    15.2   Physical Layer, 302
    15.3   MAC Layer, 303
            15.3.1   Power Save, 304
                15.3.1.1   Distribution of Power Save State Information, 306
                15.3.1.2   CCo Power Save, 307
            15.3.2   Bandwidth Sharing Between Green PHY and HomePlug AV and IEEE, 307
                15.3.2.1   Green PHY Preferred Allocation, 307
                15.3.2.2   Distributed Bandwidth Control, 308
            15.3.3   PEV–EVSE Association, 309
                15.3.3.1   PEV–EVSE Association Procedure, 310
    15.4   Summary, 311

16   HomePlug AV2   312

    16.1   Introduction, 312
    16.2   MIMO, 312
    16.3   Extended Frequency Band, 315
            16.3.1   Power BackOff, 316

16.4 Efficient Notching, 316
16.5 Short Delimiter and Delayed Acknowledgment, 316
    16.5.1 Short Delimiter, 317
    16.5.2 Delayed Acknowledgment, 320
    16.5.3 TCP and UDP Efficiency Improvements, 320
16.6 Immediate Repeating, 321
16.7 Power Save, 322
16.8 Summary, 323

Appendix A Acronyms 325

Appendix B HomePlug AV Parameter Specification 332

References 334

Index 337

LIST OF FIGURES

FIGURE 2.1.	System block diagram	13
FIGURE 2.2.	Protocol layer architecture	14
FIGURE 2.3.	HomePlug AV network architecture	15
FIGURE 3.1.	Sample powerline channel impulse response	20
FIGURE 3.2.	Sample channel frequency response	20
FIGURE 3.3.	Examples of powerline impulse noise	21
FIGURE 3.4.	OFDM symbol timing	23
FIGURE 3.5.	Windowed OFDM PSD	23
FIGURE 3.6.	Turbo Convolutional Code performance	24
FIGURE 3.7.	Turbo Convolution Code compared with capacity	25
FIGURE 3.8.	Beacon Period and Tone Map Regions	28
FIGURE 3.9.	Beacon Period and TDMA allocations	28
FIGURE 4.1.	HomePlug AV transceiver block diagram	33
FIGURE 4.2.	AV PPDU structure	34
FIGURE 4.3.	MPDU to PPDU encoding	35
FIGURE 4.4.	Hybrid Mode PPDU structure	36
FIGURE 4.5.	AV Mode PPDU structure	36
FIGURE 4.6.	OFDM symbol timing	37
FIGURE 4.7.	AV Preamble waveform	38
FIGURE 4.8.	AV Frame Control FEC Data Path	39
FIGURE 4.9.	Data scrambler	40
FIGURE 4.10.	AV Turbo Convolutional Encoder	41
FIGURE 4.11.	8-State Constituent Encoder	41
FIGURE 4.12.	PN Generator	50
FIGURE 4.13.	Spectral occupancy for semi-infinite number of carriers	55

FIGURE 4.14.	AV PRS waveform	57
FIGURE 4.15.	Spectral occupancy of set of HomePlug carriers	59
FIGURE 5.1.	General MPDU formats	62
FIGURE 5.2.	Beacon MPDU format	64
FIGURE 5.3.	Network identifier	67
FIGURE 5.4.	Example of Beacon Relocation	75
FIGURE 5.5.	Start-of-Frame MPDU format	77
FIGURE 5.6.	Measurement of FL_AV	80
FIGURE 5.7.	PHY Block formats	83
FIGURE 5.8.	SACK MPDU format	85
FIGURE 5.9.	RTS/CTS MPDU format	88
FIGURE 5.10.	Duration field in RTS/CTS when IGF is set to 0b0	90
FIGURE 5.11.	Long Sound MPDU format	91
FIGURE 5.12.	Reverse SOF MPDU format	95
FIGURE 6.1.	MSDU Format	100
FIGURE 6.2.	MAC Data Plane Overview	100
FIGURE 6.3.	MAC Frame Format	101
FIGURE 6.4.	MAC Segmentation and MPDU Generation	105
FIGURE 6.5.	Transmitter MAC Frame Stream FSM	108
FIGURE 6.6.	Receiver MAC Frame Stream FSM	110
FIGURE 6.7.	Illustration of Multinetwork Broadcast Transmission	113
FIGURE 6.8.	Example of MPDU Bursting	115
FIGURE 6.9.	Bidirectional Burst Mechanism	116
FIGURE 6.10.	Interframe Spacing during Bidirectional Burst	117
FIGURE 6.11.	Bidirectional Bursts during CSMA	117
FIGURE 7.1.	User-Appointed CCo	124
FIGURE 7.2.	Transfer of CCo function	126
FIGURE 8.1.	HomePlug AV network architecture	134
FIGURE 8.2.	Basic Beacon Period structure	134
FIGURE 8.3.	Example of Beacon Period structure in Uncoordinated Mode	137
FIGURE 8.4.	Example of Beacon schedule persistence	139
FIGURE 8.5.	Beacon Period structure in CSMA-only mode	141
FIGURE 8.6.	Example of Beacon Period structure in Uncoordinated Mode	142
FIGURE 8.7.	Example of Beacon Period structure in Coordinated Mode	143
FIGURE 8.8.	Medium States when a MPDU is transmitted or detected in Contention State	144
FIGURE 8.9.	Medium States when a station gets pre-empted in the Priority Resolution Period and detects no MPDU transmission in Contention State	145
FIGURE 8.10.	Medium States when MPDU frame control errors or collision lead to a busy state and no delimiter is detected for an EIFS_X period	145

LIST OF FIGURES

FIGURE 9.1.	Connection setup	160
FIGURE 9.2.	Global Link setup	162
FIGURE 9.3.	Connection teardown for Connections with only Local Links	163
FIGURE 9.4.	Connection teardown for Connections with Global Links	165
FIGURE 9.5.	Connection reconfiguration	166
FIGURE 9.6.	Connection Squeeze/De-Squeeze	169
FIGURE 9.7.	Global Link life cycle	170
FIGURE 10.1.	Power-on Network Discovery Procedure	174
FIGURE 10.2.	Unassociated STA behavior	175
FIGURE 10.3.	Unassociated CCo behavior	176
FIGURE 10.4.	Behavior as an STA in an AVLN	176
FIGURE 10.5.	Behavior as a CCo in an AVLN	177
FIGURE 10.6.	Getting Full AVLN information	179
FIGURE 10.7.	STA Association	180
FIGURE 10.8.	Provision NEK for a new STA (Authentication)	182
FIGURE 10.9.	AVLN formation by two unassociated STAs with matching NIDs	184
FIGURE 10.10.	AVLN formation using a DAK-Encrypted NMK	185
FIGURE 10.11.	AVLN formation using UKE by one STA in SC-Add and one STA in SC-Join	187
FIGURE 10.12.	AVLN formation using UKE by two STAs in SC-Join	188
FIGURE 10.13.	New STA joins existing AVLN with matching NID	190
FIGURE 10.14.	New STA ioins AVLN by DAK-Encrypted NMK	191
FIGURE 10.15.	New STA joins existing AVLN using UKE	192
FIGURE 10.16.	Disassociation—STA leaves AVLN	193
FIGURE 10.17.	Provision NEK for part or all of the AVLN	204
FIGURE 10.18.	Encrypted payload message when PID is between **0x00** and **0x03**	205
FIGURE 10.19.	Encrypted payload message when PID = **0x04**	206
FIGURE 11.1.	Initial channel estimation	214
FIGURE 11.2.	Dynamic channel adaptation	215
FIGURE 11.3.	HomePlug AV bridging to Ethernet Networks	220
FIGURE 11.4.	LBDAT and RBAT of HomePlug AV Stations	221
FIGURE 11.5.	Proxy Network created by Network 1	226
FIGURE 11.6.	HSTA association	228
FIGURE 12.1.	Neighbor Network Mode transitions	236
FIGURE 12.2.	Flowchart for computing INL allocation	239
FIGURE 12.3.	MSC to set up a new network in Coordinated Mode	242
FIGURE 12.4.	New CCo detects two groups of networks	243
FIGURE 12.5.	MSC to request additional bandwidth in Coordinated Mode	245
FIGURE 12.6.	MSC to shut down an AVLN in Coordinated Mode	247
FIGURE 13.1.	Management Message format	251
FIGURE 13.2.	Illustration of fragmentation of a MMENTRY	253

FIGURE 14.1.	Wavelet transmitter and receiver	275
FIGURE 14.2.	PPDU frame format	276
FIGURE 14.3.	Generator	278
FIGURE 14.4.	Scrambler	278
FIGURE 14.5.	CRC encoder	279
FIGURE 14.6.	Convolutional encoder	280
FIGURE 14.7.	Example of frame control transmitted data	284
FIGURE 14.8.	Example of TMI transmitted data	284
FIGURE 14.9.	Example of FL transmitted data	285
FIGURE 14.10.	Example of frame body diversity transmitted data	285
FIGURE 14.11.	Image of the relation between ramp process and output wave of IDWT	286
FIGURE 14.12.	Short preamble signal	287
FIGURE 14.13.	Preamble signal	287
FIGURE 14.14.	PPDU frame format with postamble	291
FIGURE 14.15.	Frame bodies using pilot signals	292
FIGURE 14.16.	Example transmission spectrum of Wavelet OFDM with notches (up to 30 MHz)	293
FIGURE 14.17.	Notch frequency characteristics	294
FIGURE 14.18.	Sync points	295
FIGURE 14.19.	ISP time Window and ISP fields concept	296
FIGURE 14.20.	Periodicity of ISP windows	296
FIGURE 14.21.	ISP fields	297
FIGURE 14.22.	ISP general TDMA structure	299
FIGURE 14.23.	Structure of TDMA	299
FIGURE 15.1.	Examples of Power Save Schedules	305
FIGURE 15.2.	Illustration of PEV–EVSE association	309
FIGURE 15.3.	Charging harness	310
FIGURE 16.1.	HomePlug AV2 MIMO transceiver block diagram	314
FIGURE 16.2.	Short delimiter	317
FIGURE 16.3.	PPDU format with Short Delimiter	318
FIGURE 16.4.	Short Delimiter efficiency improvement	319
FIGURE 16.5.	Delayed acknowledgment	320
FIGURE 16.6.	Throughput improvement for CSMA	321
FIGURE 16.7.	Throughput improvement for TDMA	321
FIGURE 16.8.	Immediate repeating channel access for CSMA	322
FIGURE 16.9.	AV2.0 SISO PHY rate with repeating	323

LIST OF TABLES

TABLE 1.1.	HomePlug Specifications Timeline	4
TABLE 4.1.	PPDU Formats	35
TABLE 4.2.	OFDM Symbol Characteristics	37
TABLE 4.3.	Rate $^1/_2$ Puncture Pattern	42
TABLE 4.4.	Rate 16/21 Puncture Pattern	42
TABLE 4.5.	Interleaver Parameters	42
TABLE 4.6.	Interleaver Seed Table for FEC Block Size of 16 Octets	43
TABLE 4.7.	Interleaver Seed Table for FEC Block Size of 136 Octets	43
TABLE 4.8.	Interleaver Seed Table for FEC Block Size of 520 Octets	43
TABLE 4.9.	Channel Interleaver Parameters	45
TABLE 4.10.	Sub-Bank Switching	45
TABLE 4.11.	ROBO Mode Parameters	46
TABLE 4.12.	Modulation Characteristics	49
TABLE 4.13.	Tone Mask Amplitude Map and Tone Map	50
TABLE 4.14.	Bit Mapping	52
TABLE 4.15.	Symbol Mapping (Except 8 QAM)	52
TABLE 4.16.	Symbol Mapping for 8 QAM	53
TABLE 4.17.	Modulation Normalization Scales	53
TABLE 4.18.	Relative Power Levels	54
TABLE 4.19.	North American Carrier and Spectral Masks	58
TABLE 4.20.	Amplitude Map	59
TABLE 5.1.	HomePlug AVMPDU Types and their Functionality	62
TABLE 5.2.	General AV Frame Control Fields	63
TABLE 5.3.	Delimiter Type Field Interpretation	63

TABLE 5.4.	HomePlug AV Beacon Payload Fixed Fields Grouped based on Functionality	66
TABLE 5.5.	Beacon Management Information Format	69
TABLE 5.6.	Beacon Entry Header Interpretation	69
TABLE 5.7.	Nonpersistent Schedule BENTRY	70
TABLE 5.8.	Session Allocation Information Format without Start Time	70
TABLE 5.9.	Session Allocation Information Format with Start Time	70
TABLE 5.10.	Persistent Schedule BENTRY	71
TABLE 5.11.	Regions BENTRY	72
TABLE 5.12.	Discovered Info BENTRY	73
TABLE 5.13.	Start-of-Frame Frame Control Fields Grouped based on Functionality	78
TABLE 5.14.	PHY Block Header Fields	84
TABLE 5.15.	SACK Frame Control Fields Grouped based on Functionality	86
TABLE 5.16.	RTS/CTS Frame Control Fields Grouped based on Functionality	89
TABLE 5.17.	Sound Frame Control Fields Grouped based on Functionality	92
TABLE 5.18.	Sound Payload Fields	94
TABLE 5.19.	Reverse SOF Frame Control Fields Grouped based on Functionality	96
TABLE 6.1.	MAC Frame Stream Command Interpretation	107
TABLE 6.2.	MAC Frame Stream Response Interpretation	107
TABLE 8.1.	Setting the VCS Timer	146
TABLE 8.2.	CW and DC as a Function of BPC and Priority	147
TABLE 8.3.	Channel Access Priority versus Priority Resolution	148
TABLE 9.1.	Recommended VLAN User Priority-to-CSMA Priority Mapping	151
TABLE 9.2.	Recommended Application Class-to-User Priority Mappings	152
TABLE 9.3.	QoS and MAC Parameter Exchanged between HLE and CM and Between CMs	153
TABLE 9.4.	Additional QoS and MAC Parameter Fields Exchanged Between Two CMs	154
TABLE 9.5.	QoS and MAC Parameter Fields Between CM and CCo	155
TABLE 9.6.	Summary of Link and Connection Identifiers	158
TABLE 10.1.	TEI Values	180
TABLE 10.2.	Security-Level Interpretation	197
TABLE 10.3.	Security Level and NMK Provisioning	197
TABLE 12.1.	Rules for Computing INL Allocation	240
TABLE 13.1.	Station–Central Coordinator Messages	254
TABLE 13.2.	Proxy Coordinator Messages	260
TABLE 13.3.	Central Coordinator-Central Coordinator Messages	260
TABLE 13.4.	Station–Station Coordinator Messages	262
TABLE 14.1.	Rate 16/18 Puncture Pattern	271
TABLE 14.2.	Puncture Patterns	280

TABLE 14.3.	Number of Transmission Bits for Tail-Bit Coding	281
TABLE 14.4.	FEC Type Parameter	282
TABLE 14.5.	Modulation Methods for Each Informational Type	283
TABLE 14.6.	Applications of Tone Maps/Tone Masks	283
TABLE 14.7.	Wavelet OFDM Major Specifications	292
TABLE 14.8.	ISP Signal Phases and their Use	295
TABLE 14.9.	Meaning of ISP Window Fields	297
TABLE 14.10.	ISP Parameter Specification	300
TABLE 14.11.	ISP Parameter Specification	301
TABLE 15.1.	ROBO Mode Parameters	303
TABLE 15.2.	Allowed PPDU Combinations	308
TABLE B.1.	HomePlug AV Parameter Specifications	332

PREFACE

Broadband powerline communication systems are continuing to gain significant market adoption worldwide for applications ranging from IPTV delivery to the Smart Grid. The suite of standards developed by the HomePlug Powerline Alliance plays an important role in the widespread deployment of broadband PLC. To date, more than 100 million HomePlug modems are deployed and these numbers continue to rise.

HomePlug AV is one of the most successful HomePlug Standards and it also forms the basis of the IEEE 1901 Standard. It was the result of several years of research by multiple companies to define a communication system that optimizes performance for the harsh powerline environment. This book is intended to provide insight into the unique design choices made in the HomePlug AV Specification. This book also contains a history of the HomePlug Powerline Alliance as well as general details on the PHY and MAC and other features of the HomePlug AV Specification. The objective is to produce a handbook that would serve as a supplement and guide to HomePlug AV while also providing background on the evolution and development of the specification. Thus, the authors sought to prepare a handbook that would be useful for designers and users of HomePlug AV compliant devices, providing a clear and simple description of the key features and capabilities of the specification.

Since HomePlug AV technologies feature prominently in the finalized IEEE 1901 Standard, the authors decided to include a discussion of the IEEE 1901 Standard as part of the HomePlug AV Handbook and hence the title of the present volume is *HomePlug AV and IEEE 1901: A Handbook for PLC Designers and Users*. The book also introduces the recently released HomePlug Green PHY and HomePlug AV 2.0 Standards.

It is the hope of the authors that this book will indeed prove useful as a PLC reference and in providing an accessible exposition of the core HomePlug and IEEE 1901 technologies that have significantly impacted high speed Powerline Communications and will continue to do so for years to come.

Haniph A. Latchman
Srinivas Katar
Lawrence W. Yonge III
Larry Yonge
Sherman Gavette

ACKNOWLEDGMENTS

It is impossible to undertake a project of the scope and magnitude encompassed in the finished volume *HomePlug AV and IEEE 1901: A Handbook for PLC Designers and Users*, without incurring a great deal of indebtedness.

The authors, having all participated in various roles in the development of the powerline specifications and standards described in this book, are intimately and acutely aware of the enormous amount of work and time that went into the development of both the HomePlug AV Specification and the IEEE 1901 Standard. Thus, it is our pleasure and duty to acknowledge the indefatigable and resolute efforts of our many colleagues in the HomePlug AV Alliance (including past members who contributed to the HomePlug specifications) as well as the work of those colleagues in the IEEE1901 working groups, who contributed to the IEEE 1901 Standard. We hope that our feeble attempts to describe their work in a format accessible to PLC designers and users, does some semblance of justice to their combined intellectual capital embodied in those standards and specifications.

The authors would also like to acknowledge the very useful feedback provided by the reviewers, Jim Allen and Scott Willy, at various stages of the project. Their insightful comments, observations, and suggestions helped us to focus the book on explaining the reasoning behind the technical decisions that were made in the development of the HomePlug AV Specification and the IEEE 1901 Standard.

We also gratefully acknowledge the support and help of the Wiley-IEEE staff in moving this project along, despite the delays occasioned by the evolution of HomePlug AV and the emergence of IEEE 1901 as the project was proceeding. The guidance, reminders, and suggestions of Mary Hatcher, the Associate Editor assigned to the project were invaluable as the project progressed.

In addition, the authors would like to thank their employers (Qualcomm Atheros and the University of Florida) for allowing them the freedom to produce this book. Thanks also to our many work and professional colleagues and students who have contributed to this work in some form and in guiding and honing our skills in this area of work. A special word of thanks goes to Anim Amarsingh and Duotong Yang, graduate students at the University of Florida, for their work in helping to proof read the final draft of the book and their assistance in refining the references and acronym definitions. We would also like to thank Peter Scarborough for designing the book cover.

Finally, the authors would like to thank their families without whose patience and tolerance, this work would not have been possible.

HANIPH A. LATCHMAN
SRINIVAS KATAR
LAWRENCE W. YONGE III
LARRY YONGE
SHERMAN GAVETTE

BIOGRAPHICAL SKETCHES OF THE AUTHORS

Dr. Haniph A. Latchman received the B.Sc. (Hons) from the University of the West Indies in 1981 and the D.Phil. from the University of Oxford in 1986. He is presently Professor of Electrical and Computer Engineering and teaches and conducts research in the areas of Control Systems, Communications, and Computer Networks. He is a Rhodes Scholar (Jamaica, St. Edmund Hall, 1983), a Senior Member of the IEEE and has published over 170 technical journal articles, conference proceedings, and three books in the general areas of Communication Networks and Control Systems. He has guided 23 Ph.D. dissertations and 37 M.S. theses. He has served as an Associate Editor and Guest Editor for several international journals, and as General Chair and member of technical program committees for conferences in the areas of communications, control systems, and networks. His teaching and research have been recognized by numerous awards, including several Best Paper Awards, the University of Florida Teacher of the Year Award, the IEEE Undergraduate Teaching Award, the Boeing Summer Faculty Fellowship, and a 2001 Fullbright Fellowship. He is an active member of the IEEE Communication Society Technical Committee on Broadband Powerline Communications and was the General Chair of the 2006 IEEE International Symposium on Powerline Communications (ISPLC) and Co-chair of the General Conference of Globecom 2006 and 2013. From 1999 to 2009 he served as a consultant to Intellon Corporation, now known as Qualcomm Atheros (PLC Technologies).

Dr. Srinivas Katar received his Bachelor of Technology degree from the Indian Institute of Technology, Kanpur in 1998 and the Ph.D. degree from the University of Florida in 2006. He is currently Principal Engineer at Qualcomm Atheros and is responsible for research and development of Medium Access Control protocols for powerline communication systems. He has more than 13 years of experience in developing PLC systems. He was a prolific contributor to the MAC Layer for several PLC standards including HomePlug AV, IEEE 1901, HomePlug AV2, and HomePlug Green PHY. He is currently an active member of HomePlug AV technical working group and Green PHY technical working group. He is an inventor of several key features in HomePlug and IEEE 1901 Standards and has more than 50 issued patents and pending patent applications. He has also authored or co-authored a number of journal and conference papers. He is a senior member of the IEEE.

Lawrence Yonge III is currently Senior Director of Technology for Qualcomm Atheros and is responsible for advanced powerline communications technology development in Ocala, Florida. He manages a research team that was a key contributor to the development of the HomePlug powerline specifications including HomePlug 1.0, HomePlug AV1.1, HomePlug Green PHY, HomePlug AV2.0, and the IEEE 1901 Standard. He is the chairman of the HomePlug AV Technical Working Group which developed the HomePlug AV1.1 and AV2.0 specifications. He joined Intellon Corporation in 1997 where he was Vice President of Research and Development. Intellon Corporation was acquired by Atheros Communication in 2009, which was acquired by Qualcomm in 2011. From 1987 to 1997 he worked as an independent engineering consultant to communication companies. He was a co-founder and President of Raydx Satellite Systems, Inc. from 1983 to 1987 and a design engineer for Microdyne Corporation from 1980 to 1983. He holds a B.S. in engineering from LeTourneau University. He is an inventor of approximately 80 US issued patents and pending applications.

Sherman Gavette received his B.S. in Mathematics from Arizona State University, Tempe and his M.S. in Information Science from the University of Chicago. He has over 47 years of experience in computers and communications, including telephony (ESS and PCS) and networking (Tymnet, UWB, and PLC). His work experience includes stints at Bell Telephone Laboratories, Tymshare (Tymnet), and Omnipoint, where he was the "guardian of the protocols" for IS661, an FCC Pioneer's Preference winner in PCS. While working at Sharp Laboratories of America (SLA), he actively participated in the development of both HomePlug 1.0 and HomePlug AV. His work on HomePlug 1.0 earned him a HomePlug Fellow Award. For HomePlug AV, he served as Chair of the Pre-Spec Working Group (PSWG), which created the AV Baseline specification, as Vice Chair of the Specification Working Group (SWG) which finalized the HomePlug AV specification, and as a member of the HomePlug Board of Directors. He holds several patents in PLC and Mobile Communications. He retired from SLA as a Principle Scientist in 2007. Since then he has served as Technical Editor on several HomePlug projects and on several IEEE Standards, including IEEE Std 1901^{TM}-2010.

1

INTRODUCTION

The explosive growth of the Internet has led to the need for ubiquitous data and multimedia communications in the twenty-first century. In-home distribution of multimedia content is still a challenge, particularly for homes not equipped with specialized wiring to support high-speed data and multimedia communications. Retrofitting buildings with new wiring is prohibitively expensive and wireless solutions do not reliably provide "whole house" coverage for multimedia applications. Hence, the need arises for a new LAN technology that enables affordable and ubiquitous connectivity within the home. The existing electrical wiring is unique in that it provides a large number of connection points throughout the home, eliminating a significant limitation of other existing wires, specifically coax cables or traditional telephone lines. However, unlike the relatively clean communication coax and phone line channel, there are many challenges in communication over the harsh powerline channel that must be overcome to make powerline communication (PLC) a viable "no new wires" home networking solution for high speed multimedia applications.

This book provides a clear and simple overview of key features of the HomePlug AV specification and the associated IEEE 1901 standard. It provides details on how the challenges associated with communicating over the electrical wiring were overcome and also the justifications and explanations of the reasoning behind the technical decisions that were made in the specification. The reader is referred to the HomePlug AV specification and the IEEE 1901 standard for the technical details needed for implementation. The book primarily focuses on HomePlug AV and discusses the IEEE 1901 standard, highlighting the features of HomePlug AV that

---

HomePlug AV and IEEE 1901: A Handbook for PLC Designers and Users, First Edition.
Haniph A. Latchman, Srinivas Katar, Larry Yonge, and Sherman Gavette.
© 2013 by The Institute of Electrical and Electronics Engineers, Inc. Published 2013 by John Wiley & Sons, Inc.

have been incorporated into the standard, as well as providing a description of key provisions of IEEE 1901 standard. The book also gives an overview of the HomePlug Green PHY specification that targets Smart Energy applications (10 Mbps) and the HomePlug AV 2.0 specification that adds MIMO, repeating, and other enhancements to HomePlug AV, thus providing operation up to 1.5 Gbps.

1.1 HomePlug AV AND ITS RELATIONSHIP TO IEEE 1901

The HomePlug AV 1.1 specification [1] (referred to throughout this book as HomePlug AV) was released in May 2007 by the HomePlug Powerline Alliance. At that time HomePlug AV was the flagship specification in an emerging PLC suite by the HomePlug Powerline Alliance, including the HomePlug 1.0.1 specification (2001). The HomePlug AV specification describes a PLC system operating at 200 Mbps and built upon an Orthogonal Frequency-Division Multiplexing (OFDM) FFT-based physical layer (PHY) protocol and a hybrid TDMA/CSMA Medium Access Control (MAC) protocol. The CSMA component is identical to that used in the HomePlug 1.0.1 CSMA technology. The OFDM FFT PHY uses the frequency band of 1.8–30 MHz with modulation up to 1024 QAM and a turbo convolutional code FEC.

The emergence of the HomePlug Powerline Alliance specifications together with a number of other incompatible industry-driven PLC specifications led to the need for a globally coordinated PLC standard. Industry organizations such as the Home-Plug Powerline Alliance with membership open to competing PLC companies made some attempts at generating consensus on design choices in specification development. However, competing corporate interests often made this a very difficult and often impossible process.

In 2005, the IEEE Communications Society (COMSOC) sponsored the IEEE P1901 project to define a global IEEE standard for high-speed PLC systems. Several competing proposals were considered from research and development groups and manufacturers of PLC equipment in Europe, Asia, and the Americas. In 2007, about the same time as the release of the HomePlug AV 1.1 specification, the IEEE 1901 working group selected a consolidated proposal by the HomePlug Powerline Alliance and the HD-PLC Alliance. The final proposal featured three technology areas or "clusters,", namely, the In-home cluster, the Access cluster, and the Coexistence cluster, and the approved IEEE 1901 standard [2] was published in December 2010. IEEE 1901 represented a standard of compromise between the FFT-based OFDM PHY in HomePlug AV and the Wavelet-based OFDM PHY used in Panasonic's HD-PLC devices. The standard specifies both PHYs as optional, with an Intersystem protocol (ISP) providing coexistence but not interoperability between the in-home FFT and Wavelet PHY realizations of IEEE 1901. This compromise with two noninteroperable PHY specifications is, in reflection, analogous to the case of the original IEEE 802.11 standard that was released with two noninteroperable PHY specifications, namely, the direct-sequence and frequency-hopping spread spectrum. In addition to enabling coexistence between these noninteroperable PHYs, the ISP is also designed to ensure coexistence between in-home IEEE 1901 system and IEEE 1901 Access or G.hn systems.

While both the HomePlug AV and the newer IEEE 1901 specifications contemplated and provided for coexistence with PLC Access systems, at present efforts to build or deploy HomePlug AV- or IEEE 1901-based PLC Access systems are at best in the very early stages. Though the technology for Access systems is available and technically viable, past experience with PLC-based systems for Internet access has been commercially unsuccessful in the United States and Europe. However, the emergence of Smart Grid and Smart Energy market drivers may portend new developments in this area in the near future.

In the present-day in-home PLC market, in the absence of interoperability, the major emphasis is on manufacturing HomePlug AV devices with IEEE 1901 certification to ensure coexistence guarantees as outlined above. At the same time, IEEE 1901 PLC devices with non-HomePlug options may still be deployed with guaranteed coexistence. The legacy HomePlug AV devices and newer IEEE 1901-certified devices with the HomePlug FFT currently have the largest market penetration and momentum globally.

1.2 FOCUS OF THE BOOK

Numerous scholarly papers, white papers, and reports have been published that present individual aspects of the HomePlug AV specification and the IEEE 1901 standard, and of course there is full technical specification of each. The IEEE 1901 standard can be purchased from the IEEE and the full HomePlug AV specification is available to members of the HomePlug Powerline Alliance. In contrast, the goal of this handbook is to provide a comprehensive yet clear and simple description of HomePlug AV and the IEEE 1901 standard that will be useful to designers of HomePlug compliant devices and also accessible and beneficial to network administrators and individual users of compliant PLC networks. From this perspective, the main focus of the book will be on HomePlug AV with relevant details from IEEE 1901 presented to clarify how IEEE 1901 is built upon and expands on HomePlug AV technologies.

In the fact, if we consider an IEEE 1901 compliant device with the FFT HomePlug AV-based FFT option for the PHY, what we have is essentially an augmented HomePlug AV system with certain extensions to the HomePlug AV 1.1 PHY and MAC, as well as the mechanisms to enable coexistence with IEEE 1901 devices with the Wavelet PHY option and with G.hn and IEEE 1901 Access systems. In summary, the extensions of the of the HomePlug AV PHY include (i) the extension of the frequency band from 1.8–30 to 1.8–50 MHz, (ii) a larger set of guard intervals—some shorter and some larger than that in HomePlug AV, (iii) a higher code rate—8/9 to complement the 1/2 and 16/21 code rates, and (iv) a higher order modulation—4096 QAM (the HomePlug AV maximum was 1024 QAM). MAC augmentations include (i) the addition of repeating, which was not present in HomePlug AV 1.1, (ii) adjustments to the Short Network IDentifiers (SNID), and (iii) the addition of the ISP for coexistence [2].

The book will not only provide a general understanding of the features and capabilities of HomePlug AV but will also give sufficient details of the PHY and MAC and other features to be helpful to PLC product designers. The book will be a supplement and guide to the HomePlug AV specification and the IEEE 1901 standard and will provide background on the evolution and development of the related specifications and standards.

1.3 THE HomePlug POWERLINE ALLIANCE

The HomePlug Powerline Alliance is a powerline networking industrial association that was formed in 2000 to promote the rapid development and adoption of powerline communications solutions. The charter of the alliance is to develop specifications and certification programs for using the powerlines for reliable home networking and Smart Grid applications. HomePlug is the largest and most established industry group for PLC, with about 65 member companies. The HomePlug specifications were designed specifically to serve a number of in-home digital entertainment and networking applications. These include easy access to services such as online video and music programming from anywhere with a power outlet, high-speed PLC broadband connections to HDTV's, Blu-ray players, DVRs, PCs, and game consoles, as well as general purpose in home computing. As of early 2012, there are a number of different HomePlug chipsets available from at least 6 vendors and close to 280 different HomePlug PLC products, with HomePlug products controlling over 90% of the broadband PLC market and over 100 million HomePlug products shipped worldwide.

1.3.1 HomePlug Specifications

Toward achieving the goals of its charter, in the last decade the HomePlug Powerline Alliance has released or has been a major contributor to a range of PLC specifications and standards operating between 14 Mbps and 1.5 Gbps. Table 1.1 shows the timeline and data rate supported by six such specifications and standards.

With this suite of specifications and associated products, the HomePlug Powerline Alliance is well poised to make a significant contribution to the converged digital and

TABLE 1.1 HomePlug Specifications Timeline

Specification/Standard	Ratification/Publication Date	Data Rate Supported
HomePlug 1.0.1	December 2001	14 Mbps
TIA-1113 (HomePlug 1.0.1)	May 2008	14 Mbps
HomePlug AV v1.1	May 2007	200 Mbps
IEEE 1901	December 2010	400 Mbps
HomePlug Green PHY v1.0	August 2010	10 Mbps
HomePlug AV v2.0	March 2012	1.5 Gbps

Smart Grid networking requirements, with products that are both compatible and interoperable. In fact, HomePlug AV is one of the four technologies (the others being Wi-Fi [3] (IEEE 802.11x [4]), Ethernet (IEEE 802.3 [5]), and MoCA [6]) that will form the new IEEE 1905.1 standard that provides a common interface for the most compelling converged home networking technologies, in support of interoperable voice, video, and data services inside the smart home.

1.3.2 How the HomePlug AV Specification Was Developed

The HomePlug Industrial Alliance uses a well-streamlined process in the development of all its specifications, somewhat paralleling the operation of other standards organizations such as the IEEE.

For the HomePlug AV specification, the process started with the HomePlug Board of Directors (BoD) appointing a committee to develop a Marketing Requirements Document (MRD) that specifies what features and characteristics HomePlug AV should have in order to address the perceived needs.

The MRD included several typical cases of multiple multimedia video sessions, IP telephony sessions, gaming sessions, and data networking applications occurring simultaneously and specified the target range of aggregate data rate that HomePlug AV should support. A Technology Evaluation Group (TEG) was also appointed by the BoD and the TEG issued a request for proposals for technologies that could meet the MRD. The TEG reviewed submissions and made the baseline selection of core technologies from multiple submissions that would form the basis of the HomePlug AV specification. Since no member company had all the requisite technologies in their portfolio, HomePlug AV was developed from a merger of ideas from several HomePlug member companies.

Finally, the HomePlug AV specification was developed through much laborious and intensive work by the HomePlug AV Technical Working Group (TWG), which produced the HomePlug AV 1.1 specification published in May 2007. Note that although HomePlug AV version 1.0 specification was published in December 2005, version 1.1 became the definitive HomePlug AV base specification since it was published about the time the first AV products were introduced to the market.

Various members of the HomePlug Powerline Alliance participated in the IEEE 1901 standardization efforts, contributing technologies and solutions to the various challenges faced in the development of this important global PLC standard.

One may now regard the present relationship between IEEE 1901 and HomePlug as being similar to the relationship between IEEE 802.11 and Wi-Fi (the Wi-Fi Alliance). Today the Wi-Fi Alliance certifies that wireless networking products conform to the 802.11x standard in much the same way as the HomePlug provides certification to PLC products as IEEE 1901 compliant. HomePlug will also provide certification for IEEE 1905.1 products that will feature HomePlug AV technology.

The *Compliance and Interoperability Working Group*(C&IWG) of the HomePlug Powerline Alliance is responsible for the processes and protocols for testing and certifying chips, devices, and products to be compliant with HomePlug specifications or IEEE standards. The C&IWG coordinates HomePlug Plugfests for multivendor interoperability and compliance testing for the HomePlug AV specification and the IEEE 1901 powerline networking standard.

1.3.3 The Regulatory Working Group

The Regulatory Working Group (RWG) of the HomePlug Powerline Alliance is responsible for the development of strategy and processes to ensure that HomePlug products conform to global regulatory requirements. The design of PLC systems such as HomePlug AV and IEEE 1901 must also deal with these regulatory constraints. For example, in the United States, PLC systems operate under FCC Part 15 [7] rules in the frequency band between 1.8 and 30 MHz. Several subbands within this range are notched out in HomePlug products to prevent interference in licensed services. Moreover, the regulatory environment in Europe is in flux: aeronautical bands may be added and power levels are under debate. Japan made amendments to their regulations in 2006 that enabled in-building PLC and is currently considering further amendments that would allow outdoor PLC. This relatively unstable international regulatory environment requires that PLC systems be flexible to adapt to changing regulations.

In sections 1.3.3.1–1.3.3.3, we consider the present PLC regulatory domains in the United States, Europe, and the rest of the world.

1.3.3.1 The United States and the FCC Regulations for powerline communications in the United States are established by the Federal Communications Commission (FCC) and are specified in Title 47 of the Code of Federal Regulations, Part 15 [7]. The FCC rules define "Access Broadband over Power Line" (Access BPL) and "In-House Broadband over Power Line" (In-House BPL).

In-House BPL is defined as "A carrier current system, operating as an unintentional radiator, that sends radio frequency energy by conduction over electric powerlines that are not owned, operated or controlled by an electric service provider. The electric powerlines may be aerial (overhead), underground, or inside the walls, floors or ceilings of user premises. In-House BPL devices may establish closed networks within a user's premises or provide connections to Access BPL networks, or both."—47 CFR Ch. I (10–1–10 Edition), § 15.3 [7].

1.3.3.1.1 FCC Compliance Testing Part 15.31 (d) [7] specifies that carrier current devices be tested for compliance with the FCC regulations in three typical installations. The ANSI C63.4-1992 [8] document gives details concerning how to install the test equipment and how to make the measurements. The equipment is installed in a typical operating scenario, usually with a transmit duty factor very close to 100%. Measurements are then made at 16 equally spaced radials around the periphery of the house in which the equipment is operating. Measurements are made with a loop antenna in the H-field, with the antenna centered 1 m above the ground and oriented to maximize the readings.

The equipment built to the HomePlug specification and providing nominally the highest signal level permitted by the HomePlug specification is tested for compliance with Part 15 requirements in three typical home installations. The powerline modem under test is plugged into an electrical outlet that is on an exterior wall. Tests are usually conducted by independent FCC-certified test labs such as Compatible Electronics of Brea, CA, for west coast locations and Product Safety Engineering of Dade City, FL, for the east coast.

The FCC Part 15 requirement for unlicensed devices, including PLC, is that the device may not cause "harmful interference." Regulatory compliance testing ensures that products bearing the HomePlug or IEEE 1901 stamp do not cause such harmful interference. If the FCC receives a report of harmful interference from a HomePlug or IEEE 1901 PLC device, the manufacturers may refer to the testing protocol and results, but will be also be required to respond to the complaint to remove the harmful interference.

Notching amateur band is not required by FCC Part 15. However, HomePlug PLC devices notch the amateur frequency bands to avoid interference in these bands. HomePlug Compliance testing evaluates both the quality and effectiveness of the notching implemented in the target devices to ensure compliance with the HomePlug specification and the adherence to FCC rules across the spectrum.

1.3.3.2 Europe, CISPR, and CENELEC Although the European Union does not have an equivalent body as the U.S. FCC, PLC products sold in Europe are required to comply with the essential requirements of "Directive 2004/108/EC of the European Parliament and of the Council" [9].

This Directive regulates the electromagnetic compatibility of equipment. The following is the essential requirement:

"*Equipment shall be so designed and manufactured, having regard to the state of the art, as to ensure that:*

(a) *the electromagnetic disturbance generated does not exceed the level above which radio and telecommunications equipment or other equipment cannot operate as intended;*

(b) *it has a level of immunity to the electromagnetic disturbance to be expected in its intended use which allows it to operate without unacceptable degradation of its intended use.*"

The broad principles set forth in the Directive are given more explicit technical expression by harmonized European standards, yet to be adopted by the various European standardization bodies such as the European Committee for Standardization (CEN), the European Committee for Electrotechnical Standardization (CENELEC), and the European Telecommunications Standards Institute (ETSI).

Thus, the standards set by these organizations become, in essence, the PLC regulations for Europe and PLC devices can be put on the market or into service in the European Union only if the manufacturers concerned have established that such devices have been designed and manufactured in conformity with the requirements of this EU Directive. Approved devices should bear the "CE" marking attesting their compliance (Conformité Européene) with the EU Electromagnetic Compatibility Directive.

To date, no formal harmonized standards for PLC have been adopted in Europe. Previous efforts in the Comité International Spécial des Perturbations Radioélectriques (CISPR) to amend CISPR-22 [10] specifying compliance testing procedures for powerline communications generated various draft standards, but these were never approved by CISPR member states. More recently, a draft standard has been created in CENELEC prEN50561-1 [11], which is largely based on the most recent draft from CISPR, and it is hoped that this draft will be ratified and approved.

CE mark certification can be obtained by having the devices tested by a "Notified Body" that is recognized and registered by the European Union. The testing organization will then apply state-of-the-art assessment, which in practice is based on the existing CISPR and CENELEC draft documents or related standards from other international organization such as the International Telecommunication Union—Radiocommunication (ITU-R). The "no harmful interference" approach allows static notching in spectral mask or even dynamic notching based on sensing the presence of interfering signals. The essential expectation is that there should be again "no harmful interference" and the manufacturers are held liable to respond to and rectify any legitimate complaints.

It should also be noted that in the European Union, individual countries are not bound by the above formative PLC regulations and can set their own independent standards for PLC.

1.3.3.3 Rest of the World Other countries apart from the European Union and the United States have a variety of stances on PLC regulations. Each country usually has the equivalent of a Public Utility Regulator that manages access to and use of radio frequency spectrum. Some countries are very open to the use of PLC technology, often adopting or adapting to the EU or U.S. standards, while others generate local standards. Each manufacturer of HomePlug or IEEE 1901 devices destined for such countries needs to ensure compliance with relevant regulations and the HomePlug Regulations Working Group ensures this for the HomePlug Powerline Alliance.

1.4 THE ROLE OF PLC IN MULTIMEDIA HOME NETWORKING AND SMART ENERGY APPLICATIONS

Powerline communication has a unique role to play in the broad areas of in-home networking and Smart Energy applications.

HomePlug AV and IEEE 1901 have as a main target multimedia high speed in home networking in support of video, voice, gaming, and data services. PLC has the advantage of near whole house coverage with a large number of convenient outlets at data rates that are usable for high-speed applications. PLC products based on IEEE 1901 and HomePlug AV offer application data rates to support one or more HDTV channels. Most of the PLC products are individual adapters that include an Ethernet interface. Several manufacturers are now embedding HomePlug AV and IEEE 1901 devices directly into multimedia set-top boxes to allow media delivery over the PLC network. Some network routers are also now being made with both wireless and PLC interfaces.

Many competing technologies have also targeted the multimedia in-home market, the most well known being the wireless networking technologies based on the IEEE 802.11 (Wi-Fi) suite of protocols. Although Wi-Fi does provide a convenient mobile solution, it faces the challenge of whole house coverage with sustained and reliable throughput when subject to typical interferences. This situation is especially challenging when Wi-Fi is used in large residences and those made of solid concrete walls through which wireless signal do not propagate well. In this regard, PLC provides an ideal complement to the Wi-Fi mobile solution by offering PLC Wi-Fi extender products that use the PLC infrastructure as the backbone connecting multiple wireless routers and thus allowing seamless mobility without the need for spectrum sharing Wi-Fi repeaters.

Of course, one could also use dedicated Ethernet and also digital communications over Coaxial cable (Multimedia over Coaxial (MoCA) [6]) or even over the telephone lines (HomePNA [12]). The emerging IEEE 1905.1 standard provides an integrated solution with seamless routing and load balancing over Ethernet, PLC (HomePlug AV), MoCA, and Wi-Fi and promises to provide a well-suited solution for the converged digital home.

PLC is also well suited for Smart Energy applications. Since the energy is electrical in nature, attached PLC devices to the electrical wires delivering the energy are ideal candidates for Smart Grid monitoring, control, and computational optimization applications. Indeed, the "simplified HomePlug AV" (aka HomePlug *Green PHY*) specification directly provides a low-cost Smart Energy solution, ensuring full interoperability with broadband HomePlug AV and IEEE 1901 PLC systems. We can now envision a world where appliances, meters, and other electrical devices have embedded Smart Energy PLC devices that enable a truly intelligent home.

1.5 BOOK OUTLINE

Following this introductory chapter, the rest of this book is organized as follows.

Chapter 2 gives an overall description of the HomePlug AV network architecture, delineating the protocol layers in HomePlug AV and how these relate to the standard OSI protocol layers. Special attention is given to the function and role of the HomePlug AV Convergence layer. This chapter also introduces the HomePlug AV network topology, identifying the various station roles and defining the HomePlug AV Local Network (AVLN). With this infrastructure in place, the chapter then proceeds to outline the essentials of peer-to-peer communication, bridging, network membership, and channel access in HomePlug AV networks.

In Chapter 3, the authors take a step back and examine the overall philosophy and reasoning that guided the technology selection in overcoming the challenges of high-speed multimedia communications in one of the harshest communication channels known to man. The frequency and time characteristics of the typical PLC channel are reviewed with snapshots of real measurements taken to illustrate the enormity of the challenge. The chapter then gives the details of the frequency band selection, the selection of windowed FFT-based OFDM with some commentary on comparisons with other possible technologies, the use of Turbo Convolution Codes (TCC) again with observations about the performance of rival approaches and providing the guiding principles in the selections made. An entire section is dedicated to a discussion of the intelligent channel adaptation schemes used in HomePlug AV, giving details of the bit-loading adaptive modulation scheme adopted and the exploitation of the cyclostationary noise behavior in the AC line cycle-based adaptation and Beacon synchronization especially with a focus on enabling TDMA allocations in HomePlug AV. The chapter also discusses how the two-level segmentation and reassembly scheme used in HomePlug AV yields higher overall efficiency and how this is implemented in the Data Plane. The chapter concludes with an explanation on persistent and nonpersistent schedules for TDMA operation.

Chapter 4 presents the details of the HomePlug AV physical layer protocol, including the preamble used for synchronization as well as the structure of the PHY Protocol Data Unit (PPDU), Frame Control, and Payload. The adaptive Tone Maps used to achieve high-throughput in the PLC channel as well as the associated parameters used for robust (ROBO) communication for critical control-related messages are described. The chapter concludes with a discussion of the Tone Mask used to mitigate interference, the Amplitude Map used to maintain acceptable power levels in each subband, and the overall functional transceiver block diagram.

Chapter 5 is devoted to the MAC Protocol Data Unit (MPDU) and discusses the various types of delimiters and associated variant fields used in HomePlug AV. The structure of the payloads for the Beacon, Data, and Sound frames are presented in detail.

Chapter 6 discusses the HomePlug AV Data Plane, giving specific attention to the segmentation and reassembly strategies used to convert between the MAC Service Data Unit (MSDU) and the MPDU via a PHY Block (PB) used as a basic unit of encryption. The chapter also explains the HomePlug AV queuing strategy for management, broadcast, connectionless and connection-oriented queues, as well as the operation of HomePlug AV Data Plane structures in a multinetwork infrastructure.

Chapter 7 takes up the important topic of the operation of the Central Coordinator (CCo) in HomePlug AV, including the CCo selection, backup, and failure recovery, as well as the CCo discovery process. Since the CCo is central to the operation of HomePlug AV, CCo functions are also discussed as needed in other chapters of the book.

Chapter 8 explains the hybrid HomePlug AV CSMA/TDMA channel access mechanisms. It shows how the operation of the Beacon and the Beacon period structure enable the coordination of CSMA access and TDMA access with admission control and persistent and nonpersistent scheduling.

Chapter 9 begins with an overview of the packet classification mechanism that enables HomePlug AV to distinguish the QoS needs of various MSDUs. This is followed by details on connection specification (CSPEC) and the associated connection setup, modification, and teardown procedures that enable the provisioning of parameterized QoS. The role of the CCo in bandwidth management is also explained.

Chapter 10 examines the question of security and network formation in HomePlug AV, from power-on to association, authentication, and authorization. The chapter discusses the various security keys used in HomePlug AV, including the various key entry modes (direct, remote, and push button).

The next three chapters present useful details on the practical functioning of HomePlug AV. Chapter 11 discusses the key HomePlug AV operations of channel adaptation, bridging, coexistence with HomePlug 1.0.1, and Proxy Networking. Chapter 12 covers bandwidth sharing in neighbor networks in HomePlug AV, including the specific cases of CSMA-Only, Uncoordinated, and Coordinated modes. Chapter 13 presents a summary of key management messages used in HomePlug AV.

Chapter 14 provides an overview of the IEEE 1901 standard, with separate sections devoted to the PHY and MAC of the FFT-based realization and the Wavelet-based option. The chapter also discusses the functional elements of the Inter System Protocol (ISP) used to ensure coexistence between FFT and Wavelet IEEE 1901 devices as well as between any IEEE 1901 in-home device and an IEEE 1901 Access device or a G.hn device.

The concluding chapters give an overview of the HomePlug Green PHY specification (Chapter 15) as a simplified HomePlug AV incarnation and the HomePlug AV2 specification (Chapter 16) as an enhancement of HomePlug AV, with MIMO, power save, repeating, delayed acknowledgments, and larger bandwidth, all of which combine to yield a PHY rate of 1.5 Gbps.

2

THE HomePlug AV NETWORK ARCHITECTURE

2.1 INTRODUCTION

This chapter presents an overall description of the HomePlug AV network and its associated architecture. Special attention is given to the function and role of the HomePlug AV PHY, Medium Access Control (MAC), and Convergence layers as well as the grouping of protocol entities into the Data Plane and the Control Plane. The chapter also defines the HomePlug AV network topology, station roles, and the HomePlug AV Local Network (AVLN) and its associated Central Coordinator (CCo). The chapter then uses these definitions to outline the essentials of peer-to-peer communication, bridging, network membership, and channel access in HomePlug AV networks.

2.2 PROTOCOL LAYERS

At the highest level of abstraction, the HomePlug AV system consists of the protocol layers shown in Figure 2.1. The functions at the transmitter are also implemented in reverse order at the receiver. The PHY layer performs forward error correction (FEC), mapping data onto OFDM symbols, and the generation of requisite time-domain waveforms. The MAC layer determines the correct position of transmission, formats the data frames into fixed-length entities for transmission on the channel, and

---

HomePlug AV and IEEE 1901: A Handbook for PLC Designers and Users, First Edition.
Haniph A. Latchman, Srinivas Katar, Larry Yonge, and Sherman Gavette.
© 2013 by The Institute of Electrical and Electronics Engineers, Inc. Published 2013 by John Wiley & Sons, Inc.

PROTOCOL LAYERS

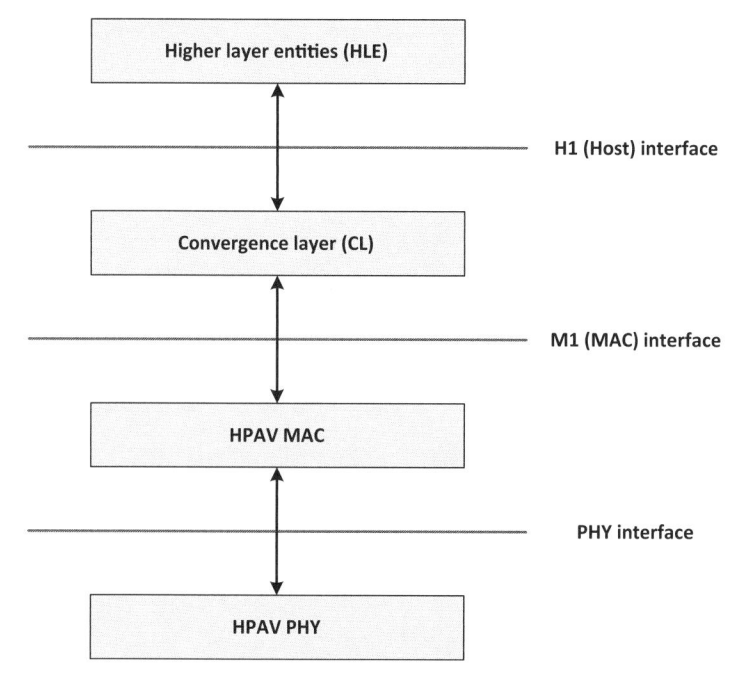

FIGURE 2.1 System block diagram.

ensures timely and error-free delivery through Automatic Repeat Request (ARQ). The MAC and PHY layers are separated by a logical PHY interface. The Convergence layer performs bridging, classification of traffic into Connections, and data delivery smoothing functions. The Convergence and MAC layers are separated by a logical M1 (MAC) interface. The logical H1 (Host) interface exposes the services provided by HomePlug AV to higher layer entities (HLE).

In relation to the International Standards Organization's (ISO) Open System Interconnect (OSI) model, the HomePlug AV specification covers the lower two layers, namely, the PHY layer and the data link layer.

Figure 2.2 shows the protocol entities defined in the HomePlug specification. Protocol entities that are directly involved in the transfer of user Payload make up the Data Plane of the protocol stack, while protocol entities that are involved in creating, managing, and terminating the flow of data are defined in the Control Plane. The HomePlug AV specification further divides the Control Plane into a Central Coordinator component and a Connection Manager (CM) component. In each AV Logical Network (AVLN), defined in greater detail shortly, one station (STA) is designated as the CCo. The CCo is responsible for setting up and maintaining the logical network, managing the communication resource on the wire and coordinating with neighbor networks (NNs). The control functions associated with the CCo are treated as part of the CCo component of the Control Plane, while functions associated with each local station fall within the CM component of the Control Plane.

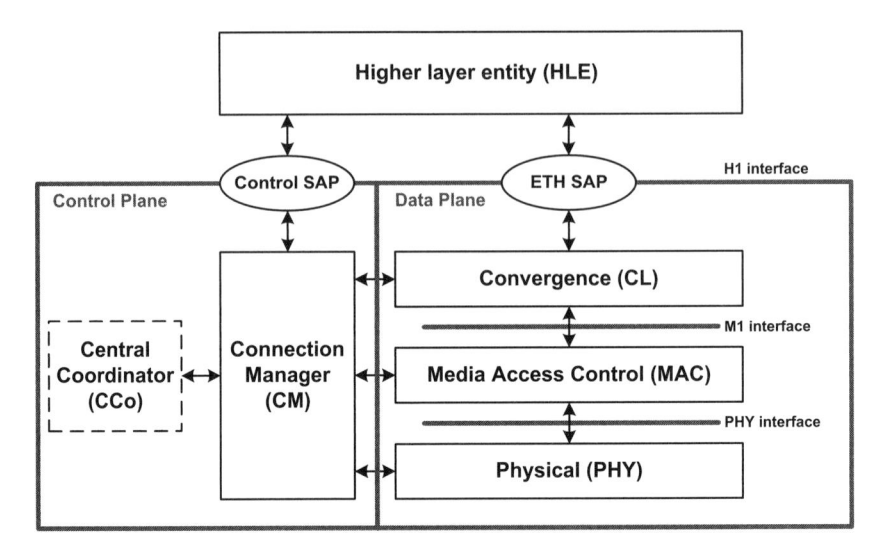

FIGURE 2.2 Protocol layer architecture.

2.3 NETWORK ARCHITECTURE

A HomePlug AV Powerline Network consists of a set of HomePlug stations connected to the AC powerline. From the physical layer perspective, stations in one dwelling might be able to communicate with stations in another dwelling. However, HomePlug AV enables stations to be logically separated by a privacy mechanism based on a 128-bit AES encryption scheme associated with a unique Network Encryption Key (NEK). An AV Logical Network is the set of STAs, typically used in a home environment, that possess the same Network Identifier (NID) and Network Membership Key (NMK). In certain situations, the CCo may deploy multiple NEKs (possibly using multiple NMKs), thus forming several logical subnetworks of the AVLN. These are called sub-AVLNs. Coordination, clock reference, and scheduling are performed on the basis of an AVLN. Cryptographic isolation is provided at the level of the sub-AVLN.

Each AVLN is managed by a single controlling station, the CCo, introduced earlier (Figure 2.3). The CCo performs network management functions such as authentication and association of new stations joining the AVLN, AC line cycle synchronization of transmission intervals, and admission control and scheduling for Time Division Multiple Access (TDMA), and Carrier Sense Multiple Access (CSMA) sessions and allocations.

The authentication of new stations is based on the knowledge of a shared secret, namely, the NMK. A user can provide the new station with the NMK or use a push-button approach for enabling it to join the AVLN. Successful authentication will enable the station to associate with the AVLN. During the association process, the CCo provides the new station with a Terminal Equipment Identifier (TEI). The TEI is used to identify a station uniquely within the AVLN. All transmissions in HomePlug

AV Logical Network

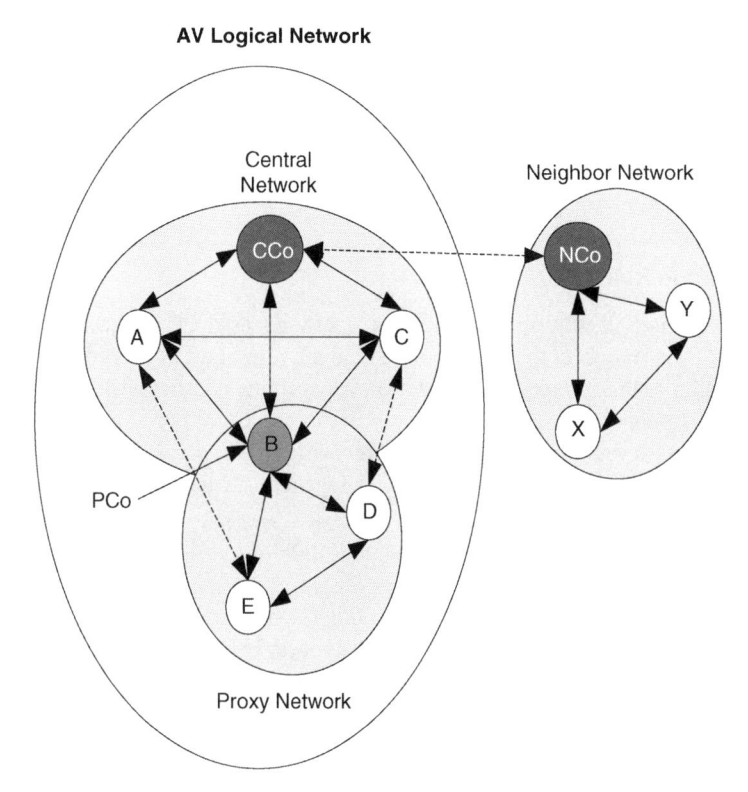

FIGURE 2.3 HomePlug AV network architecture.

AV carry the source and destination TEIs for addressing. It should be noted that although the CCo is used to manage the AVLN as describer earlier, HomePlug AV stations normally communicate directly with one another without having to go through the CCo. Communication with the CCo is needed only to manage TDMA allocations and certain other infrequent control functions. This is in contrast with popular technologies like Wi-Fi in which all transmissions go through the Access point in Infrastructure mode.

Figure 2.3 shows the organization of HomePlug AV devices into different classes of networks. The CCo and the devices in the logical network that can directly communicate with it form the Central Network (CN). The attenuation and noise characteristics on the powerline channel may give rise to situations where certain devices that belong to the same home network may not be able to communicate with the CCo. A Proxy Network (PN) is instantiated in such scenarios to allow control of the "hidden devices" through a relay of communications between a Proxy Coordinator (PCo) and a CCo. Direct peer-to-peer communications are still enabled between devices in a PN and devices in the CN with which the PN is associated. The PN concept improves coverage by enabling communications for hidden devices. Due to the robust physical layer used by HomePlug AV, proxy networks are very rare.

While PNs always depend on and are associated with a CN, a Neighbor Network (NN) is an entirely autonomous association of HPAV devices. Neighbor networks are independent networks that can exist in neighboring homes. The HPAV system provides for coordination among the neighboring networks so that access to the medium is shared fairly by the various networks and Quality of Service (QoS) is preserved for communications within a network.

2.3.1 Station Roles

Each HomePlug AV station is capable of operating as a CCo of the AVLN. The selection of the station in the AVLN to assume the role of a CCo is typically done in an automated manner based on the station capabilities and network topology. The user may also appoint a specific station to act as the CCo. For example, it may be prudent to assign the Home Gateway/Router with HPAV capabilities as the default CCo. HomePlug AV defines three different levels of CCo capability:

- Level-0 CCo: A basic CCo that can only support CSMA-based channel access (i.e., supports CSMA-Only mode).
- Level-1 CCo: A CCo that can support TDMA-based channel access when there are no neighbor networks (i.e., supports Uncoordinated mode).
- Level-2 CCo: A CCo that can support TDMA-based channel access even in the presence of neighbor networks (i.e., supports Coordinated mode).

All HomePlug AV stations are required to, at a minimum, support Level-0 CCo functionality. Furthermore, all HomePlug AV stations are required to be capable of operating under Level-0, Level-1, and Level-2 CCos.

HomePlug AV stations may also assume the role of a Proxy Coordinator to enable hidden stations to join the AVLN. Proxy Coordinators are selected by the CCo of the AVLN.

All other HomePlug AV stations in the AVLN operate as normal stations and rely on stations that assumed the role of CCo and PCo for managing the AVLN.

2.3.2 Bridging

One or more stations in the AVLN may also act as bridges to other networks. The bridge is responsible for routing traffic between the AVLN and other networks based on the list of MAC addresses of devices it is bridging for. The bridge also provides this list to other stations in the AVLN so that other stations can efficiently deliver traffic within the AVLN using unicast transmissions.

2.3.3 Channel Access

HomePlug AV is designed to provide high-quality multimedia streaming within the home network. To enable strict guarantees on QoS (i.e., bandwidth, latency, jitter, and packet loss probability), the MAC layer is based on a hybrid TDMA and CSMA protocol. The CCo is responsible for coordinating medium access and providing TDMA allocations based on QoS requirements by periodically generating a Central Beacon with information on TDMA and CSMA allocations.

2.4 SUMMARY

The HomePlug AV network architecture and protocols introduced in this chapter are at the core of the HomePlug AV specification and the associated IEEE 1901 standard. The detailed operation of the various elements in this architecture will be developed in the rest of the book.

Chapter 3 examines the overall philosophy and reasoning that guided the technology selection in overcoming the challenges of high-speed multimedia communications over the powerline channel.

3

DESIGN APPROACH FOR POWERLINE CHANNELS

3.1 INTRODUCTION

The HomePlug AV PHY and MAC Layers were jointly designed to address the unique properties and challenges of powerline channels. Various problems had to be addressed to communicate reliably on the powerline medium at data rates relatively close to capacity and with Quality of Service (QoS) guarantees. For example, the framing and segmentation process in HomePlug AV was tied directly to the Forward Error Correction coding (FEC) and channel interleaving so that only the portion of frames that experienced decoding failure need to be retransmitted. Another example is that the HomePlug AV MAC Beacon Period was defined to be synchronous to the 50- or 60-Hz powerline frequency to facilitate, among other things, being able to support different bit-loading maps for different phases of the AC powerline cycle since noise is often synchronous with the powerline cycle.

The rest of this chapter will provide further details on the power line communication (PLC) channel characteristics and the solutions adopted in HomePlug AV to address them. The following is a list of key HomePlug AV Physical Layer features that will be discussed from a philosophical perspective in this chapter and more technically in the rest of the book:

- Windowed OFDM:
    ○ provides simple spectral notching for preamble, Frame Control, and payload,

---

HomePlug AV and IEEE 1901: A Handbook for PLC Designers and Users, First Edition.
Haniph A. Latchman, Srinivas Katar, Larry Yonge, and Sherman Gavette.
© 2013 by The Institute of Electrical and Electronics Engineers, Inc. Published 2013 by John Wiley & Sons, Inc.

- ○ uses a total of 917 OFDM carriers (excluding Amateur bands) from 1.8 to 30 MHz,
- ○ adopts a 40.96-μs OFDM symbol length.
- Bit-loaded modulation with constellations from Binary Phase Shift Keying (BPSK) to 1024 Quadrature Amplitude Modulation (QAM):
  - ○ Bit-loading is unique between a transmitter and receiver and can be unique for different regions of the AC line cycle.
- Turbo Convolutional Code FEC for Frame Control, beacon, and payload
  - ○ supports 16-, 136-, and 520-byte block sizes,
  - ○ achieves near Shannon Capacity performance.
- Channel interleaver for impulse noise and other powerline impairments
  - ○ Turbo Convolutional Code FEC block is interleaved.
- Diversity coding for reliable Frame Control, Beacon, and ROBO
  - ○ Copy code spread in time and frequency.
- Coexistence mode uses HomePlug 1.0 Frame Control
  - ○ AV preamble can be detected by 1.0 devices.
- 200-Mbps PHY channel rate
  - ○ 150-Mbps PHY information rate.

3.2 CHANNEL CHARACTERISTICS

The powerline channel is characterized by variable attenuation with frequency as a result of physical attenuation, delay spread (multipath), and impedance mismatches. The PLC channel response is attributable to the AC wiring topology, losses in the wires and the loads presented by various devices, and appliances attached to the AC power wiring. While a typical channel may exhibit an average attenuation of about 40 dB, it is not uncommon for portions of the frequency band to experience greater than 80 dB of attenuation. Similarly, while most powerline channels have a delay spread between 1 and 2 μs, it is not uncommon for some channels to exhibit a delay spread larger than 5 μs. Figures 3.1 and 3.2 show a sample powerline channel impulse response and its associated frequency response, respectively.

One of the unique characteristics of powerline channels is the synchronization of noise to the AC line cycle. Electric appliances may turn on and off and/or draw electric power as a function of the AC line cycle. While this may change both the channel's frequency response and noise profile, it is more common to see changes only in the noise profile.

The major sources of noise on the powerline are from electrical appliances, which generate noise components that extend well into the high-frequency spectrum. Induced radio frequency signals from broadcast, commercial, military, citizen band, and amateur stations also impair certain frequency bands. Worse, there are often impulse noise sources that may be further categorized as either periodic or continuous based on whether or not it occurs relative to the underlying AC line cycle.

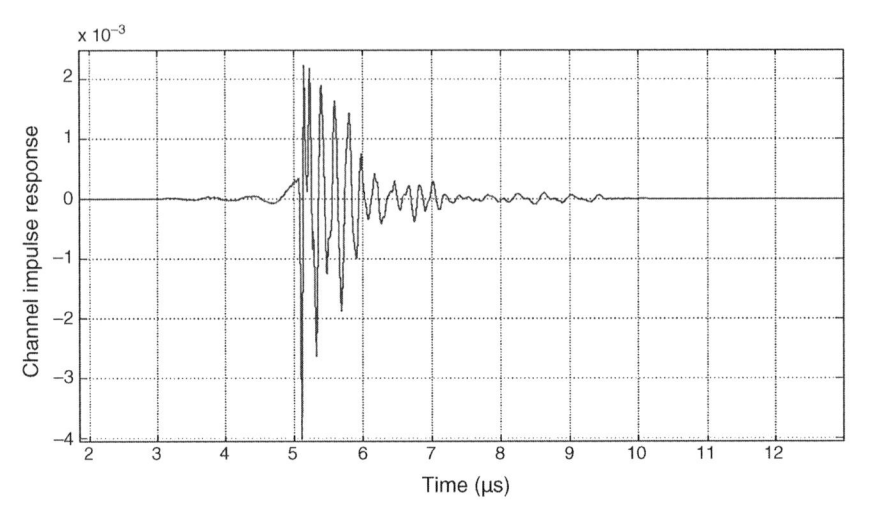

FIGURE 3.1 Sample powerline channel impulse response.

Triac dimmers used for lighting will typically generate one large impulse at the phase in the line cycle where the triac begins to conduct while switching power supplies may output a continuous stream of impulse at the frequency of the switcher. Impulse noise may also be quite large, especially at the point where a triac begins to conduct power, resulting in impulse noise amplitude greater than several volts peak-to-peak. Figure 3.3a–d shows periodic (cyclostationary) impulse noise from a halogen lamp with a triac dimmer, a yard light with switching power supply, a light dimmer switch of unknown and a hair dryer, respectively. All four figures also include the HomePlug AV PPDU signal (rectangular shape). Arcing occurs in motors with brush

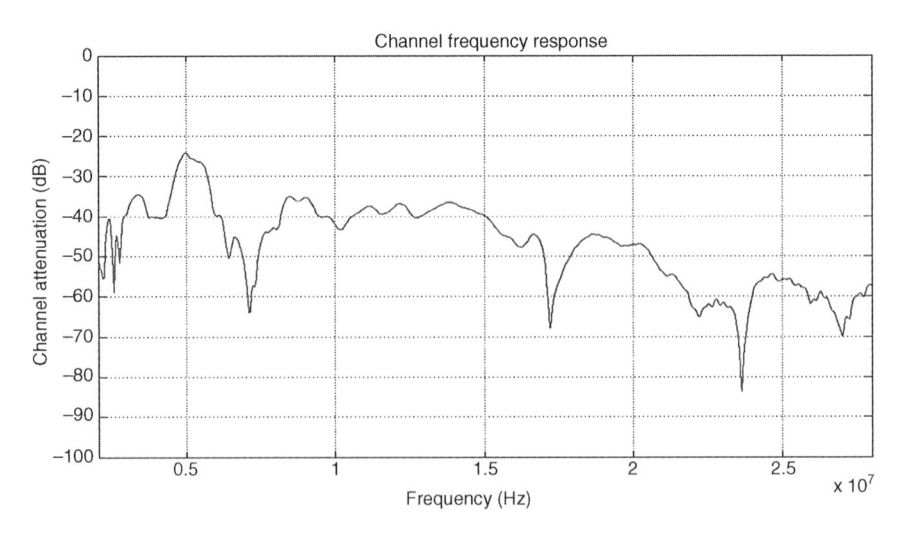

FIGURE 3.2 Sample channel frequency response.

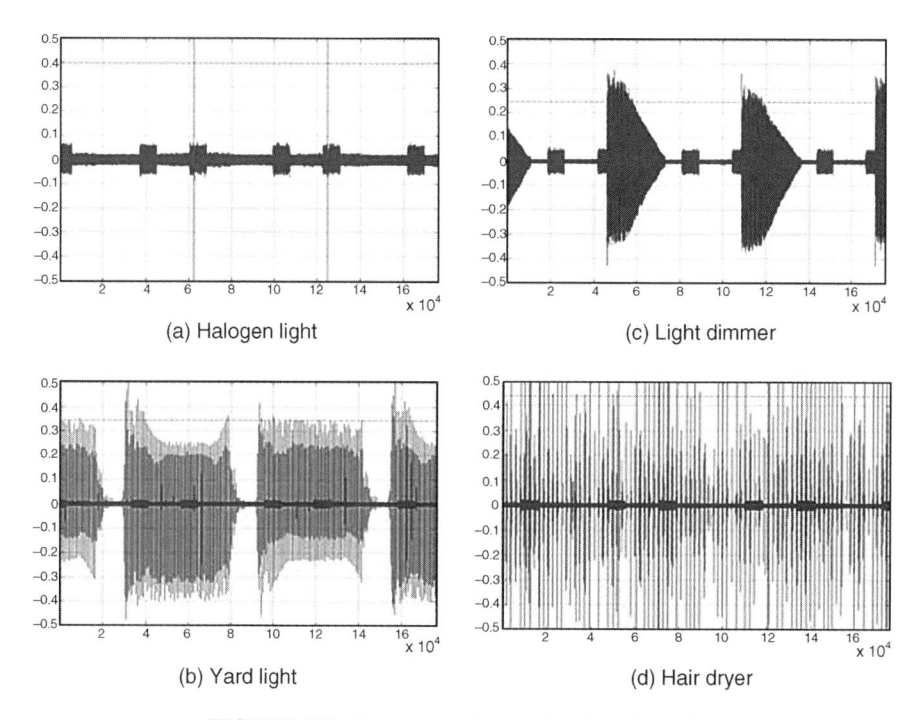

FIGURE 3.3 Examples of powerline impulse noise.

armatures creating a relatively random stream of impulse noise. Hair dryers (Figure 3.3d), vacuum cleaners, blenders, and power tools are common household appliances that contain brush motors. Because much of the noise experienced by each node may be highly localized due to attenuation, the noise profile seen by each PLC device may be significantly different. Therefore, powerline channels are not typically symmetric.

3.3 FREQUENCY BAND

The frequency band selected for HomePlug AV is from 1.8 to 30 MHz. Several reasons influenced this choice including throughput and reliability objectives. A significant factor was the issue of regulations, discussed in Chapter 1. In the United States, the Federal Communications Commission (FCC) regulations for PLC are specified in 47 CFR Chapter I, Part 15. 15.209 and sets the radiation limit for PLC within the range from 1.705–30.0 MHz to 30 μV/m at a distance of 30 m, measured with a quasi-peak detector and a measurement bandwidth of 9 KHz.

At either end of this frequency range, there is a significant reduction in the transmit power allowed. 15.107(c)(2) specifies a conducted limit in the Amplitude Modulation (AM) radio band for PLC of "1000 μV within the frequency band of 535–1705 kHz, as measured using a 50 μH/50 Ω Line Impedance Stabilized Network (LISN)." 15.109(a) sets the radiated limits for in-home PLC within the range from 30–88 MHz to 100 μV/m at a distance of 3 m, measured with a quasi-peak detector and a measurement bandwidth of 120 KHz. The combination of field strength, distance, measurement bandwidth, and other factors results in approximately 30 dB lower transmit power above 30 MHz than for the range of 1.705–30.0 MHz.

Combining the regulations, which limits transmit power versus frequency, with powerline noise and attenuation characteristics and the need for relatively wide bandwidth required to support greater than 100 Mbps data rates, the 1.705–30.0 MHz frequency band is the best choice. 1.8 MHz was chosen as the lower limit to provide a 95-KHz guard band to meet the conducted limits for the AM radio band.

Since the Amateur Radio Bands are licensed spectrum that is commonly used in residential areas, a number of 30 dB deep notches were added to the 1.8–30.0 MHz band to minimize the likelihood of causing harmful interference to these licensed users. The original decision to notch the Amateur Radio Bands was made for HomePlug 1.0, the HomePlug specification that preceded HomePlug AV. Joint testing was performed in December 2000 with the Amateur Radio Relay League (ARRL) which confirmed that the 30-dB notches provided sufficient protection against harmful interference when HomePlug device were installed in the same home as the Amateur Radio. The report from this testing can be found in the ARRL website [13].

3.3.1 Tone Mask

The Tone Mask defines the set of Orthogonal Frequency Division Multiplexing (OFDM) carriers used by a HomePlug AV device. Encoding of Frame Control and ROBO Mode are dependent on the Tone Mask. Thus, it is necessary for all devices to use the same Tone Mask for Frame Control and ROBO Mode to interoperate. The Tone Mask is typically set during manufacturing of devices.

HomePlug AV specifies a Default Tone Mask for North America, which has resulted in its being widely used worldwide. The Default Tone Mask specifies 917 out of 1155 carriers between 1.8 and 30.0 MHz, removing 238 carriers for the 30 dB deep notches for Amateur Radio Bands.

3.3.2 Amplitude Map

The Amplitude Map defines the signal level of each carrier in the Tone Mask. This provides the ability to specify nonuniform PSDs for regions such as Japan, where there is a 10 dB step change in the transmit power allowed above 15 MHz [14]. Also, since one of the values for signal level in the Amplitude Map is zero, additional notches may be added to meet changes in regulation or regulations in different regions without adversely affecting interoperability. This is due to the Tone Mask defining the encoding of Frame Control and ROBO Mode, which are encoded with significant redundancy so that the receiver can decode the information bits when some carriers are lost due to the channel or when they are not transmitted due to the Amplitude Map setting.

3.4 WINDOWED OFDM

HomePlug AV defines a time domain window that is applied to each OFDM symbol and corresponding guard interval (or cyclic extension). As shown in Figure 3.4, the windowed OFDM symbol is overlapped with the adjacent symbols to maintain constant power in time and reduce overhead. Note that a portion of the guard interval is used for the overlapped interval *RI*.

The selection of the OFDM symbol length combined with the window for the AV specification provides spectral properties for the OFDM signal that enable 30 dB deep notches (e.g., for the Amateur Radio Bands) simply by turning carriers off. Thus, spectral notches can be implemented through configuration, for example, selection of the Tone Mask and Amplitude Map. Typically, four additional carriers on each side of a notch need to be turned off to achieve a 30 dB notch depth, resulting in about 200 KHz of guard band overhead for each notch.

Figure 3.5 shows the HomePlug AV spectral mask and windowed OFDM PSD for a HomePlug AV signal with the Default Tone Mask.

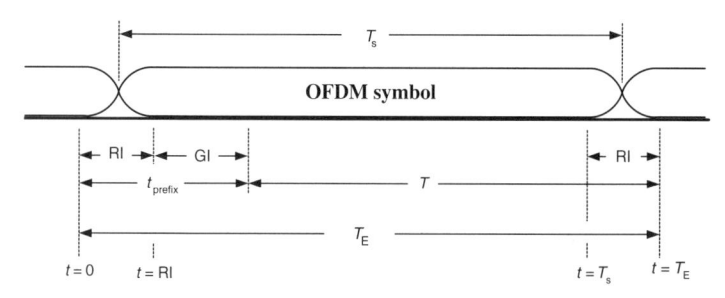

FIGURE 3.4 OFDM symbol timing.

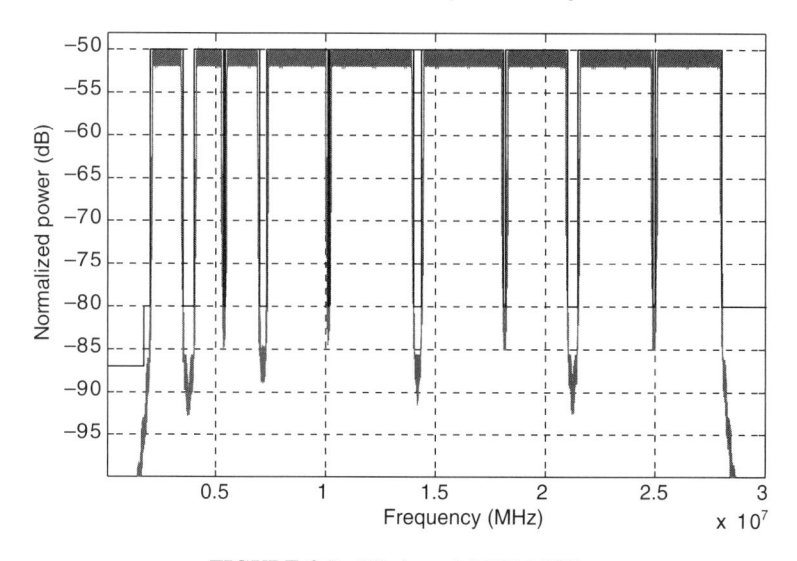

FIGURE 3.5 Windowed OFDM PSD.

3.5 TURBO CONVOLUTIONAL CODE

The forward error correction code for the HomePlug AV specification was selected based on extensive analysis and simulation on measured powerline channel impulse responses and captured noise samples. A number of codes were considered, but the final choice was between Low Density Parity Check Codes (LDPCCs) and Turbo Convolutional Codes (TCCs) because they provided 30–50% throughput gain due to the approximately 3 dB of coding gain compared with the more conventional forward error correction codes such as the concatenated code specified in HomePlug 1.0.

Based on the analysis on the Additive White Gaussian Noise (AWGN) channel, LDPCCs had a small advantage over TCCs with respect to the error floor at very low block error rates, but TCCs showed about 1 dB better performance at the lower signal-to-noise ratios (SNRs), which is important to achieve good performance on the poorest channels, where improvement is most needed. Simulation studies on powerline channels also showed that the lower error floor provided by LDPCCs provided no performance benefit on the powerline channel, most likely due to the powerline noise characteristics, which are impulse noise dominated.

The TCC specified in the AV specification is a duo-binary code that operates within 1.2 dB of capacity at 1 bit/s/Hz spectral density (Quadrature Phase Shift Keying (QPSK) with $\frac{1}{2}$ rate code). Figure 3.6 shows the block (or packet) error rate versus Eb/No performance for the TCC decoder based on a specific hardware implementation. Each trace represents different combinations of code rate and constellations from BPSK to 1024 QAM. Figure 3.7 shows the performance

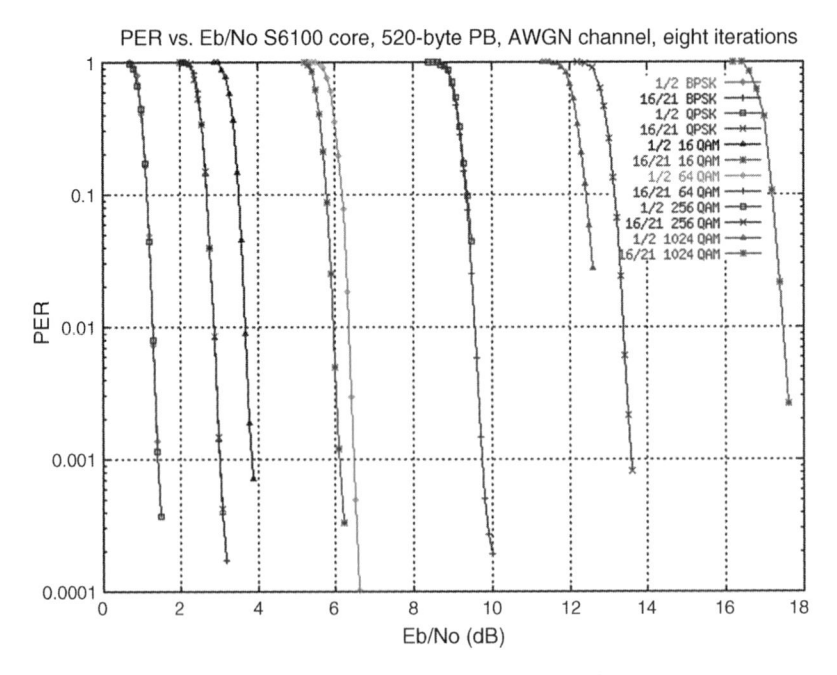

FIGURE 3.6 Turbo Convolutional Code performance.

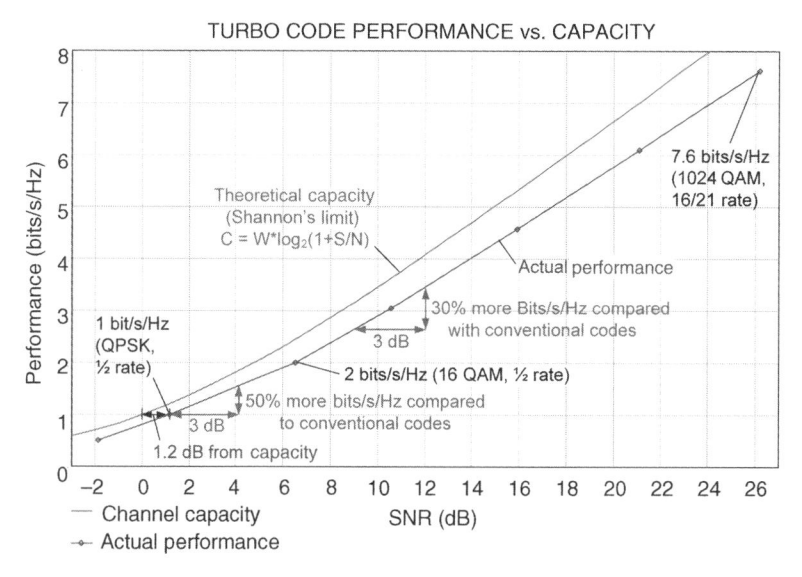

FIGURE 3.7 Turbo Convolution Code compared with capacity.

compared with capacity in terms of bits/s/Hz versus signal-to-noise ratio (SNR) at a block error rate of 0.01, which is a typical operating point on the powerline channel. This figure also shows that, for the approximately 3 dB coding gain that the TCC provides over a conventional concatenated convolutional plus Reed-Solomon code (such as was used in HomePlug 1.0.1), the throughput performance gain is quite substantial, and that this gain is higher where needed on poorer channels (e.g., 50%) than on typical channels (e.g., 30%).

3.6 CHANNEL ADAPTATION

The channel between each pair of powerline devices is unique in its attenuation versus frequency transfer function. The noise seen at each powerline device is also often unique. Furthermore, the attenuation and noise characteristics can also vary as a function of AC line cycle. Thus, the throughput capacity of each link in each direction may be significantly different. To provide maximum network throughput, one or more Tone Maps are generated for each link. Note that in this context, a link is the connection from one transmitter to another receiver. The reverse direction is a different link.

The channel adaptation process is initiated by the transmitter sending a number of Sound PHY Protocol Data Units (PPDUs) to the receiver. The receiver processes these PPDUs to estimate channel properties such as the signal-to-noise ratio of each carrier, noise characteristics, and channel delay spread. From the channel analysis the receiver performs, it selects the modulation methods that are optimized for the channel in the form of a Tone Map that is sent to the transmitter together with a Tone Map Index to uniquely identify the Tone Map from other possible Tone Maps that may be used on the same link. The Tone Map contains the information about the constellation assigned to each carrier, the code rate selected for the Turbo Convolutional Code, and the guard interval length. The transmitter will apply the selected Tone Map on future transmissions to the receiver. The Tone Map Index is communicated in the Frame Control and thus is used by the receiver to identify the Tone Map for decoding.

Once the initial Tone Map is in use, the receiver will continuously monitor the Tone Map-encoded PPDU and will send an updated Tone Map when a change in the channel is detected. In this event, the channel changes significantly for the worse such that communication fails using the existing Tone Map. The transmitter will identify this condition from the large number of failed segments indicated in the selective acknowledgement and will restart the sounding process.

One critical characteristic of the powerline channel is that the noise, both impulse and nonimpulse noise, is usually synchronous with the AC line cycle. Thus, the noise level and noise characteristics in one portion of the AC line cycle phase may be distinctly different from the noise level and noise characteristics in another. In general, the noise level is typically lowest near the zero-crossings and highest near the peaks of the 50 Hz or 60 Hz cycle. To avoid having to adapt to the worst-case noise in the worst noise level of the AC line cycle, the HomePlug AV specification supports defining multiple Tone Maps for different segments (or regions) of the AC line cycle. The number of Tone Map Regions and the start and end points of each region is selected by the receiver based on the noise and channel characteristics. The specification allows up to a maximum of seven Tone Maps for different phases of the AC line cycle.

Another key characteristic of the PLC channel is that the attenuation versus frequency transfer function, and its associated impulse response, may change synchronously with the AC line cycle. Thus, the impulse response in one portion of the AC line cycle phase may be distinctly different from the impulse response in another. The changes are usually step changes, that is, from one impulse response to another at a certain segments of the line cycle. This is due to devices or appliances conducting power during portions of the line cycle and thus presenting a load that affects the impulse response for a portion of the line cycle. For example, there may be a higher attenuation when the device or appliance is conducting power. This is not as common as noise being synchronized to the AC line cycle; however, it occurs frequently enough that it needs to be addressed to maximize the total throughput. Fortunately, the receiver can address this by specifying the appropriate Tone Map Regions, which is the same as the approach to address the case of synchronous noise. The only difference in the solution is additional mechanisms that may be needed in the receiver to detect this condition.

Dividing the line cycle into different regions with different Tone Maps for the purpose of channel adaptation provides significant performance gain. By isolating particularly noisy portions of the line cycle, the data rate in the good regions can often be 50% higher or more than in the poor regions. Even if a source transmits with equal probability at all times in the line cycle, this provides a significant performance gain. This synchronization of noise is unique to the PLC environment and is fully exploited in HomePlug AV.

3.6.1 Bit-Loading

Bit-loading is a part of the channel adaptation process. The receiver estimates the signal-to-noise ratio for each OFDM carrier and determines the highest order constellation each carrier can support. It should be noted that bit-loading in HomePlug AV is distinctly different from bit-loading in Digital Subscriber Line (DSL). In DSL, the total transmit power is limited and thus the bit-loading processing includes a water-filling algorithm to set the power level on each OFDM carrier within the total transmit power constraint to achieve the maximum throughput. Thus for DSL, bit-loading is performed using the estimated signal-to-noise ratio for each carrier after the water-filling algorithm is applied.

On the powerline medium, the regulations for the 1.705–30.0-MHz band limit the Power Spectral Density (PSD) measured in a 9-KHz bandwidth, thus the transmit power of each OFDM carrier, which is about 24.4 KHz in bandwidth in HomePlug AV, is effectively limited individually. Thus, unlike DSL, the water-filling algorithm is not used on the powerline.

3.7 BEACON PERIOD SYNCHRONIZED TO AC LINE CYCLE

The HomePlug AV MAC protocol specifies a Beacon Period that is synchronized to the AC line cycle and that is two AC line cycles in length. Careful synchronization of the Beacon Period in HomePlug AV provides a solution to a number of problems, given that the noise and impulse response can be different for different regions of the AC line cycle. The first problem solved is reliable reception of the Central Beacon, which is transmitted by the Central Coordinator and provides information necessary for the network to function properly. The Central Coordinator can select a location in the AC line cycle that provides the most reliable reception in the network. All other stations in the network indicate to the Central Coordinator (CCo) in the Frame Control of every Start-Of-Frame (SOF) whether they detected the beacon in the current Beacon Period. Based on this information, the CCo can move the Central Beacon (and Beacon Period) to another location in the AC line cycle. It is common for the CCo to select the zero-cross of the AC line cycle for the Central Beacon since the powerline channel typically exhibits the lowest noise and attenuation at this location and thus usually provides the highest reliability.

The second problem solved by Beacon Period synchronization is the timing and synchronization of the Tone Map Regions. The receiver communicates the location of each Tone Map Region in the Tone Map and it sends to the transmitter based on the beginning of the Beacon Period. Note that it is common to define a set of Tone Maps and respective Tone Map Regions for the first half line cycle interval (the first quarter of the Beacon Period) and repeat the Tone Maps and Regions for the next three half line cycle interval since the channel changes are typically identical for the positive and negative half line cycles. Figure 3.8 shows an example of this. When the two halves of the AC line cycle are not identical, unique Tone Maps and Regions may be defined for a full AC line cycle. Unique Tone Maps and Regions may also be

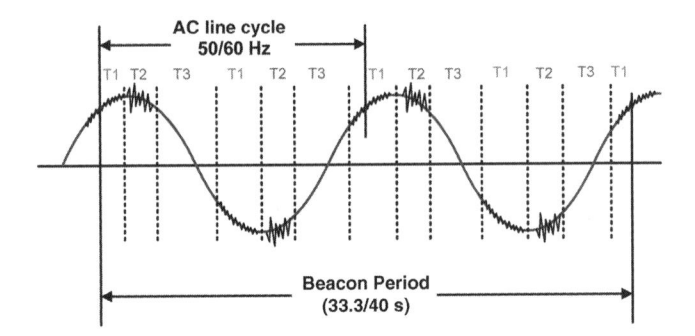

FIGURE 3.8 Beacon Period and Tone Map Regions.

specified for the entire Beacon Period. This may be useful when there are certain QoS requirements for traffic in a Carrier Sense Multiple Access (CSMA) or Time Division Multiple Access (TDMA) allocation. For example, the Tone Map could be more conservative for a particular TDMA allocation to achieve lower latency rather than achieve maximum throughput.

A third problem solved by synchronizing the Beacon Period to the AC line cycle is providing stable TDMA allocations, and this is described next.

3.7.1 AC Line Cycle Synchronization for TDMA Allocations

If TDMA allocations were not synchronized to the AC line cycle, the data rate supported in the TDMA allocation of successive Beacon Periods will likely be significantly different, making efficient scheduling of the length of the TDMA allocation difficult. Figure 3.9 shows an example of a Beacon Period with three TDMA allocations.

As mentioned in the previous section, more than one Tone Map and Tone Map Region may be specified by the receiver for the TDMA allocation such that the

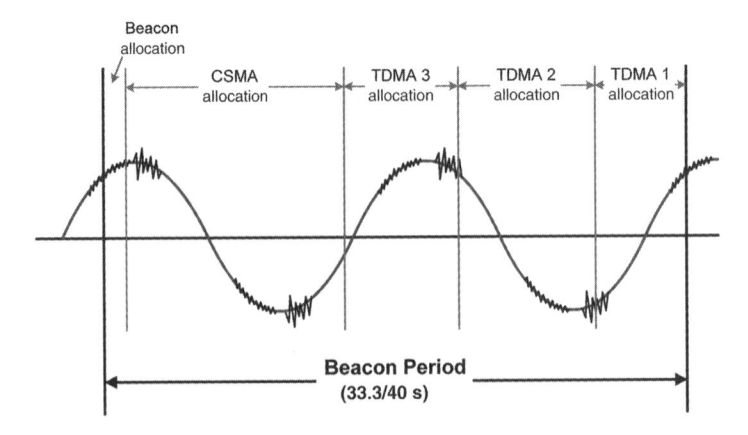

FIGURE 3.9 Beacon Period and TDMA allocations.

Tone Maps are optimized for the QoS required for the traffic in the TDMA allocation [15].

3.8 TDMA WITH PERSISTENT AND NONPERSISTENT SCHEDULES

One of the challenges in managing TDMA allocations on the powerline is reliably providing the schedule for each Beacon Period. Noise conditions on the powerline can result in decoding failure of the Central Beacon for some AV stations in the network, thus these stations will not know the schedule for this Beacon Period [16]. HomePlug AV addresses this problem by supporting persistent schedules for allocations communicated in the Central Beacon. When an allocation schedule is communicated in the Central Beacon, the number of Beacon Periods it is valid for is also communicated. For example, if the persistence is set to 2, then the start time and duration of the allocation will be valid for the current Beacon Period plus the following 2. Thus, if the transmitter and/or the receiver that are using this allocation fail to receive the next two Central Beacons, they will still be able to use the allocation. When the persistence is set to 0, then the start time and duration of the allocation is only valid for the current Beacon Period. This is called nonpersistent allocation.

Another challenge in managing TDMA allocations is changing channel conditions. The powerline channel can change when a device or appliance is turned on or off, or when a new device or appliance is plugged into the powerline. A further challenge is that the data rate for the TDMA allocation may be variable. Both of these create a challenge in providing a schedule for the allocation, while maintaining high efficiency. The goal is to provide an allocation that is the minimum necessary to meet the QoS requirements for the traffic in this allocation. Much of the allocation would be unused if the allocation were fixed to the duration necessary to accommodate the worst-case channel conditions together with the peak traffic data rate.

To optimize the efficiency of a TDMA allocation, fields were defined in the Start of Frame (SOF) delimiters of PPDUs transmitted in a TDMA allocation that communicate to the CCo the data rate of the channel based on the tone maps derated by the PHY Block (PB) error rate and the number of PHY Block pending in the queue for the allocation. This information, combined with the Central Beacon reliability information also communicated in the SOF and the QoS requirements for the traffic communicated to the CCo during the TDMA connection setup, provides the CCo the information needed to precisely schedule the TDMA allocation. A typical approach by the CCo would be to provide a persistent allocation to meet most of the bandwidth requirement for the traffic, as well as an additional nonpersistent allocation occasionally when the PHY Block pendings in the queue become high. The CCo would typically change the size of the persistent allocation based on the changes to the data rate supported in the allocation. Additionally, the transmitter may optionally send some of the traffic for the allocation in the CSMA allocation when the PHY Block pendings become high, if the allocations provided by the CCo are occasionally insufficient to support the traffic.

3.9 DATA PLANE: TWO-LEVEL FRAMING, SEGMENTATION, AND REASSEMBLY

Impulse noise on the powerline channel creates a challenge for reliable communication. One mechanism in HomePlug AV that helps address this problem is matching the segmentation process to the Turbo Convolutional-encoded blocks in the PPDU. This results in a one-to-one relationship between decoding failures on the channel and the data that need to be retransmitted. To maintain high efficiency and minimize padding, a two-level framing and segmentation mechanism is used. MAC Service Data Units (MSDUs) from the host have a header and a cyclic redundancy check added to create a Medium Access Control (MAC) Frame. MAC Frames are concatenated together into a MAC Frame Queue. This is the first level of framing. The MAC Frame Queue is then segmented into 512-byte blocks. Each block is encrypted and a header and cyclic redundancy check is added to generate a 520-byte PHY Block (PB). PBs are individually encoded with the Turbo Convolutional encoder and one or more encoded PBs are transmitted in a PPDU. Typically, a large number of PHY Blocks are transmitted in a single PPDU. This is the second level of framing.

The acknowledgment transmitted by the receiver indicates the success or failure of each PB in the PPDU, using a selective acknowledgment scheme. Thus, only the PBs that fail Turbo Convolutional Decoding need to be retransmitted. The 512-byte size of the segment and its corresponding 520-byte PHY Block was selected to optimize the overall throughput, Turbo Convolutional Code performance, implementation complexity, latency, and overhead requirements.

This two-level framing approach gives the implementer of a HomePlug AV product the ability to optimize the Tone Map for throughput. When a large impulse noise is present, such as from a triac dimmer, higher throughput is usually achieved by generating a Tone Map for noise other than impulse noise and retransmitting the PBs that fail decoding due to the impulse noise. This enables the majority of the PBs to be transmitted at a much higher Physical layer (PHY) rate, yielding a significantly higher overall throughput considering the retransmission, than would be the case if the Tone Map had been generated to decode successfully with the impulse noise present. Interestingly, adapting the Tone Map to a higher data rate when impulse noise is present may actually reduce the PHY Block error rate. The reason for this is that at the higher data rate, a PHY Block spans a smaller amount of time on the channel and therefore has a lower probability of being affected by the impulse noise.

3.10 PHY CLOCK SYNCHRONIZATION

The OFDM symbol length and its corresponding carrier spacing results in a requirement that the clock accuracy between a transmitter and receiver be within a few parts-per-million (ppm) to avoid significant intercarrier interference between adjacent OFDM carriers. To meet the cost requirements for HomePlug AV products, the HomePlug AV specification allows the clock crystals to have a clock tolerance of ±25 ppm, resulting in possible clock error of 50 ppm. To address this problem, among others, a Network Time Base is generated by the CCo and communicated in the Central Beacon. All other stations in the network track the Network Time Base and adjust their transmit and receive clocks to match with the Network Time Base frequency. This provides a means to keep the clock accuracy for transmitting and receiving PPDU signals between any stations in the network less than 1 ppm using a low-cost crystal. The clock adjustment is typically implemented using digital resampling of the transmit and receive signals.

3.11 SUMMARY

This chapter examined the overall philosophy and reasoning that guided the technology selection in overcoming the communications challenges over the noisy powerline channel. Details were given about the frequency band selection and about the selection of windowed FFT-based OFDM with some commentary on comparisons with other possible technologies. Justification was presented for the use of Turbo Convolution Codes (TCCs) again with observations about the performance of rival approaches and providing the guiding principles in the selections made. This chapter also provided a discussion of the intelligent channel adaptation schemes used in HomePlug AV, including the bit-loading adaptive modulation methodology adopted, the exploitation of the cyclostationary noise behavior in the AC line cycle-based adaptation and beacon synchronization especially with a focus on enabling TDMA allocations in HomePlug AV. This chapter also discussed how the two-level segmentation and reassembly scheme used in HomePlug AV yields higher overall efficiency.

Equipped with these philosophical guiding principles, Chapter 4 now presents a more in-depth view of the HomePlug AV Physical layer (PHY) protocol.

4

PHYSICAL LAYER

4.1 INTRODUCTION

The HomePlug AV Physical layer is designed to operate as close as possible to the channel capacity of the powerline channel and is based on bit-loaded Orthogonal Frequency Division Multiplexing (OFDM) modulation. Each carrier can be coherently modulated at up to 10 coded bits/carrier. The length of the Guard Interval (GI) is also adaptable. Channel adaptation in HomePlug AV is synchronized to the AC line cycle, which is divided into multiple Tone Map (TM) Regions. Each Tone Map Region for a link uses a GI and a Tone Map tailored to this link and to the noise characteristics of this part of the AC line cycle for this receiver.

The forward error correction (FEC) used in HomePlug AV is Turbo Convolutional Coding (TCC), which is widely known to provide performance close to the theoretical channel throughput limits with moderate complexity. The standard block size for payload transmission is 520 information bytes. A larger block could have provided a theoretically higher coding gain, but other variables, such as decoding latency and overhead when the number of bits to be transported is small, favor a smaller size. Impulse noise is handled in HomePlug AV by a combination of channel adaptation and efficient retransmission of uncorrectable FEC blocks using Automatic Repeat ReQuest (ARQ) at the Medium Access Control (MAC) level. In general, the channel adaptation mechanism is used to reduce the Physical (PHY) layer errors to a range where they can be effectively corrected by the MAC layer. Impulse noise channels will produce temporal effects in channel capacity. Between

---

HomePlug AV and IEEE 1901: A Handbook for PLC Designers and Users, First Edition.
Haniph A. Latchman, Srinivas Katar, Larry Yonge, and Sherman Gavette.
© 2013 by The Institute of Electrical and Electronics Engineers, Inc. Published 2013 by John Wiley & Sons, Inc.

INTRODUCTION

two impulse noise events, the channel capacity can be very high, while during the impulse noise the capacity is very low. To obtain a low FEC block error rate, the channel has to be adapted close to the lower end (i.e., capacity near impulse noise events) of its capacity and hence is undesirable. A better approach is to adapt aggressively and use the MAC ARQ mechanism to correct errors. HomePlug AV uses selective retransmission of FEC blocks received in error. This approach enables a high overall throughput even under high FEC block error rates (up to 20% by design).

4.1.1 Transceiver Block Diagram

A block diagram of a HomePlug AV transceiver is shown in Figure 4.1, based on a 75-MHz sampling clock. On the transmitter side, the PHY layer receives its inputs from the Medium Access Control (MAC) layer. Three separate processing chains are shown because of the different error correction codings required for HomePlug 1.0.1 control information, HomePlug AV control information, and HomePlug AV packet body data or payload. The control information carries information such as destination, source addresses, and frame length, while the payload contains the actual information to be transported. AV control information is processed by the AV Frame Control FEC Encoder block, which uses a Turbo Convolutional Code (TCC) and diversity copier, while the HomePlug AV data stream passes through a scrambler, a TCC encoder, and a

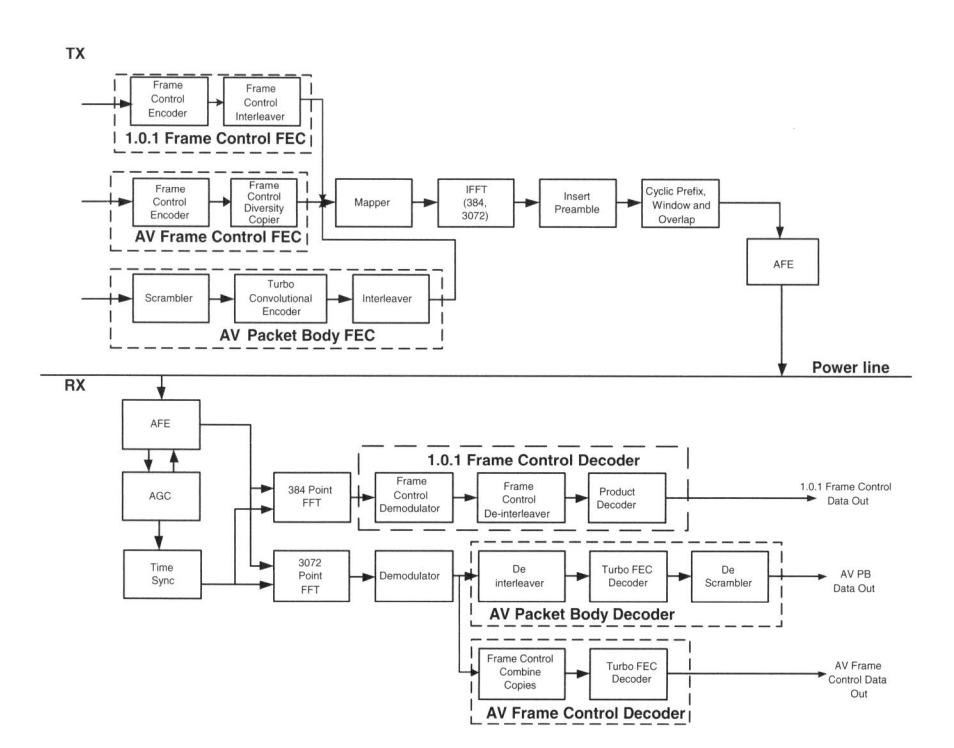

FIGURE 4.1 HomePlug AV transceiver block diagram.

channel interleaver. The HomePlug 1.0.1 Frame Control (FC) information passes through a separate HomePlug 1.0.1 FEC unit, which is based on a turbo product code. The outputs of the three FEC Encoders lead to a common OFDM modulation structure. This consists of a mapper, an Inverse Fast Fourier Transform (IFFT) processor, preamble, and cyclic prefix insertion, and windowed overlapping. The output feeds the Analog Front End (AFE) module that couples the signal onto the powerline.

At the receiver, the AFE operates with an Automatic Gain Controller (AGC) and a time synchronization module to feed separate control and data information recovery circuits. Assuming the digitized waveform out of the AFE is sampled at 75 MS/s, the Frame Control is recovered by processing the received sampled stream through a 384-point fast Fourier Transform (FFT) (for HomePlug 1.0.1 delimiters) and a 3072-point FFT (for HomePlug AV). It then passes through separate Frame Control Demodulators and Frame Control Decoders for each mode. The sampled packet body or payload stream contains only HomePlug AV-formatted symbols. It is processed through a 3072-point FFT, a demodulator with SNR estimation, a channel deinterleaver, a turbo FEC decoder, and a descrambler to recover the transmitted information.

4.2 PPDU

The term "PHY Protocol Data Unit" (PPDU) refers to the physical entity that is transmitted over the powerline. PPDUs are generated by the PHY for transmission on the powerline at the PHY interface.

The MAC layer provides the physical Layer (PHY) with a MAC Protocol Data Unit (MPDU) to encode and transmit. The MPDU consists of a 128-bit Frame Control (FC), possibly followed by one or more payload PHY Blocks (PBs). At the PHY, the FC is encoded in a single OFDM symbol and is prepended with a preamble, with which the FC forms a delimiter. In networks containing the existing 1.0.1 nodes, a 1.0.1-compatible Frame Control may be inserted between the preamble and AV Frame Control. The PBs form the PHY payload. Each PB is encoded to generate an FEC block. A number of OFDM symbols sufficient to transmit the FEC blocks are encoded after the delimiter; these symbols constitute the PHY body. The delimiter and the PHY body (if present) form the PHY Protocol Data Unit (PPDU) as shown in Figure 4.2. All samples in this section are based on 75-MHz sampling clock. Additional details of the encoding process from MPDU to PPDU is shown in Figure 4.3.

There are two PPDU types: long and short. Short PPDUs consist solely of a delimiter (preamble and FC), whereas long PPDUs include a PHY body. The latter

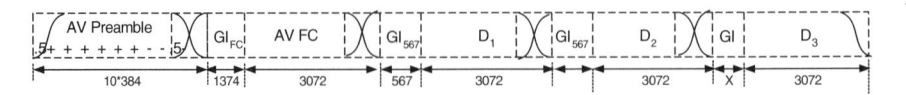

FIGURE 4.2 AV PPDU structure.

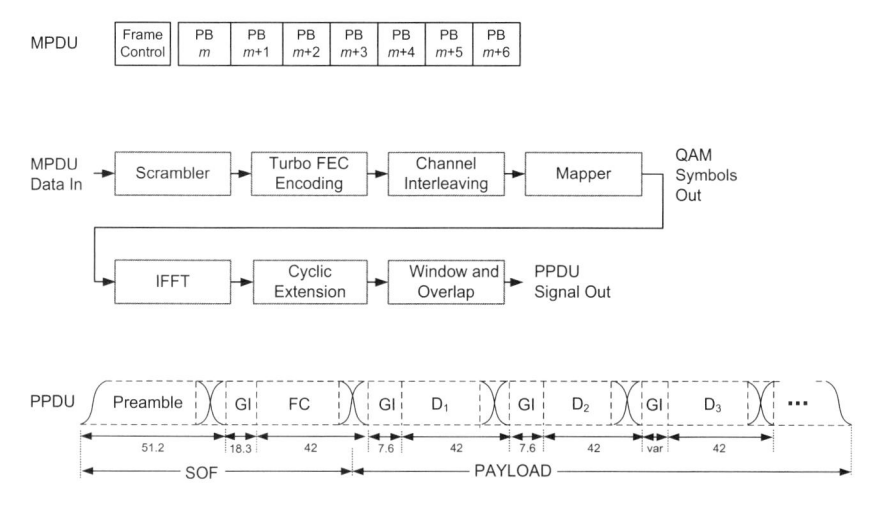

FIGURE 4.3 MPDU to PPDU encoding.

are used for data transmissions, whereas the former are used for control purposes (e.g., MAC-level acknowledgments).

4.2.1 PPDU Formats

The HomePlug AV specification supports four PPDU formats as shown in Table 4.1 that match their corresponding MAC Protocol Data Unit (MPDU) formats defined in Section 5.3.2.

The PPDU payload may be encoded using one of the following formats:

- one or more 520-octet FEC blocks modulated using negotiated Tone Maps (TMs),
- one 520-octet FEC block modulated using Standard ROBO Modulation,
- one, two, or three 520-octet FEC blocks using High-Speed ROBO Modulation,

TABLE 4.1 PPDU Formats

PPDU Format	Preamble Type	HP1.0.1 Frame Control (FC)	HPAV Frame Control (FC)	Payload
AV-Only Long PPDU	HPAV	No	Yes	Yes
Hybrid Long PPDU	Hybrid	Yes	Yes	Yes
AV-Only Short PPDU	HPAV	No	Yes	No
Hybrid Short PPDU	Hybrid	Yes	Yes	No

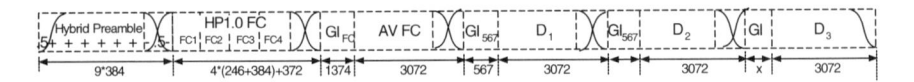

FIGURE 4.4 Hybrid Mode PPDU structure.

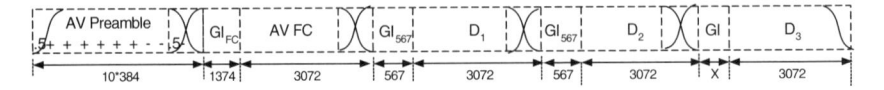

FIGURE 4.5 AV Mode PPDU structure.

- one 136-octet FEC block modulated using negotiated TMs,
- one 136-octet FEC block modulated using Mini-ROBO Modulation.

4.2.2 PPDU Structure

There are two types of PPDU structures, depending on whether the PHY is operating in HomePlug 1.0.1-compatible Mode or AV-Only Mode. In the following paragraphs, these modes are referred to as Hybrid and AV Modes, respectively.

When the PHY operates in Hybrid Mode, the PPDU structure consists of a HomePlug 1.0.1-compatible Preamble followed by a 1.0.1-compatible FC, AV FC, and optionally a PPDU payload, as shown in Figure 4.4. When the PHY operates in AV Mode, the PPDU structure consists of an AV Preamble, an AV FC, and optionally a PPDU payload, as shown in Figure 4.5. The combination of a preamble, HomePlug 1.0.1 Frame Control (if any), and HomePlug AV Frame Control is referred to as a "Delimiter."

For both Hybrid and AV Modes, the first and second payload symbols (D1 and D2) have a fixed guard interval of 567 samples. Starting with the third payload symbol (D3), one of the three payload symbol guard intervals listed in Table 4.2 may be used for the remaining PPDU. The guard interval length is specified in the Tone Map used for the PPDU.

4.2.3 Symbol Timing

The OFDM time domain signal is based on a 75-MHz sampling clock and is generated as follows. For AV Frame Control and payload symbols, a set of data points from the mapping block is modulated onto the subcarrier waveforms using a 3072-point IFFT, resulting in 3072 time samples (IFFT interval). A fixed number of samples from the end of the IFFT are taken and inserted as a cyclic prefix at the front of the IFFT interval to create an extended OFDM symbol. Figure 4.6 shows the OFDM symbol timing, with interval durations identified in Table 4.2.

TABLE 4.2 OFDM Symbol Characteristics

Symbol	Description	Time Samples	Time (μs)
T	IFFT interval	3072	40.96
t_{prefix}	Cyclic prefix interval	RI + GI	4.96 + GI
T_E	Extended symbol interval (T + t_{prefix})	$T + t_{prefix}$	45.92 + GI
RI	Rolloff Interval	372	4.96
T_S	Symbol Period	3072 + GI	40.96 + GI
GI_{FC}	Frame Control Guard Interval	1374	18.32
GI	Payload symbol guard interval, generically	417, 567, 3534	5.56, 7.56, 47.12
GI_{SR}	STD-ROBO_AV Payload Symbol(s) Guard Interval	417	5.56
GI_{HR}	HS-ROBO_AV Payload Symbol(s) Guard Interval	417	5.56
GI_{MR}	MINI-ROBO_AV Payload Symbol(s) Guard Interval	567	7.56
GI_{417}	Guard Interval, length = 417 samples	417	5.56
GI_{567}	Guard Interval, length = 567 samples	567	7.56
GI_{3534}	Guard Interval, length = 3534 samples	3534	47.12

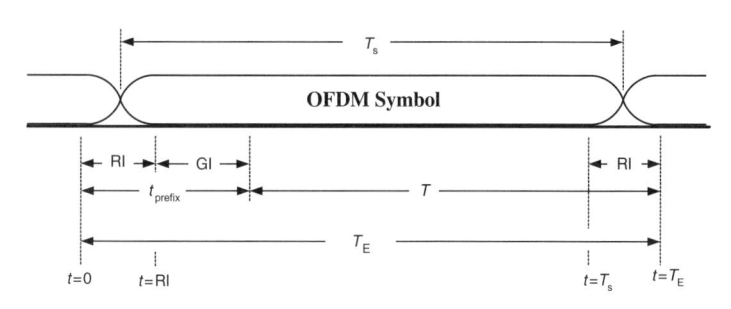

FIGURE 4.6 OFDM symbol timing.

4.3 PREAMBLE

The AV Preamble structure was picked such that it can be robustly detected by both 1.0.1 and AV nodes. The same phases are transmitted on the 76 standard HomePlug 1.0.1 carriers, though extra carriers have been added in the AV Preamble to reflect AV's use of extended bandwidth (1.8–30 MHz) versus 1.0.1 (4.5–20.7 MHz). As these extra carriers are orthogonal, they have no impact on the preamble detection performance of 1.0.1 nodes. The preamble's time domain waveform consists of a repeating pattern of 7.5 "plus synchronization" (SYNCP) symbols followed by 1.5 (effective) "minus synchronization" (SYNCM) symbols. The SYNCP symbols are "chirps" or frequency sweeps, spanning the used bandwidth (though excluding

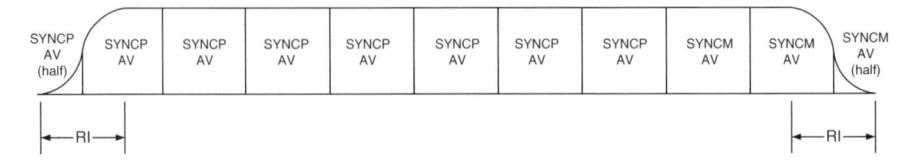

```
| SYNCP | SYNCP | SYNCP | SYNCP | SYNCP | SYNCP | SYNCP | SYNCP | SYNCM | SYNCM | SYNCM |
| AV    | AV    | AV    | AV    | AV    | AV    | AV    | AV    | AV    | AV    | AV    |
| (half)|       |       |       |       |       |       |       |       |       | (half)|
```

FIGURE 4.7 AV Preamble waveform.

masked carriers), and are 5.12 μs each in duration. The SYNCM symbol is the same as the SYNCP symbol, except it is phase shifted by 180°. The collection of SYNCP and SYNCM symbols provides a repeating pattern, as well as a transition point that can be detected robustly using a simple correlation technique for synchronization. It also provides a training sequence for channel estimation and equalization, as the preamble is an a priori known signal. A receiver may, for example, average SYNCP symbols (reducing noise) to estimate the channel state information (CSI) for equalization of the following AV Frame Control symbol. Similar to the AV Frame Control and PHY body OFDM symbols, described in the later sections, the basic preamble pattern may also be generated with the 3072-IFFT. The IFFT output can then be extended, and shaped so that it also meets the spectral mask requirements without additional filtering. Figure 4.7 details the components of the extended AV Preamble.

The parameter RI refers to Rolloff Interval, which is 4.96 μs and is the portion of the time domain waveform that is windowed. The last RI samples of the preamble are shaped and overlapped with the first RI samples of the FC. When the extended preamble is created with the 3072-IFFT, the SYNCP and SYNCM symbols are not the exact replicas of each other, but similar enough so as to have essentially no effect on 1.0.1 or AV synchronization performance.

4.4 FRAME CONTROL

The AV Frame Control field consists of 128 information bits that are encoded and modulated over one OFDM symbol. The AV Frame Control symbol has an IFFT interval of 40.96 μs and an effective (nonoverlapped) guard interval (GI) of 18.32 μs. The entire Frame Control signal is transmitted at 0.8 dB higher power than the payload to increase robustness and improve the channel state information estimate. As the Frame Control duty cycle is relatively low in typical traffic, this extra power boost does not result in any measurable effect on radiated emissions.

The 128 information bits are encoded using the $\frac{1}{2}$ rate Turbo Convolutional Code. The 256 encoded bits are interleaved; then passed through an outer repetition code that copies each bit as many times as possible, depending on the number of enabled carriers in the Tone Mask.

In contrast to the HomePlug 1.0.1 Frame Control's coherent BPSK modulation, AV uses coherent Quadrature Pulse Shift Keying (QPSK) modulation to provide twice the number of copies per encoded bit. While in an Additive White Gaussian

PAYLOAD 39

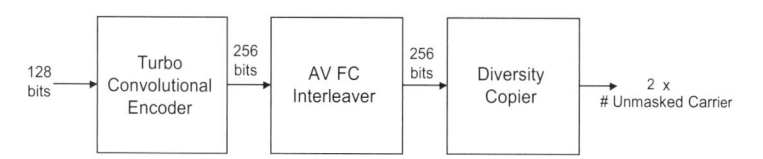

```
128 bits → [Turbo Convolutional Encoder] →256 bits→ [AV FC Interleaver] →256 bits→ [Diversity Copier] → 2 x # Unmasked Carrier
```

FIGURE 4.8 AV Frame Control FEC Data Path.

Noise (AWGN) channel, sending x bits with BPSK versus sending $2x$ bits with QPSK is equivalent in terms of coding gain (assuming equal gain combining); the greater number of copies provides more diversity gain in the fading channel typical of the powerline medium. For the standard 917 carriers of the North American Tone Mask, each encoded bit is copied approximatedly 7.2 times in the outer code. The bit copies are mapped onto the frame control symbol with maximum spreading for time and frequency diversity. As there are 256 bits before copying, the bit copies are transmitted consecutively in frequency (with wrapping) with a bit index offset of 128 between the in-phase (I) and quadrature (Q) channels. The entire AV Frame Control FEC data path is shown in Figure 4.8.

For decoding, the channel state information may be estimated from the known preamble pattern.

The FC contains information that is needed by both the PHY and by the MAC. PHY-related content consists of delimiter type, length of PHY body and Tone Map Identifier (TMI). Delimiter type is needed to determine the presence of payload symbols and the length, and TMI are required to demodulate the PHY body symbols, if present. The TMI is an index to the Tone Map for the transmitter used to modulate the OFDM symbols of the PHY body. It is chosen by the receiver during channel adaptation and is sent along with the Tone Map to the transmitter. The PHY payload length indicates to the PHY how many symbols to demodulate. The FC is also used to estimate the channel state information to decode the payload from the known FC information (after successful FC decoding).

4.5 PAYLOAD

For payload symbols, the information bits from the PHY Blocks are first passed through a scrambler. The scrambler serves to "whiten" the data, resulting in a fairly even distribution of 0s and 1s at the output, so as not to cause significant variations in average power when mapped. The scrambled bits are then passed to the Turbo Convolutional Code encoder. The output of the encoder is subsequently interleaved, so that any channels errors will be spread relatively uniformly through the encoded PHY Block during decoding.

The OFDM time domain signal, based on a 75-MHz system clock, is then created as follows. The (interleaved) bits out of the TCC encoder are passed through the mapper resulting in baseband constellation symbols, which are then modulated onto the carrier waveforms using a 3072-point IFFT. The result is 3072 time samples (referred to as the IFFT interval). A fixed number of samples from the end of the IFFT output are copied to the front of the IFFT interval as a cyclic prefix, to create an extended OFDM symbol. The first and last 372 samples are then pulse shaped for the Rolloff Interval and finally, the first 372 samples are overlapped with the last 372 samples of the preceding OFDM symbol or preamble. Similarly, the last 372 samples will be overlapped with the following OFDM symbol.

Note that three guard interval (GI) lengths are available for use during the PHY body. GI_{417} and GI_{567} provide two "standard" guard intervals of 5.56 and 7.56 μs, respectively. The smaller GI may be selected on channels with shorter delay spread to reduce GI overhead. GI_{3534} is 47.12 μs in length and is included as a special case to handle particularly harsh channels with severe impulse noise. GI_{3534} provides two full copies of the OFDM symbol that even with the higher overhead can result in higher throughput. By detecting impulse noise, the receiver can intelligently combine segments from the two copies and generate an OFDM symbol for decoding with most or all the impulse noise removed. Better bit-loading and higher throughput can be achieved than with the shorter GIs on channels with large impulse noise with a high repetition rate, such as from some switching power supplies.

The PHY payload uses a 520-byte or a 136-byte PHY Block information frame size (PB520 or PB136, respectively), and may be encoded at either rate $1/2$ or rate 16/21. The PHY Blocks of a PPDU are concatenated for modulation and a pseudo-random pattern is used to generate pad bits to fill any unused portions of the last OFDM symbol. Rate $1/2$ typically provides better throughput performance on relatively poor channels, whereas rate 16/21 typically provides better throughput performance on relatively good channels.

4.5.1 Scrambler

The data scrambler block helps give the data a random distribution. The data stream is "XOR-ed" with a repeating Pseudo Noise (PN) sequence using the following generator polynomial (Figure 4.9):

$$S(x) = x^{10} + x^3 + 1$$

The bits in the scrambler are initialized to all ones at the start of processing each MPDU.

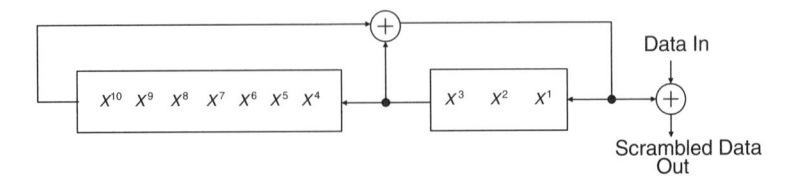

FIGURE 4.9 Data scrambler.

PAYLOAD 41

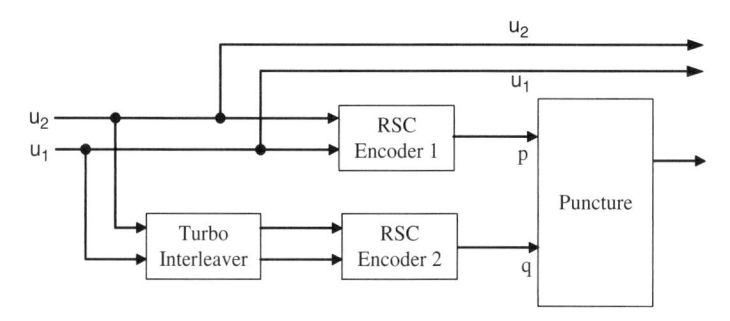

FIGURE 4.10 AV Turbo Convolutional Encoder.

4.5.2 Turbo Convolutional Encoder

In contrast to HomePlug 1.0.1's concatenated code (variable, high rate Reed-Solomon outer code and $1/2$ or $3/4$ rate convolutional inner code), AV uses a state-of-the-art turbo convolutional code TCC. This provides an additional coding gain of about 2.5 dB versus the FEC of HomePlug 1.0.1, and thus significantly increases channel throughput. The TCC is also used for the AV Frame Control, providing an additional coding gain of about 1 dB over HomePlug 1.0.1's turbo product code.

The TCC encoder consists of two rate 2/3, 8-state recursive systematic convolutional (RSC) constituent codes and one turbo interleaver as shown in Figure 4.10. The TCC supports sizes of 520, 136, and 16 bytes. The encoded output is then punctured to a code rate of either $1/2$ or 16/21.

4.5.2.1 Constituent Encoders Figure 4.11 shows the 8-state encoder used for Encoder 1 and Encoder 2.

The first bit of the PB is mapped to **u1**, the second to **u2**, and so on. Only output **x0** is passed to the puncture circuit.

4.5.2.2 Termination Tail-bitten termination is used in each constituent encoder. Two passes of each encoder are required. First, the encoder is initialized to the all-zeros state $S_0 = [s1\ s2\ s3] = [0\ 0\ 0]$, and then the entire FEC block is passed through

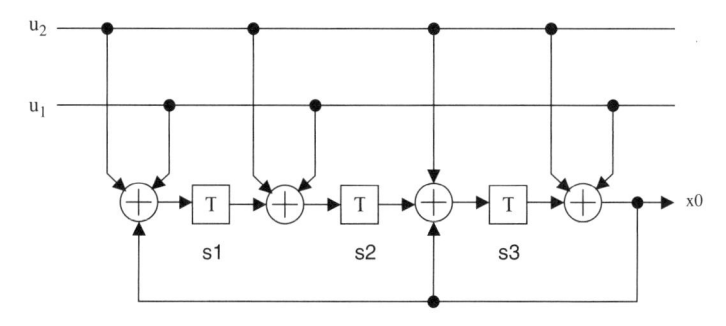

FIGURE 4.11 8-State Constituent Encoder.

TABLE 4.3 Rate ½ Puncture Pattern

p	1111111111111111
q	1111111111111111

TABLE 4.4 Rate 16/21 Puncture Pattern

p	1001001001001000
q	1001001001001000

TABLE 4.5 Interleaver Parameters

PB Size (Octets)	N Value	M Value	Interleaver Length, L
16	8	8	64
136	34	16	544
520	40	52	2080

the encoder (the outputs are unused). The final state S_N is used to determine the starting state for the second pass (i.e., $S'_0 = F[S_N]$, where the prime (′) indicates states of the second pass). The function F[.] is chosen so that the final state S'_N will equal the initial state S'_0 at the conclusion of the second pass. The entire FEC block is passed through the encoder for the second time, this time outputting to the puncture circuit.

4.5.2.3 Puncturing The u1, u2 systematic bits are never punctured. They are written to the data output buffer in natural order.

The p and q parity bits from Encoder 1 and Encoder 2, respectively, are punctured and written to the parity output buffer in natural order.

Table 4.3 shows the puncture pattern for rate ½ (i.e., no puncturing).

Table 4.4 shows the puncture pattern for rate 16/21.

4.5.2.4 Turbo Interleaving A Turbo Interleaver is required to interleave the original sequence for the second constituent code. The Turbo Interleaver interleaves the PB in dibits, not bits, thus keeping the original bit pairs together. As a result, the Interleaver length (that is, the length of the Interleaver input and output sequences) equals the FEC dibit block size. The Turbo Interleaver output may be generated algorithmically, although a seed table is required for each FEC block size supported.

Table 4.5 lists the parameters used for each FEC block size, with its corresponding seed tables provided in Tables 4.6–4.8. It is the seed table S, the corresponding seed table length **N**, and the interleaver length **L** that specify each interleaver mapping I(.). The Turbo Interleaver mapping is defined through the following interleaver equation:

$$I(x) = [S(x \bmod N) - (x \, div \, N) * N + L] \bmod L \text{ for } x = 0, 1, \ldots, (L-1),$$

where *div* is an integer division operation and *mod* is the modulo operation.

PAYLOAD

TABLE 4.6 Interleaver Seed Table for FEC Block Size of 16 Octets

x	0	1	2	3	4	5	6	7
S(x)	54	23	61	12	35	2	40	25

TABLE 4.7 Interleaver Seed Table for FEC Block Size of 136 Octets

x	0	1	2	3	4	5	6	7
S(x)	369	235	338	436	169	200	397	59
x	8	9	10	11	12	13	14	15
S(x)	298	20	265	429	294	466	16	48
x	16	17	18	19	20	21	22	23
S(x)	525	461	187	86	216	387	41	142
x	24	25	26	27	28	29	30	31
S(x)	247	314	79	486	512	103	476	345
x	32	33						
S(x)	4	105						

The interleaver mapping **I(x)** is then used to interleave the bits as shown below.

Note: When the output index **x** is even, the corresponding interleaved information bit pairs (bit 0 and bit 1 of the pair) are swapped.

$$\begin{aligned}
&\text{if } x \bmod 2 == 0 \\
&\qquad IntData(2 \cdot x) = Data(2 \cdot I(x) + 1) \\
&\qquad IntData(2 \cdot x + 1) = Data(2 \cdot I(x)) \\
&\text{if } x \bmod 2 == 1 \\
&\qquad IntData(2 \cdot x) = Data(2 \cdot I(x)) \\
&\qquad IntData(2 \cdot x + 1) = Data(2 \cdot I(x) + 1)
\end{aligned} \right\} \text{for } x = 0, 1, 2, \ldots, L-1$$

where **Data** and **IntData** refer to the pre and postinterleaver bit sequences, respectively.

TABLE 4.8 Interleaver Seed Table for FEC Block Size of 520 Octets

x	0	1	2	3	4	5	6	7
S(x)	1558	1239	315	1114	437	956	871	790
x	8	9	10	11	12	13	14	15
S(x)	833	1152	147	506	589	388	1584	265
x	16	17	18	19	20	21	22	23
S(x)	981	220	1183	102	1258	1019	1296	737
x	24	25	26	27	28	29	30	31
S(x)	694	1495	612	453	1049	1450	531	47
x	32	33	34	35	36	37	38	39
S(x)	1368	645	166	322	1323	1404	0	881

4.5.3 Channel Interleaver

The natural bit order at the encoder output starts with all the data bits in the same order as at the input to the encoder, followed by all the parity bits in the same order as generated by the encoder. The two parity bits generated by the encoder (**p** and **q** from Figure 4.10) are interlaced, with the **p** bit coming first. An entire turbo encoded PHY Block is interleaved by the Channel Interleaver before mapping.

To facilitate parallel turbo decoding by as many as four decoder engines, the interleaver was designed to allow simultaneous reads by the turbo decoders from four sub-banks of the deinterleaver buffer.

In the following example, **k** represents the number of information bits and **(n–k)** represents the number of parity bits. The information bits are divided into four equal sub-blocks of **k/4** bits, and the parity bits are divided into four equal sub-blocks of **(n–k)/4** bits. For both PB520 and PB136, for both code rates (1/2 and 16/21), and for the FC case, the number of both information bits and parity bits is divisible by 4. In the encoder, the output buffer is split into four information sub-banks of **k/4** bits and four parity sub-banks of **(n–k)/4** bits. The encoder writes the first **k/4** information bits (in natural order) to the first information sub-bank, the next **k/4** bits to the second information sub-bank, and so on. Then it writes the first **(n–k)/4** parity bits (in natural order) to the first parity sub-bank, the next **(n–k)/4** bits to the second parity sub-bank, and so on.

Each of the information sub-blocks is interleaved in an identical manner by the nature in which the bits are read out of the sub-banks. The four information sub-banks of length **k/4** may be thought of as one matrix consisting of **k/4** rows and four columns, with column 0 representing the first sub-bank, column 1 representing the second sub-bank, and so on. Groups of four bits on the same row (one bit from each sub-block) are read out from the matrix at a time, starting with row 0. After a row has been read out, the row pointer is incremented by StepSize before performing the next row read (**0, StepSize, 2∗ StepSize, . . . , [k/4]–StepSize**). After **(k/4)/ StepSize** row reads, the end of the matrix has been reached. The row pointer is then initialized to 1 (rather than starting at 0) and the process is repeated—reading rows, incrementing the row pointer by StepSize, and incrementing the starting row by 1 after **(k/4)/StepSize** row reads. For example the second pass will read rows (**1, 1+StepSize, 1+2∗StepSize, . . . , [k/4]–StepSize+1**) and the third pass will read rows (**2, 2+StepSize, 2+2∗StepSize, . . . , [k/4]–StepSize+2**). As a result, StepSize passes of **(k/4)/StepSize** row reads are required to read all bits from the matrix.

The parity bits for rate $1/2$ are interleaved similarly to the information sub-blocks. However, instead of reading out of the **([n–k]/4)x4** parity matrix starting with row 0, the parity reads begin at a predetermined offset and wrap around when the end of the matrix is reached until arriving back at the starting row. Let **t** represent the sub-bank length for the parity sub-banks, **t = [n–k]/4**. The first rows that are read are (**offset, (offset+ StepSize) mod t, . . . , [(offset+t–StepSize) mod t]**). Then the starting row pointer is incremented by 1 and the process is repeated **StepSize–1** more times for a total of StepSize passes of **((n–k)/4)/StepSize** row reads, thus reading rows

TABLE 4.9 Channel Interleaver Parameters

PB Size	Code Rate	Offset (Parity)	Step Size
16 Octets	$1/2$	16	4
520 Octets	$1/2$	520	16
136 Octets	$1/2$	136	16
520 Octets	16/21	170	16
136 Octets	16/21	40	8

(**offset+1**, (**offset+1+StepSize**) **mod t**, ..., [(**offset+1+t−StepSize**) **mod t**]) during the second pass, for example.

The parity bits for rate 16/21 are interleaved similarly to the rate $1/2$ parity. The reading out of the parity matrix starts with an offset and wraps, as for rate $1/2$ parity, except the row pointer is not re-initialized for each of the successive StepSize−1 passes. Again, let **t** represent the sub-bank length for the parity sub-banks, where **t = [n−k]/4**. After each row read, StepSize is added to the row pointer and a modulo-t is performed (**offset**, [**offset+StepSize**] **mod t**, [**offset+2∗StepSize**] **mod t**, ...). This process continues without row pointer reinitialization, until all **n−k** parity bits have been read.

Table 4.9 lists the parity offsets used for each PB size and code rate combination. Interleaved bits are read from each of the sub-banks in turn.

- For rate 1/2, the first four bits to be transmitted on the channel are information bits, the next four bits are parity bits, and so on.
- For rate 16/21, the first three 4-bit outputs are information bits, followed by four parity bits out. This is repeated for a total of five times (resulting in 20 × 4-bit outputs) before extra information 4-bit read is output. This pattern repeats until all bits are transmitted.

In addition to the above process, sub-bank switching is performed to further interleave the bit-stream. The switching reorders the 4-bit outputs, regardless of whether the nibble contains information or parity bits. Table 4.10 shows the switching, where **b0, b1, b2,** and **b3** represent bit outputs from information or

TABLE 4.10 Sub-Bank Switching

Output Nibble Number	Switched Bit Order
1 or 2	**b0 b1 b2 b3**
3 or 4	**b1 b2 b3 b0**
5 or 6	**b2 b3 b0 b1**
7 or 8	**b3 b0 b1 b2**
9 or 10	**b0 b1 b2 b3**

parity sub-bank 0, 1, 2, and 3, respectively. The leftmost bit in Table 4.10 is transmitted first (i.e., it has the smallest time index). The switched bit order changes after every two nibbles read, independent of whether the nibbles contained information, parity, or both.

4.5.4 ROBO Modes

As discussed earlier, the PHY adapts to each simplex path using the bit-loaded tone maps to optimize throughput for individual, simplex data paths between two nodes. However, the system must also support more reliable and hence lower-throughput communications over unknown channels for initial communication between two nodes and for broadcast and multicast communications to multiple nodes. Three robust signaling schemes, referred to as for ROBust mOdulation (ROBO-AV) modes, are supported for these purposes. These modes are used for beacon, control, and data broadcast, multicast communication, session setup, and initial unicast communication until a set of Tone Maps have been exchanged. All ROBO modes use QPSK modulation, along with the $\frac{1}{2}$ rate turbo convolutional code. Special interleavers for the ROBO modes introduce further redundancy in the form of an outer repetition code (i.e., each coded bit is represented with multiple copies at the output of the ROBO interleaver). The encoding spreads the copy of each coded bit in both frequency and time to maximize the probability of successful decoding for transmission over an unknown fading channel with impulse noise. The parameters for the three ROBO modes are summarized in Table 4.11.

The Standard ROBO mode (STD-ROBO_AV) is normally used with 520-byte PHY Blocks. The transmitter may use High-Speed ROBO mode (HS-ROBO_AV) if reliable communication can be achieved with fewer copies. Mini-ROBO mode (MINI-ROBO_AV) is used with 136-byte PHY Blocks, and is for cases when a small payload needs a high degree of reliability.

4.5.4.1 ROBO Interleaver
The definition of the ROBO Interleaver requires that the number of used carriers be a multiple of the number of ROBO copies. Thus, in Step 5 of the following procedure, a special Tone Map is defined for ROBO that ensures that a feasible number of carriers are modulated.

If $V_{\text{int}}(i)$ denotes the sequence of bits at the output of the Channel Interleaver, the bit sequence at the output of the ROBO Interleaver $V_{\text{robo\_int}}(i)$ is determined as follows. The notation $x = \lfloor a \rfloor$ indicates that x is the largest integer less than

TABLE 4.11 ROBO Mode Parameters

ROBO Mode	Number of Copies (N_{copies}) After TCC	PHY Rate (Default Carrier Mask)	PB Block
STD-ROBO_AV	4	4.9226 Mbps	PB520
HS-ROBO_AV	2	9.8452 Mbps	PB520
MINI-ROBO_AV	5	3.7716 Mbps	PB136

PAYLOAD

or equal to a, and $x = \lceil a \rceil$ indicates that x is the smallest integer greater than or equal to a.

- Define the following:
  - N_{raw}: the number of bits per PB (information and parity) at the output of the Channel Interleaver
  - $N_{carrier}$: the number of carriers turned on (in Tone Mask)
  - N_{copies}: the number of redundant copies of the data (Table 15.1)
  - BPC: the number of bits per carrier (2 for QPSK)
- Determine the number of bits to pad:

$$N_{carrier\_robo} = N_{copies} \left\lfloor \frac{N_{carrier}}{N_{copies}} \right\rfloor$$

$$CarriersInSegment = N_{carrier\_robo} / N_{copies}$$

$$BitsPerSymbol = BPC \cdot N_{carrier\_robo}$$

$$BitsInSegment = BPC \cdot CarriersInSegment$$

$$BitsInLastSymbol = N_{raw} - BitsPerSymbol \left\lfloor \frac{N_{raw}}{BitsPerSymbol} \right\rfloor$$

if $BitsInLastSymbol == 0$
   $BitsInLastSymbol = BitsPerSymbol$
   $BitsInLastSegment = BitsInSegment$
else
   $BitsInLastSegment = BitsInLastSymbol - BitsInSegment$

$$\times \left\lfloor \frac{BitsInLastSymbol - 1}{BitsInSegment} \right\rfloor$$

end
$N_{pad} = BitsInSegment - BitsInLastSegment$

- Determine the cyclic shift:

  if $N_{copies} == 2$
    if $BitsInLastSymbol <= BitsInSegment$
      $CyclicShift(0, 1) = (0, 0)$;
    else
      $CyclicShift(0, 1) = (0, 1)$;
    end
  elseif $N_{copies} == 4$
    if $BitsInLastSymbol <= BitsInSegment$
      $CyclicShift(0, 1, 2, 3) = (0, 0, 0, 0)$;

$\quad$ elseif $BitsInLastSymbol <= 2 \cdot BitsInSegment$
$\quad\quad CyclicShift(0, 1, 2, 3) = (0, 0, 1, 1);$
$\quad$ elseif $BitsInLastSymbol <= 3 \cdot BitsInSegment$
$\quad\quad CyclicShift(0, 1, 2, 3) = (0, 0, 0, 0);$
$\quad$ else
$\quad\quad CyclicShift(0, 1, 2, 3) = (0, 1, 2, 3);$
$\quad$ end
elseif $\;N_{copies} == 5$
$\quad$ if $BitsInLastSymbol <= 4 \cdot BitsInSegment$
$\quad\quad CyclicShift(0, 1, 2, 3, 4) = (0, 0, 0, 0, 0);$
$\quad$ else
$\quad\quad CyclicShift(0, 1, 2, 3, 4) = (0, 1, 2, 3, 4);$
$\quad$ end
end

- Assign output of the ROBO Interleaver:

for $\;k = 0 : N_{copies} - 1$
$\quad$ if $CyclicShift(k) == 0$
$\quad\quad$ for $\;i = 1 : N_{raw}$
$\quad\quad\quad V_{robo\_int}(i + k(N_{raw} + N_{pad})) = V_{int}(i)$
$\quad\quad$ end
$\quad\quad$ for $\;i = 1 : N_{pad}$
$\quad\quad\quad V_{robo\_int}(N_{raw} + i + k(N_{raw} + N_{pad})) = V_{int}(i)$
$\quad\quad$ end
$\quad$ end
$\quad$ if $CyclicShift(k) > 0$
$\quad\quad NumberBitsShifted = (CyclicShift(k)-1) \cdot BitsInSegment + BitsInLastSegment$
$\quad\quad$ for $\;i = 1 : NumberBitsShifted$
$\quad\quad\quad V_{robo\_int}(i + k(N_{raw} + N_{pad})) = V_{int}(N_{raw} - NumberBitsShifted + i)$
$\quad\quad$ end
$\quad\quad$ for $\;i = 1 : N_{pad}$
$\quad\quad\quad V_{robo\_int}(i + NumberBitsShifted + k(N_{raw} + N_{pad})) = V_{int}(i)$
$\quad\quad$ end
$\quad\quad$ for $\;i = 1 : N_{raw} - NumberBitsShifted$
$\quad\quad\quad V_{robo\_int}(i + NumberBitsShifted + N_{pad} + k(N_{raw} + N_{pad})) = V_{int}(i)$
$\quad\quad$ end
$\quad$ end
end

PAYLOAD 49

- Set ROBO Tone Map
  Set Tone Map to indicate that the first $N_{carrier\_robo}$ carriers are used with QPSK modulation, and that any remaining carriers ($N_{carrier} - N_{carrier\_robo}$) are encoded with BPSK using the PN generator defined in Section 4.5.5.1.

4.5.5 Mapping and Tone Maps

The mapping function distinguishes between FC information, which is mapped into coherent QPSK only, and regular data, which generally is mapped into coherent Quadrature Amplitude Modulation (BPSK, QPSK, 8 QAM, 16 QAM, 64 QAM, 256 QAM, or 1024 QAM). Table 4.12 shows the modulation and bits per carrier for FC and payload information. Except for ROBO-AV Mode, in which all unmasked carriers use QPSK modulation, a mixture of the modulations (also known as "bit-loading") may be used for different carriers in the mask when creating the OFDM payload symbol(s). As in DSL, bit-loading can be supported on the powerline channel since the frequency response generally does not change significantly with time. When the channel does change, the channel adaptation process provides a means to rapidly readapt the bit-loading for the channel.

The mapping block is responsible for assuring that the transmitted signal conforms to the given Tone Map and Tone Mask.

Tone Masks are defined system wide (for all transmitters) and specify which carriers are used by the system. Tone Masks are obeyed by the transmitter during the Priority Resolution Symbols, Preamble, FC Symbols, and all types of data modulation. On the other hand, the Tone Map contains a list of Modulation Types for all unmasked carriers (or tones) that are to be used on a particular unicast communication link between two stations. For example, carriers that are experiencing fades may be avoided, and no information may be transmitted on these carriers. The Tone Map is obeyed by the data modulation modes and ignored when transmitting Frame Control, ROBO_AV, Preamble, and Priority Resolution Symbols. Table 4.13 shows

TABLE 4.12 Modulation Characteristics

Information Type	Bits Per Carrier	Modulation Type
Frame Control	2	Coherent QPSK
ROBO_AV (STD-ROBO_AV, HS-ROBO_AV, MINI-ROBO_AV)	2	Coherent QPSK
Data	1	Coherent BPSK
	2	Coherent QPSK
	3	Coherent 8 QAM
	4	Coherent 16 QAM
	6	Coherent 64 QAM
	8	Coherent 256 QAM
	10	Coherent 1024 QAM

TABLE 4.13 Tone Mask Amplitude Map and Tone Map

Symbols	Tone Mask	Amplitude Map	Tone Map
PPDU payload: BPSK, QPSK, 8 QAM, 16 QAM, 64 QAM, 256 QAM, 1024 QAM	Comply	Comply	Comply
Frame Control, ROBO-AV, Preamble, Priority Resolution Symbol	Comply	Comply	Ignore

the signaling types that obey or ignore the Tone Map/Tone Mask. Table 4.13 also shows the impact of the Amplitude Map on various signaling types.

4.5.5.1 Empty Tone Filling When the Tone Map indicates that a particular carrier is not used for information transmission, the mapping function uses coherent BPSK, modulated with a binary value from the PN sequence defined below. The PN sequence is generated using the following generator polynomial (see also Figure 4.12):

$$S(x) = x^{10} + x^3$$

The bits in the PN sequence generator are all initialized to all ones at the start of the first OFDM payload symbol of each PPDU. When a carrier is encountered that is nonmasked and not used for information, the existing value in the X^1 register is used for coherent BPSK modulation and the sequence is advanced. The sequence is advanced only when used as described above.

4.5.5.2 Last Symbol Padding In general, the encoded PB bits to be mapped on to OFDM Symbols will not exactly fill all carriers of the last symbol in the PPDU. The PN sequence above is also used (without reinitialization) to fill the last PPDU OFDM symbol. For a bit-loaded PPDU, a variable number of bits will be used from

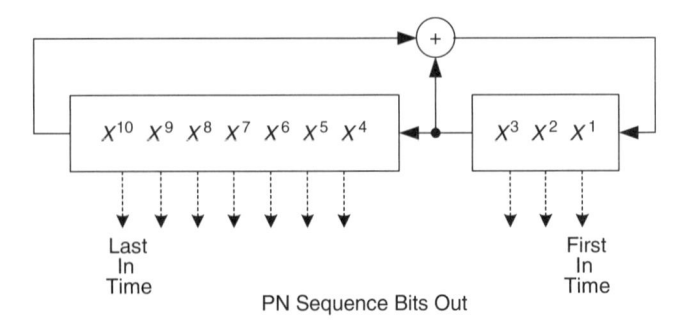

FIGURE 4.12 PN Generator.

the X^1 to X^{10} registers, beginning with X^1, depending on how many bits are required to produce a constellation symbol for each carrier's modulation type. Similar to the case of the empty tone filling, the sequence is only advanced after each time it is used (i.e., once per carrier, when this carrier did not have enough remaining CI bits to entirely fill one constellation symbol).

4.5.5.3 Mapping Reference The phase reference ϕ is used as the initial phase of the FC and payload symbol. The AV Specification defines a set of phase angle references for each of the 1155 carriers. The phase reference is described in equation form below. First, create the floating-point phase reference, ϕ_{FLT}, as

$$for\ c = 74 : 1228,$$
$$\Phi_{FLT}(c) = (\Phi_{FLT}(c-1) - (c-74) * 2\pi/(1228 - 74 + 1))\ Mod\ 2\pi;$$
$$end$$

where

$$\phi_{FLT}(c-1) = 0\ \text{when}\ c = 74.$$

ϕ is then created by quantizing ϕ_{FLT} as follows:

$$\varphi(c) = \left\lfloor \frac{\varphi_{FLT}(c)}{(\pi/4)} \right\rfloor \cdot \frac{\pi}{4} \quad \text{for} \quad 74 \leq c \leq 1227$$

$$\varphi(c) = \frac{7\pi}{4} \quad \text{for} \quad c = 1228$$

In the above equation, indices 74 and 1228 correspond to 1.8 and 30 MHz, respectively. These values are the minimum and maximum frequencies that the HomePlug-AV system can use.

4.5.5.4 Mapping for BPSK, QPSK, 8 QAM, 16 QAM, 64 QAM, 256 QAM, 1024 QAM Data bits are mapped for coherent QAM modulation. Mapping is performed on both the I and Q channels in the following way:

1. The mapper takes 1, 2, 3, 4, 6, 8, or 10 bits, depending on the constellation of the current symbol and maps them into the I and Q values of a symbol. Table 4.14 shows how the bits are mapped to a symbol. In all cases, the LSB x_0 has the earliest time index.
   Note: BPSK has nothing transmitted in the Q channel.
2. These bits are then mapped to the values in Tables 4.15 and 4.16, resulting in I and Q values for each symbol.
3. The symbols are scaled to produce a unity average power symbol. The I and Q values are multiplied by the power scale value in Table 4.17.

TABLE 4.14 Bit Mapping

Modulation Scheme	Bits from Channel Interleaver	I Channel	Q Channel
BPSK	x0	x0	–
QPSK	x1x0	x0	x1
8 QAM	x2x1x0	x1x0	x2
16 QAM	x3x2x1x0	x1x0	x3x2
64 QAM	x5x4x3x2x1x0	x2x1x0	x5x4x3
256 QAM	x7x6x5x4x3x2x1x0	x3x2x1x0	x7x6x5x4
1024 QAM	x9x8x7x6x5x4x3x2x1x0	x4x3x2x1x0	x9x8x7x6x5

TABLE 4.15 Symbol Mapping (Except 8 QAM)

Mapped Value	1024 QAM $(x_4x_3x_2x_1x_0)$ $(x_9x_8x_7x_6x_5)$	256 QAM $(x_3x_2x_1x_0)$ $(x_7x_6x_5x_4)$	64 QAM $(x_2x_1x_0)$ $(x_5x_4x_3)$	16 QAM (x_1x_0) (x_3x_2)	QPSK (x_0) (x_1)	BPSK (x_0) $(-)$
+31	11000					
+29	11001					
+27	11011					
+25	11010					
+23	11110					
+21	11111					
+19	11101					
+17	11100					
+15	10100	1100				
+13	10101	1101				
+11	10111	−1111				
+9	10110	1110				
+7	10010	1010	110			
+5	10011	1011	111			
+3	10001	1001	101	11		
+1	10000	1000	100	10	1	1
−1	00000	0000	000	00	0	0
−3	00001	0001	001	01		
−5	00011	−0011	011			
−7	00010	0010	010			
−9	00110	0110				
−11	00111	0111				
−13	00101	0101				
−15	00100	0100				
−17	01100					
−19	01101					
−21	01111					
−23	01110					
−25	01010					
−27	01011					
−29	01001					
−31	01000					

PAYLOAD

TABLE 4.16 Symbol Mapping for 8 QAM

Mapped Value I	Mapped Value Q	8 QAM ($x_2 x_1 x_0$)
-1	-1.29	000
-3	-1.29	001
$+1$	-1.29	010
$+3$	-1.29	011
-1	$+1.29$	100
-3	$+1.29$	101
$+1$	$+1.29$	110
$+3$	$+1.29$	111

The mapping function for QAM modulation will obey the Tone Mask, that is

- Carriers that are masked (refer to Section 4.5.5.2) are not assigned I and Q constellation symbols and
- The amplitude is set to zero.

Additionally, all non-ROBO-AV PPDU payload mapping obey the Tone Map of a given link (refer to Section 9.4).

4.5.5.5 Mapping for ROBO-AV For ROBO-AV modulation, the majority of nonmasked carriers are mapped with coherent QPSK modulation. A small number of carriers may become unusable in a particular ROBO Mode, since the ROBO

TABLE 4.17 Modulation Normalization Scales

Modulation	Power Scale
BPSK	$\frac{1}{\sqrt{1}}$
QPSK	$\frac{1}{\sqrt{2}}$
8 QAM	$\frac{1}{\sqrt{5 + 1.29^2}}$
16 QAM	$\frac{1}{\sqrt{10}}$
64 QAM	$\frac{1}{\sqrt{42}}$
256 QAM	$\frac{1}{\sqrt{170}}$
1024 QAM	$\frac{1}{\sqrt{682}}$

modes require the number of carriers to be an integer multiple of the number of redundant copies in the interleaver.

4.5.6 Payload Symbols

The generation of the Payload Symbols complies with the following equations. The time domain discrete waveform for one payload symbol is defined below.

$$S_{Data}[n] = \frac{w[n] \cdot 10^{2.2/20}}{\sqrt{T}} \sum_{c \in M} \alpha(d) \cdot \cos\left(\frac{2 \cdot \pi \cdot c \cdot (n - GI - RI)}{T} + \varphi(c) + \gamma(d)\right)$$
$$\text{for} \quad 0 \leq n \leq RI + GI + T - 1$$

where

$$w[n] = \begin{cases} w_{rise}[n] & \text{for } 0 \leq n \leq RI - 1 \\ 1 & \text{for } RI \leq n \leq T + GI - 1 \\ w_{fall}[n - (T + GI)] & \text{for } T + GI \leq n \leq T + GI + RI - 1 \end{cases}$$

The equation parameters RI, T, and GI are defined in Table 4.2. Window functions w_{rise} and w_{fall} are of length RI and reduce out of band and notch band energy. The scaling factor of $10^{2.2/20}$ is due to the PPDU Payload Symbol power boost defined in Table 4.18.

The set M is the subset of all carriers in the Tone Map for non-ROBO symbols and the modified Tone Mask used by ROBO symbols (refer to Section 4.5.4); c is the carrier index with values in M and $c = 0$ corresponds to D.C., $\phi(c)$ is the set of phase references, and γ and α are defined as the set of phases and amplitudes respectively of the rectangular symbols out of the mapper. For example, if the mapper output for a particular 16 QAM carrier is:

$$\frac{1 + 3i}{\sqrt{10}}$$

TABLE 4.18 Relative Power Levels

Waveform Type	Average Subcarrier Power Boost	
	HomePlug 1.0.1	HomePlug AV
Preamble Symbols	3 dB	3 dB
Frame Control Symbols	0 dB	3 dB
PPDU Payload Symbols	0 dB	2.2 dB
Priority Resolution Symbols	3 dB	3 dB

Note: The boost in average power applies to both the IFFT interval and cyclic prefix (if applicable).

then $\alpha(d) = 1$ and $\gamma(d) = 71.57°$. The first value of d for each symbol (corresponding to the minimum value of **c**) is equal to the number of carriers in the set M multiplied by the symbol number m, with $m = 0$ for the first payload symbol in the PPDU. The index d is incremented by one for each successive value of c.

4.5.7 Windowed OFDM and Symbol Shaping

An essential requirement of the AV PHY is to support flexible adaptation of its Power Spectral Density (PSD) for easy deployment in different countries, each with its own regulations regarding RF emissions of powerline devices. Moreover, the regulatory environment in many areas is in flux, further emphasizing the need for cost-effective programmability of the system PSD. The AV system uses symbol shaping to avoid costly programmable notch filters. A specifically designed window is applied to each time domain OFDM symbol (**w[n]**, **w**$_{rise}$, and **w**$_{fall}$ mentioned in Section 4.5.6), causing reduced bandwidth occupancy of the sidelobes of each carrier. A few carriers must be turned off (masked) to create a deep spectral notch. Figure 4.13 shows the semi-infinite power roll-off for a single carrier for the three GIs. The PSD drops to about 30 dB below the peak at a frequency gap of approximately 115 kHz. Conversely, to create a spectral notch all carriers within approximately 115 kHz of a desired notched band are masked. Note that even using notch filters, a guard band on either side of the frequency notch would be required.

To reduce the overhead due to the required cyclic prefix length for shaping and delay spread mitigation, the shaped portions of each OFDM symbol are overlapped with the shaped portions of the preceding and following OFDM symbols, as shown in Figure 4.6.

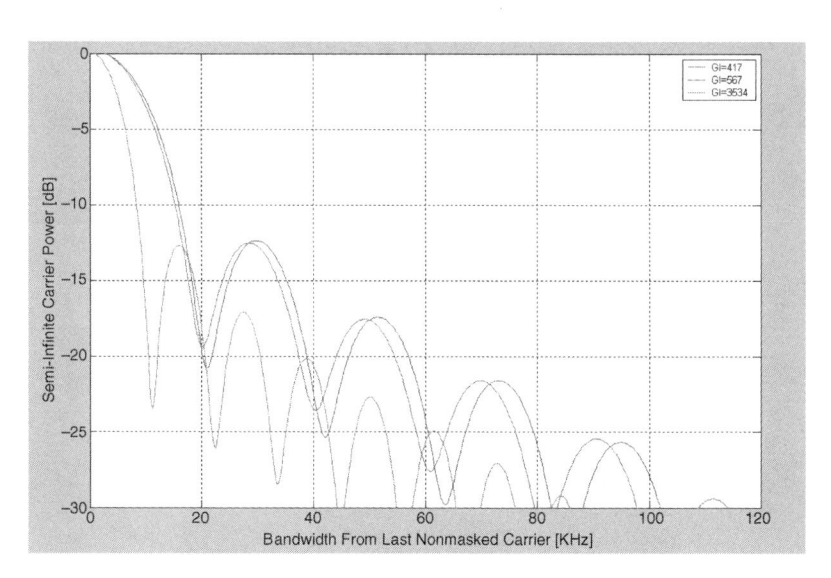

FIGURE 4.13 Spectral occupancy for semi-infinite number of carriers.

4.6 PRIORITY RESOLUTION SYMBOL

The Priority Resolution Symbol is used during the Priority Resolution Slots (PRS). The Priority Resolution Symbol is derived from the PRS waveform defined in the HomePlug 1.0.1 specification by:

- including the additional carriers defined for the AV Preamble,
- reversing the sign of the phases used in the preamble,
- affixing an extra half of a subsymbol to the beginning and end of the standard six HomePlug 1.0.1 PRS subsymbols,
- pulse-shaping the first and last RI samples of the resulting waveform,
- beginning transmission of the waveform a half of a subsymbol (2.56 μs) before the start of the actual slot time.

The nominal PRS waveform is defined as:

$$S_{NomPRSMasked}[n] = \frac{10^{3/20}}{\sqrt{384}} \sum_{c \in C_{HP1.0-ES}} \cos\left(\frac{2 \cdot \pi \cdot c \cdot n}{384} - \psi(c)\right) \quad for\ 0 \leq n \leq T-1$$

where $C_{HP1.0-ES}$ is the subset of all unmasked HomePlug 1.0.1 Extended Set carriers, c is an index with values in $C_{HP1.0-ES}$, $c = 0$ corresponds to D.C., and $\psi(c)$ denotes the phase angle number multiplied by $\pi/8$. The scaling factor of $10^{3/20}$ is due to the Priority Resolution Symbol scaling defined in Table 4.18.

The PRS waveform is then created by windowing the first and last symbols, effectively shrinking the length of the waveform by one subsymbol (5.12 μs)

$$S_{PRS\_AV}[n] = w_{PRS\_AV}[n] \cdot S_{NomPRSMasked}[n + 192] \quad for\ 0 \leq n \leq 7 \cdot 384 - 1$$

where

$$w_{PRS\_AV}[n] = \begin{cases} w_{rise}[n] & for\ 0 \leq n \leq RI - 1 \\ 1 & for\ RI \leq n \leq 7 \cdot 384 - RI - 1 \\ w_{fall}[n - (7 \cdot 384 - RI)] & for\ 7 \cdot 384 - RI \leq n \leq 7 \cdot 384 - 1 \end{cases}$$

The equation parameters T and RI are defined in Table 4.2. w_{rise} and w_{fall} are mentioned in Section 4.5.6. To minimize the effect of the symbol shaping on priority detection reliability, the transmission of the resulting PRS waveform, S_{PRS_AV}, begins one-half a subsymbol, or 2.56 μs, before the start of the actual PRS, as shown in Table 4.18.

4.7 TRANSMIT POWER, TONE MASK, AND AMPLITUDE MAP

4.7.1 Relative Power Levels

Since radiation compliance is measured using quasi-peak (not mean) power, AV traffic must have the same peak PSD profile as HomePlug 1.0.1 traffic. However, due

TRANSMIT POWER, TONE MASK, AND AMPLITUDE MAP 57

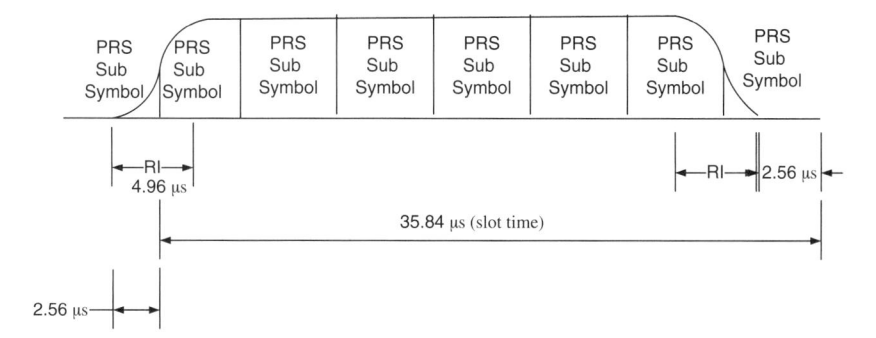

FIGURE 4.14 AV PRS waveform.

to the closer carrier spacing (24.414 kHz) of AV Frame Control and PPDU payload symbols compared with those of HomePlug 1.0.1 (195.31 kHz), the resulting AV PSD has less ripple. This translates into a boost of about 2.2 dB in average subcarrier power for AV Frame Control and PPDU payload symbols for HomePlug 1.0.1 and AV to have the same peak power levels.

Additionally, there are some waveforms with relatively low duty cycles. They can therefore be transmitted at higher average power than other types of higher duty cycle traffic, without affecting the measured peak PSD.

Table 4.18 summarizes the relative subcarrier average power levels for HomePlug 1.0.1 and HomePlug AV, with the HomePlug 1.0.1 PPDU payload symbols (called "packet body" symbols in HomePlug 1.0.1) as a reference. HomePlug AV stations adjust their average power to comply with this table.

4.7.2 Tone Mask

A Tone (or Carrier) Mask defines the set of tones that can be used in a given regulatory jurisdiction or a given application of the HomePlug AV system. Certain tones need to be turned off to comply with the spectral mask requirements of the region or application. Table 4.19 defines the default Tone Mask that will comply with current North American regulations. The spectral mask is depicted in Figure 4.15. To meet other regulations or applications that were uncertain at the time that the HomePlug AV specification was developed, the HomePlug AV specification required that stations be capable of supporting Tone Masks of any combination of unmasked carriers, from a minimum of 275 to a maximum of 1155 unmasked carriers, in the frequency range of 1.8–30 MHz (carrier numbers 74–1228). While the default Tone Mask specified for North America was based on the FCC Part 15 rules, the default Tone Mask is now in use worldwide. The only notable difference is different PSD limits. For example, the PSD Limit is 5 dB lower for Europe and there is a 10-dB reduction in PSD at 15 MHz and above in Japan.

TABLE 4.19 North American Carrier and Spectral Masks

Frequency (MHz)	PSD Limit (dBm/Hz)	Carrier On/Off	Notes
$F \leq 1.71$	-87	Carriers 0–70 are OFF	AM broadcast band and lower
$1.71 < F < 1.8$	-80	Carriers 71–73 are OFF	Between AM and 160-m band
$1.8 \leq F \leq 2.00$	-80	Carriers 74–85 are OFF	160 m amateur band
$2.00 < F < 3.5$	-50	Carriers 86–139 are ON	HomePlug carriers
$3.5 \leq F \leq 4.00$	-80	Carriers 140–167 are OFF	80 m amateur band
$4.000 < F < 5.33$	-50	Carriers 168–214 are ON	HomePlug carriers
$5.33 \leq F \leq 5.407$	-80	Carriers 215–225 are OFF	5 MHz amateur band
$5.407 < F < 7.0$	-50	Carriers 226–282 are ON	HomePlug Carriers
$7.0 \leq F \leq 7.3$	-80	Carriers 283–302 are OFF	40 m amateur band
$7.3 < F < 10.10$	-50	Carriers 303–409 are ON	HomePlug carriers
$10.10 \leq F \leq 10.15$	-80	Carriers 410–419 are OFF	30 m amateur band
$10.15 < F < 14.00$	-50	Carriers 420–569 are ON	HomePlug carriers
$14.00 \leq F \leq 14.35$	-80	Carriers 570–591 are OFF	20 m amateur band
$14.35 < F < 18.068$	-50	Carriers 592–736 are ON	HomePlug carriers
$18.068 \leq F \leq 18.168$	-80	Carriers 737–748 are OFF	17 m amateur band
$18.168 < F < 21.00$	-50	Carriers 749–856 are ON	HomePlug carriers
$21.000 \leq F \leq 21.45$	-80	Carriers 857–882 are OFF	15 m amateur band
$21.45 < F < 24.89$	-50	Carriers 883–1015 are ON	HomePlug Carriers
$24.89 \leq F \leq 24.99$	-80	Carriers 1016–1027 are OFF	12 m amateur band
$24.99 < F < 28.0$	-50	Carriers 1028–1143 are ON	HomePlug Carriers
$F \geq 28.0$	-80	Carriers 1144–1535 are OFF	10 m amateur band

4.7.3 Amplitude Map

In addition to the Tone Mask, each carrier will obey its Amplitude Map. The Amplitude Map specifies a possible transmit power-reduction factor for each subcarrier. For example, if the Amplitude Map entry for a particular carrier is **0b0010**, that carrier is transmitted at 4 dB less power than the normal PSD limit (-50 dBm/Hz for unmasked carriers for the North American spectral mask). Table 4.20 defines the power reduction for each value of the Amplitude Map entry.

The Amplitude Map affects only the amplitudes of the corresponding unmasked carriers set by the mapper and therefore does not in any way affect the encoding used for FC, Payload, or ROBO symbols and/or the number of unmasked carriers. Carriers with an Amplitude Map entry of **0b1111**, referring to "Off (No Signal)," are not considered or processed as masked by any element of the transmit chain, even though they have zero amplitude. The Amplitude Map is applied to all PHY waveforms,

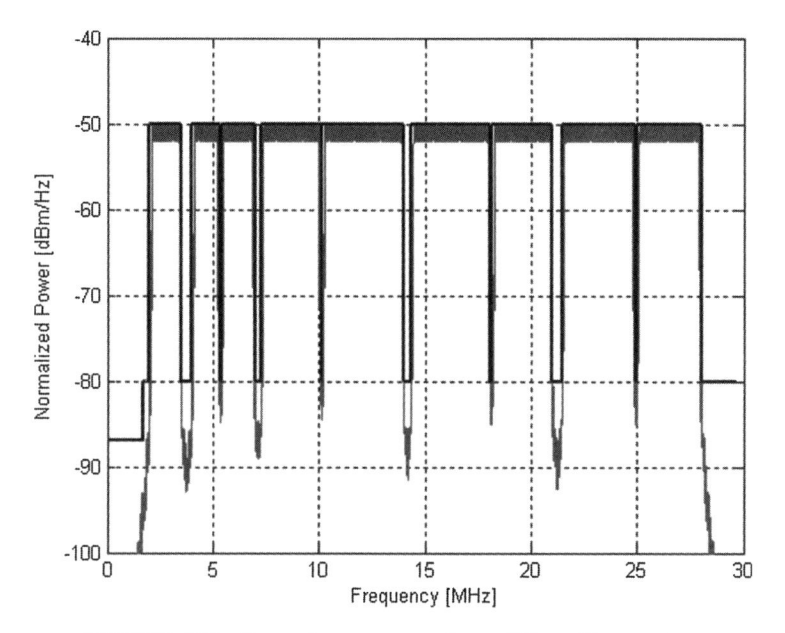

FIGURE 4.15 Spectral occupancy of set of HomePlug carriers.

including the Preamble, PRS symbols, 1.0.1 Frame Control, AV Frame Control, and Payload Symbols. Power-reduction values will be met with an accuracy of ±1 dB for AMDATA values from **0b0000** through **0b1010** and an accuracy of ±2 dB for AMDATA values from **0b1011** through **0b1110**.

An Amplitude Map with no reduction in the transmit power on all used carriers (as defined in Tone Mask) will be used as the default Amplitude Map in North America. The Amplitude Map is the mechanism used to implement the 10-dB PSD reduction at 15 MHz and above required in Japan mentioned in Section 4.7.2.

TABLE 4.20 Amplitude Map

AMDATA	TX Power Reduction (dB)	AMDATA	TX Power Reduction (dB)
0b0000	0 (no reduction—default)	**0b1000**	16
0b0001	2	**0b1001**	18
0b0010	4	**0b1010**	20
0b0011	6	**0b1011**	22
0b0100	8	**0b1100**	24
0b0101	10	**0b1101**	26
0b0110	12	**0b1110**	28
0b0111	14	**0b1111**	Off (no signal)

4.8 SUMMARY

This chapter provided an in-depth description of the HomePlug AV PHY protocol, including a discussion of the PPDU, and the use of Tone Maps, Tone Masks, and Amplitude Maps to optimize overall performance and to adhere to regional regulatory constraints. The issue of synchronization and the definition and use of reliable ROBO modes of communications in the PLC channel were also discussed.

In Chapter 5, the HomePlug AV MAC protocol is presented and the details of the protocol structures used for MPDU and Beacon, Data, and Sound frames, as well as the associated frame delimters are examined.

5

MAC PROTOCOL DATA UNIT (MPDU) FORMAT

5.1 INTRODUCTION

The MAC Protocol Data Unit (MPDU) consists of information that is exchanged between the MAC and the Physical (PHY) layers. When the MAC Layer needs to communicate information with one or more peer MAC Layers, it generates an MPDU. The MPDU generated by the MAC Layer is then converted into a PHY Protocol Data Unit (PPDU) by the PHY layer and is transmitted on the powerline medium. Any PPDU received on the powerline by the PHY layer is converted into an MPDU and provided to the MAC Layer. There is always a one-to-one relationship between an MPDU and a PPDU. The general format of an MPDU is shown in Figure 5.1.

The HomePlug 1.0 Frame Control contains 25-bits of broadcast information that is used to defer HomePlug 1.0 stations from accessing the medium during a HomePlug AV transmission. An MPDU that includes HomePlug 1.0 Frame Control is referred to as Hybrid Mode MPDU. An MPDU that does not include HomePlug 1.0 Frame Control is referred to as AV-only mode MPDU. AV-only mode MPDUs can only be used when there are no HomePlug 1.0 stations. Section 11.4 provides details on HomePlug 1.0.1 coexistence.

The AV Frame Control contains 128-bits of broadcast information that is used for control information.

The AV payload contains variable length unicast or broadcast information. AV payload is only present in some MPDUs. MPDUs that include AV payload are referred to as long MPDUs and MPDUs that do not include AV payload are referred

---

HomePlug AV and IEEE 1901: A Handbook for PLC Designers and Users, First Edition.
Haniph A. Latchman, Srinivas Katar, Larry Yonge, and Sherman Gavette.
© 2013 by The Institute of Electrical and Electronics Engineers, Inc. Published 2013 by John Wiley & Sons, Inc.

| HomePlug 1.0 Frame Control (25 bits) | HomePlug AV Frame Control (128 bits) | AV Payload |

FIGURE 5.1 General MPDU formats.

to as short MPDUs. The presence of AV payload and its format is communicated using the AV Frame Control.

HomePlug AV uses six different MPDU Types for exchange of data, management, and control information. Table 5.1 shows the MPDU Types and their functionality.

TABLE 5.1 HomePlug AV MPDU Types and their Functionality

MPDU Type	Functionality
Beacon	Beacon MPDU is used by the Central Coordinator to manage the network and control channel access for stations within the AVLN. The CCo transmits a Central Beacon once every Beacon Period. When hidden nodes are present, Proxy Coordinators repeat the information contained in each Central Beacon using Proxy Beacons. This enables hidden nodes to obtain the information contained in the Central Beacon. All stations in the network also periodically send Discover Beacons to enable topology discovery. All Beacons share the same Beacon MPDU format, but can have different content based on the Beacon Type. Each Beacon includes a Beacon Frame Control followed by a 136-octet Beacon payload that is modulated using Mini-ROBO.
Start-of-Frame (SOF)	Start-of-Frame is used to transmit data and management information to one or more stations. The SOF MPDU includes a SOF Frame Control followed by one or more PHY Blocks that contains data and/or management information. The payload of the Start-of-Frame is typically transmitted using tone maps that are tailored to the channel conditions between the transmitter and receiver.
Selective Acknowledgment (SACKs)	Selective Acknowledgments (SACKs) are used to acknowledge the reception status of the PHY Blocks from the preceding SOF or RSOF MPDUs.
Request to Send (RTS)/Clear to Send (CTS)	Request to Send (RTS) and Clear to Send (CTS) MPDU are primarily used to handle hidden nodes. RTS/CTS can also be used for other functions such as participation in multiple networks and multinetwork broadcasting. RTS and CTS MPDUs do not include any MPDU payload.
Sound	Sound MPDU is used for channel estimation. A long Sound MPDU includes a Sound Frame Control followed by either one 136 octet Mini-ROBO payload or one 520 octet Std-ROBO payload. A short Sound MPDU is used by the receiver to acknowledge reception of a long Sound MPDU.
Reverse Start-of-Frame (RSOF)	Reverse SOF is used as part of bidirectional bursting to increase the efficiency for bidirectional traffic (e.g., TCP). RSOF allows piggybacking selective acknowledgment information along with data or management information. The RSOF MPDU includes a RSOF Frame Control followed by one or more PHY Blocks. The payload of the RSOF is typically transmitted using Tone Maps that are tailored to the channel condition between the transmitter and receiver.

INTRODUCTION

TABLE 5.2 General AV Frame Control Fields

Field	Field Size (Bits)
Delimiter Type (DT_AV)	3
ACCESS	1
Short Network Identifier (SNID)	4
Variant field (VF_AV)	96
Frame Control Block Check Sequence (FCCS_AV)	24

The AV Frame Control is used to indicate the MPDU Type. The general format of AV Frame Control is described Section 5.1.1. Details of the various MPDU Types are described in Sections 5.2–5.7.

5.1.1 General AV Frame Control

The general AV Frame Control format is shown in Table 5.2. Delimiter Type (DT_AV) is a 3-bit field that identifies the delimiter (i.e., MPDU Type). Table 5.3 shows the various types of delimiters. The format of HomePlug AV variant fields (VF_AV) as well as the MPDU payload depends on DT_AV.

ACCESS is a single-bit field that indicates whether the Frame Control is transmitted by a device that is part of an in-home network or an Access network (i.e., network connecting in-home devices to the internet). HomePlug AV devices are widely deployed for in-home networking and their usage in Access networks has been fairly limited.

Short Network Identifier (SNID) is a 4-bit field that is used to distinguish between MPDUs transmitted by different networks on the same powerline medium. It is a shorthand representation of the Network Identifier. The correspondence between a SNID and a Network Identifier is established when both are transmitted in a Beacon MPDU. The Central Coordinator (CCo) of an AVLN selects a unique SNID for its AVLN. If the CCo determines that its SNID is being used by Neighboring CCo, it randomly selects a new SNID from the list of SNIDs that are not used by neighboring CCos. The CCo uses change SNID BENTRY to notify STAs in its network of the new BENTRY before it transitions to the new SNID.

TABLE 5.3 Delimiter Type Field Interpretation

DT Value	Interpretation
000	Beacon
001	Start-of-Frame (SOF)
010	Selective Acknowledgment (SACK)
011	Request to Send (RTS)/ Clear to Send (CTS)
100	Sound
101	Reverse Start-of-Frame (RSOF)
110–111	Reserved

Frame Control Check Sequence (FCCS_AV) is a 24-bit field in the HomePlug AV Frame Control Block, and is CRC-24 computed over the fields of HomePlug AV Frame Control block. FCCS_AV is used to check the integrity of the AV Frame Control information. CRC-24 is computed using the following standard generator polynomial of degree 24.

$$G(x) = x^{24} + x^{23} + x^6 + x^5 + x + 1$$

5.2 BEACON

Beacon MPDUs play an important role in network management, controlling channel access and topology discovery. Beacons are always transmitted in Hybrid Mode. This selection was made to simplify the procedure for new stations (STAs) to join the network. It should be noted that stations operating in AV-only mode will not be able to decode Hybrid Mode transmissions and vice versa. Since Beacons are always transmitted in Hybrid Mode, new stations need to search only for existing networks in Hybrid Mode thus simplifying and expediting the procedure for joining an AVLN. This selection also simplifies the coordination between neighboring networks (refer to Section 12.1.3).

Figure 5.2 shows the Beacon MPDU format. This includes an HP1.0 Frame Control followed by a 128-bit Beacon Frame Control followed by a 136-octet Beacon payload. Contents of Beacon Frame Control are described in Section 5.2.1. A Beacon payload is always transmitted using Mini-ROBO modulation and includes some fixed fields (i.e., fields present in all Beacons) followed by variable Beacon Management Information. Beacon Management Information includes one or more Beacon Entries (BENTRIES). The exact set of BENTRIES that is included in the Beacon can be varied based on the need. For example, Encryption Key Change BENTRY is only included when the CCo is changing the network encryption key. This flexible formatting also allows adding new BENTRIES in the future to add or enhance existing functionality. The Beacon payload format is described in detail in Section 5.2.2.

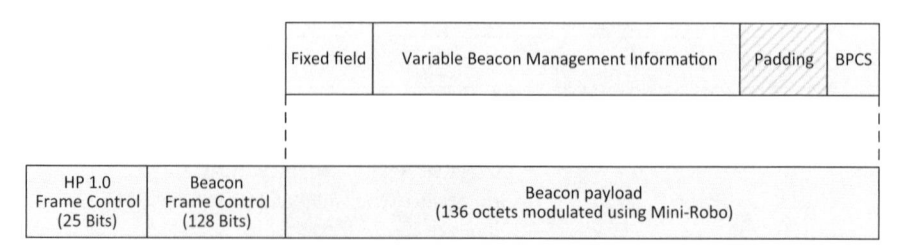

FIGURE 5.2 Beacon MPDU format.

5.2.1 Beacon Frame Control

Beacon Frame Control includes a Beacon Time Stamp (BTS) and Beacon Transmission Offset (BTO).

5.2.1.1 Beacon Time Stamp (BTS) AV stations use ±25 ppm (parts per million, PPM) local clocks for signal processing. Beacon Time Stamps are used for Network Time Base synchronization that enables stations in the network to reduce the clock error to within a fraction of a PPM. Beacon Time Stamp contains the 32-bit value of the Network Time Base at the start of Beacon PPDU transmission. The Network Time Base is based on the 25 MHz local clock of the Central Coordinator (refer to Section 7.5.1).

5.2.1.2 Beacon Transmission Offset (BTO) The Beacon Period in AV networks is synchronized to the underlying AC line cycle and is twice the AC line cycle period. There are variations in the phase and frequency of the AC line cycle from the power generating plant; hence the Beacon Period can vary as the AC line cycle period varies. For example, in North America the AC line cycle frequency is on an average of 60 Hz. However, during some times the exact AC line cycle period can be 59.9 Hz while they may be 60.1 Hz at other times. Such variation results in changes to the Beacon Period. The CCo tracks the AC line cycle periods and provides the offset of future Beacon Period Start times from their expected locations using the Beacon Transmission Offset field. This enables all station in the network to determine the exact location of the start of future Beacon Periods. The CCo can include Beacon Transmit Offsets for up to four future Beacon Periods (i.e., BTO[0], BTO[1], BTO[2], and BTO[3]).

5.2.2 Beacon Payload

The Beacon payload carries 136 octets of payload modulation using the Mini-ROBO Modulation. The duration of time required to transmit the Beacon payload depends on the number of Orthogonal Frequency Division Multiplexing (OFDM) carriers in the active Tone Mask. HomePlug AV stations use the active Tone Mask information to implicitly determine the Beacon payload duration. When default tone mask is used (refer to Section 4.7.2), the Beacon payload consists of six OFDM symbols. Beacon payload includes a set of fixed fields followed by variable length Beacon Management Information. Beacon payload fixed fields are always present in each Beacon, however, the contents of Beacon Management Information can change from one Beacon to another.

Table 5.4 shows the contents of the Beacon payload fixed fields based on their functionality. Section 5.2.2.8 provides details of the contents of Beacon Management Information.

5.2.2.1 Beacon Type Beacon Type (BT) is a 3-bit field that indicates the type of Beacon. All Beacon types share the same Beacon structure, but differ in functionality

TABLE 5.4 HomePlug AV Beacon Payload Fixed Fields Grouped based on Functionality

Functionality	Fields	Field Size (Bits)
Beacon Type (BT)	Beacon Type	3
Addressing	Network Identifier (NID)	54
	Source Terminal Equipment Identifier (STEI)	8
Neighbor Network Coordination	Noncoordinating Networks Reported (NCNR)	1
	Number of Beacon Slots (NumSlots)	3
	Beacon Slot Usage (SlotUsage)	8
	Beacon Slot ID (SlotID)	3
	AC Line Cycle Synchronization Status (ACLSS)	3
Network Operation Mode	Hybrid Mode (HM)	2
	Network Mode (NM)	2
	Network Power Save Mode (NPSM)	1
CCo Capability	CCo Capability (CCoCap)	2
Participation in Multiple Network	RTS Broadcast Flag (RTSBF)	1
CCo Handover	Handover-In-Progress (HoIP)	1

and the Beacon Management Information they carry. The various beacon types are as follows:

- *Central Beacon.* The Central Beacon is generated by the CCo of each AVLN during each Beacon Period (refer to Section 8.1). One of the main functions of Central Beacon is to carry medium allocation (or scheduling) information.
- *Discover Beacon.* The Discover Beacon is transmitted by all associated and authenticated STAs periodically to aid in network-topology discovery, where other STAs update their Discovered STA Lists and Discovered Network Lists (refer to Section 7.5.2). The Discover Beacon also allows hidden STAs (i.e., STAs that cannot decode the Central Beacons from the CCo) to ascertain the Beacon Period structure and to associate with the CCo.
- *Proxy Beacon.* The Proxy Beacon is used to manage hidden terminals (i.e., Proxy Networking) within an AVLN (refer to Section 11.5).

5.2.2.2 Addressing The addressing information in the Beacon payload includes the 54-bit Network Identifier (NID) of the network of the STA that transmitted the Beacon and its 8-bit source Terminal Equipment Identifier (TEI).

5.2.2.2.1 Network Identifier (NID) Each AVLN has a 54-bit Network ID (NID) that uniquely identifies the network. The NID is generated by combining the Security Level (2 bits) with the NID Offset (52 bits) as shown in Figure 5.3. The default NID Offset is generated by hashing the NMK using SHA-256 as the underlying hash

2 bits	52 bits
Security level	NID offset

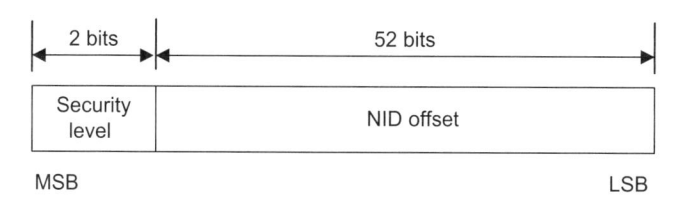

MSB · LSB

FIGURE 5.3 Network identifier.

algorithm. The Security Level is set to **0b00** when push-button-based authentication is used. In all other cases, Security Level is set to 0b1.

5.2.2.3 Neighbor Network Coordination Neighbor Network Coordination fields in the Beacon facilitate AVLN operation in the presence of Neighboring Networks (refer to Chapter 12).

- Noncoordinating Network Reported (NCNR) is a 1-bit field that indicates the absence or presence of networks that are not coordinating with the CCo. A noncoordinating network does not respect the TDMA allocation within the local network, thus causing interference during TDMA allocations.
- Number of Beacon Slots (NumSlots) is a 3-bit field that indicates the number of Beacon Slots in the Beacon Region. The maximum number of Beacon Slots in the Beacon Region is limited to 8.
- Beacon Slot Usage (SlotUsage) is an 8-bit bit-mapped field. Each bit in the field indicate whether the corresponding Beacon Slot is free or occupied.
- Beacon Slot ID (SlotID) is a 3-bit field that indicates the Beacon Slot number used by the CCo Beacon.
- AC Line Cycle Synchronization Status (ACLSS) is a 3-bit field. The ACLSS will be set to the current Beacon Slot ID if the CCo is locally tracking the AC line cycle. If the CCo is tracking the AC line cycle synchronization of a Central Beacon of another CCo in a Group of networks, ACLSS will be set to the Beacon Slot ID of that Central Beacon.

5.2.2.4 Network Operation Mode Network Operation Mode field indicates the mode of operation of the network with regards to HomePlug 1.0 Coexistence (i.e., Hybrid Mode field), TDMA operation (i.e., Network Mode field) and Network Power Save (using Network Power Save Mode field).

- Hybrid Mode (HM) is a 2-bit field that indicates the HomePlug 1.0.1 Coexistence operating mode of the AVLN (refer to Section 11.1). The Hybrid Mode (HM) can be
  o AV-only mode;
  o shared CSMA Hybrid Mode;
  o fully Hybrid Mode; and
  o fully Hybrid Mode with unrestricted frame lengths.

- Network Mode (NM) is a 2-bit field that indicates the network mode of operation of the AVLN (refer to Section 8.3). Various network modes are as follows:

  o Uncoordinated Mode;

  o Coordinated Mode; and

  o CSMA-only mode.

- Network Power Saving Mode (NPSM) is a 1-bit field that indicates whether the CCo has placed the network into Network Power Saving Mode. In practice, this field is not used. HomePlug Green PHY and HomePlug AV2.0 uses advanced power save mechanism described in Section 15.3.1.

5.2.2.5 CCo Capability The CCo Capability (CCoCap) is a 2-bit field that indicates the capability of the CCo of the AVLN (refer to Section 7.2.2.1). This can be

- Level-0 CCo Capable—does not support QoS and TDMA.
- Level-1 CCo Capable—supports QoS and TDMA but only in Uncoordinated Mode.
- Level-2 CCo Capable—supports QoS and TDMA in Coordinated Mode.
- Level-3 CCo Capable—future CCo capabilities.

5.2.2.6 Participation in Multiple Networks The RTS Broadcast Flag (RTSBF) is a 1-bit field that indicates to all stations in the network that broadcast MPDUs must use RTS/CTS when RTSBF is set to 0b1.

The CCo maintains a list of all stations in the network (defined by the SNID), where the Different CP PHY Clock Flag (DCPPCF) is set in a CC_DCPPC.IND message and removes from the list any previously listed stations where DCPPCF is cleared in a CC_DCPPC.IND message. If the list contains one or more stations, the CCo will set RTSBF to 0b1. Otherwise, it will be set to 0b0 (refer to Section 7.5.1.2).

5.2.2.7 CCo Handover A CCo may handover CCo functionality to another station in the network. This functionality is referred to as CCo Handover. Handover-In-Progress (HOIP) is a 1-bit field that indicates whether a handover is in progress. If a handover is in progress, STAs and neighbor CCos wait before sending association or bandwidth requests to the CCo.

5.2.2.8 Beacon Management Information (BMI) Beacon Management Information is a variable-length field that contains Beacon Management Messages as shown in Table 5.5.

The interpretation for various fields in the Beacon Management Information is as follows:

- Number of Beacon Entries (NBE) indicates the total number of Beacon entries present in the Beacon payload.

TABLE 5.5 Beacon Management Information Format

Field	Field Size (Octets)
Number of Beacon Entries (NBE)	1
Beacon Entry Header (BEHDR[1])	1
Beacon Entry Length (BELEN[1])	1
Beacon Entry (BENTRY[1])	N[1]
...	
Beacon Entry Header (BEHDR[L])	1
Beacon Entry Length (BELEN[L])	1
Beacon Entry (BENTRY[L])	N[L]

- Beacon Entry Header (BEHDR) indicates the type of Beacon entry. The various types of BENTRIES are shown in Table 5.6.
- Beacon Entry Length (BELEN) indicates the length of the Beacon entry, in octets.
- Beacon Entry (BENTRY) field depends on the Beacon Entry Header. A brief description of the functionality provided by each Beacon Entry is presented in the following sections.

5.2.2.8.1 Nonpersistent Schedule BENTRY Nonpersistent Schedule BENTRY contains the schedule that is valid only in the current Beacon Period (Table 5.7).

TABLE 5.6 Beacon Entry Header Interpretation

BEHDR Value	Interpretation
0x00	Nonpersistent Schedule BENTRY
0x01	Persistent Schedule BENTRY
0x02	Regions BENTRY
0x03	MAC Address BENTRY
0x04	Discover BENTRY
0x05	Discovered Info BENTRY
0x06	Beacon Period Start Time Offset BENTRY
0x07	Encryption Key Change BENTRY
0x08	CCo Handover BENTRY
0x09	Beacon Relocation BENTRY
0x0A	AC Line Sync Countdown BENTRY
0x0B	Change NumSlots BENTRY
0x0C	Change HM BENTRY
0x0D	Change SNID BENTRY
0x0E – 0xFE	Reserved for future use
0xFF	Vendor-specific BENTRY

TABLE 5.7 Nonpersistent Schedule BENTRY

Field	Field Size (Bits)
Number of Sessions (NS)	6
Session Allocation Information (SAI [0])	24 or 32
...	
Session Allocation Information (SAI [L-1])	24 or 32

- Number of Sessions (NS) is a 6-bit field that indicates the number of sessions (L) for which Nonpersistent session allocation information is contained in the Beacon payload.
- Session Allocation Information (SAI) provides the basic unit of medium allocation. The length of the SAI field is 3 or 4 octets, depending on whether the Start Time field is present. The format of SAI is shown in Tables 5.8 and 5.9.

    o Start Time Present Flag (STPF) is a 1-bit field. This field is set to 0b1 to indicate that the Start Time of the session is explicitly specified.
    o Global Link Identifier field contains the lower seven bits of the GLID for which this allocation is provided. Note that the lower seven bits of the GLID can be used to uniquely determine the 8-bit GLID of the session since the high-order bit is always 1.
    o Start Time (ST) is a 12-bit field that specifies the time at which the session associated with a particular GLID will start. This field indicates the time in

TABLE 5.8 Session Allocation Information Format without Start Time

Field	Field Size (Bits)
Start Time Present Flag (STPF)	1
Global Link Identifier (GLID)	7
ET	12
Reserved	4

TABLE 5.9 Session Allocation Information Format with Start Time

Field	Field Size (Bits)
Start Time Present Flag (STPF)	1
Global Link Identifier (GLID)	7
Start Time (ST)	12
End Time (ET)	12

TABLE 5.10 Persistent Schedule BENTRY

Field	Field Size (Bits)
Preview Schedule Countdown (PSCD)	3
Current Schedule Countdown (CSCD)	3
Number of Sessions (NS)	6
Session Allocation Information (SAI [0])	Var
...	
Session Allocation Information (SAI [L-1])	Var

multiples of Allocation Time Unit (10.24 μs). Start Time is measured relative to the Start Time of the corresponding Beacon Period. When ST is not present in SAI, then the ET of the previous session is implicitly used as the Start Time.

o End Time (ET) is a 12-bit field that specifies the time at which the session associated with a particular GLID will end. This field indicates the time in multiples of Allocation Time Unit and is measured relative to the Start Time of the session.

5.2.2.8.2 Persistent Schedule BENTRY Persistent Schedule BENTRY describes the schedules that are valid for multiple Beacon Periods. Table 5.10 shows the format of Persistent Schedule BENTRY.

The interpretation of various fields in Persistent Schedule BENTRY is as follows:

- Preview Schedule Countdown (PSCD) is a 3-bit field that indicates the Beacon Period to which the schedule information is applicable.
  o A value of **0b000** indicates that the schedule pertains to the current Beacon Period. A value of **0b001** indicates that the schedule will become current in 1 Beacon Period (i.e., in the next Beacon Period). A value of **0b010** indicates that the schedule will become current in 2 Beacon Period, and so on.
- Current Schedule Countdown (CSCD) is a 3-bit field and its interpretation depends on PSCD.
  o When PSCD is set to zero, CSCD field defines the number of Beacon Periods for which the current schedule is valid. A value of **0b000** indicates that the current schedule is valid only for the current period. Values from **0b001** to **0b110** indicate that the current schedule is valid for the current and next periods, and so on, respectively. A value of **0b111** indicates that the current schedule is valid indefinitely. The current schedule should be considered valid until new information superseding it is received. Indefinite current schedules are only used in CSMA-only mode.
  o When PSCD is nonzero the CSCD field provides a preview of the value that the published schedule will have in the first Beacon Period in which it is the current schedule.

TABLE 5.11 Regions BENTRY

Field	Field Size (Bits)
Number of Regions (NR)	6
Region Type (RT[1])	4
Region End Time (RET[1])	12
...	
Region Type (RT[L])	4
Region End Time (RET[L])	12

- Number of Sessions (NS) is a 6-bit field that indicates the number of sessions (L) for which persistent session allocation information is contained in the Beacon payload. The maximum number of sessions supported in a network is limited by the size of Beacon payload.
- Session Allocation Information (SAI) provides the basic unit of medium allocation. The format of this field is same as described in Section 5.2.2.8.1.

5.2.2.8.3 Regions BENTRY Regions BENTRY describe the top-level structure of the Beacon Period. This information is used to coordinate the sharing of bandwidth among Neighbor Networks. Regions BENTRY is only interpreted by the CCos. All Beacons include a Regions BENTRY. Furthermore, Regions BENTRY covers the entire Beacon Period, excluding the Beacon Regions. Table 5.11 shows the format of Regions BENTRY.

- Number of Regions (NR) is a 6-bit field that contains a count of the number of Regions in the Beacon Period.
- Region Type (RT) is a 4-bit field that identifies the type of Region for which allocation is made using the Regions BENTRY. A brief description of various Region types is presented here.
    - o Reserved Region indicates Regions where the CCo provides scheduled access to stations within the AVLN. The Persistent and Nonpersistent Schedule BENTRIES provide granularity on how the Reserved Region is to be used in each Beacon Period.
    - o Shared CSMA Region indicates Regions where CSMA channel access is used for sharing between stations in the local network as well as for stations in the neighboring AVLN.
    - o Local CSMA Region indicates Regions where CSMA channel access is used for sharing between stations belonging to the local AVLN.
    - o Stayout Region indicates Regions in which one or more neighboring AVLNs in the Interfering Network List (INL) have specified Reserved or Protected Regions.
    - o Protected Region indicates Regions where another group of AVLNs with a different Beacon Period Start Time have specified a Beacon Region.
    - o Beacon Region indicates Regions where one or more AVLNs transmit Beacons.

o Region End Time is a 12-bit field that defines the end time of a Region within the Beacon Period in multiples of Allocation Time Unit. Region End Time is measured relative to the Start Time of the corresponding Beacon Period.

5.2.2.8.4 MAC Address BENTRY MAC Address BENTRY specifies the 48-bit MAC address of the STA that transmits the Beacon MPDU. This BENTRY must be present in a Discover Beacon. It is optional in other types of Beacons.

5.2.2.8.5 Discover BENTRY carries the TEI of the STA that is designated to transmit the Discover Beacon. There is only one Discover BENTRY in each Central Beacon.

In Uncoordinated and Coordinated Modes, the CCo provides Nonpersistent TDMA allocation for transmission of the Discover Beacons. In CSMA-only mode, the STA should contend using CSMA/CA channel access at Channel Access Priority-2 (CAP2) for transmission of the Discover Beacon.

5.2.2.8.6 Discovered Info BENTRY specifies the following information related to the station transmitting the Beacon (Table 5.12):

- CCo Capability—indicates the CCo Capability level (refer to Section 5.2.2.5).
- Proxy Networking Capability—indicates whether the STA supports Proxy CCo capability.
- Backup CCo Capability—indicates whether the STA supports Backup CCo capability.
- CCo Status—indicates whether the STA is acting as a CCo.
- PCo Status—indicates whether the STA is acting as a Proxy CCo.
- Backup CCo Status—indicates whether the STA is acting as a Backup CCo.
- NumDisSTA—indicates the number of entries in the Discovered STA list.

TABLE 5.12 Discovered Info BENTRY

Field	Field Size (Bits)
Updated	1
CCo Capability	2
Proxy Networking Capability	1
Backup CCo Capability	1
CCo Status	1
PCo Status	1
Backup CCo Status	1
NumDisSTA	8
NumDisNet	8
Authentication Status	1
User-appointed CCo Status	1

- NumDisNet—indicates the number of entries in the Discovered Network list.
- Authentication Status—indicates whether the STA is authenticated or not with an AVLN.
- User-appointed CCo Status—indicates whether the STA is a user appointed CCo.

This BENTRY is always present in a Discover Beacon, and optional in other types of Beacons.

5.2.2.8.7 Beacon Period Start Time Offset BENTRY contains a 3-octet Beacon Period Start Time Offset (BPSTO) that indicates the offset of the start of the Preamble of this Beacon PPDU from the Start Time of the most recent Central Beacon Period of the corresponding AVLN in multiples of 40 ns. This BENTRY is always present in Proxy Beacons and Discover Beacons. BPSTO BENTRY is also present in Central Beacons when the AVLN is operating in CSMA-only mode.

5.2.2.8.8 Encryption Key Change BENTRY indicates that one of the AVLN's fundamental encryption keys is going to change. This BENTRY contains the following fields:

- A 6-bit Key Change Countdown (KCCD) field that indicates the number of Beacon Periods after which the new key will become effective. When this BENTRY is present, KCCD is decremented by one in each Beacon Period until the countdown is complete.
- A 1-bit Key Being Changed (KBC) field that identifies whether the Key Being changed is the Network Encryption Key (NEK) or the Network Membership Key (NMK).
- A 4-bit NewEKS field. If KBC = 0b0, NewEKS is the EKS of the new NEK that has been recently distributed to the active STAs. If KBC = 0b1, NewEKS = **0b0011**, the Payload Encryption Key Select (PEKS) value for NMK Encryption Key, which is the only Payload Encryption Key that requires a countdown in the Beacon. A new NMK and NID pair has been recently distributed to the active STAs.

5.2.2.8.9 CCo Handover BENTRY The Central Coordinator Handover BENTRY indicates an ongoing transfer of CCo function from the current station to a new station in the network. This BENTRY contains the following fields:

- A 6-bit Handover Countdown (HCD) field that indicates the number of Beacon Periods after which the CCo handover will occur. When this BENTRY is present, HCD will decrement by one in each Beacon Period until the countdown is complete.
- An 8-bit New CCo TEI (NCTEI) field that indicates the Terminal Equipment ID of the New CCo.

5.2.2.8.10 Beacon Relocation BENTRY indicates relocation of Beacon to a different part of the AC line cycle. This BENTRY contains the following fields:

- A 6-bit Relocation Countdown (RCD) field that indicates the number of Beacon Periods after which the Beacon relocation will occur. When this BENTRY is present, RCD will decrement by one in each Beacon Period until the countdown is complete. The Beacon Relocation BENTRY will become effective and the BENTRY removed from the Beacon MPDU payload in the Beacon immediately following the Beacon Period when RCD equals 1.
- A 1-bit Relocation Type (RLT) field used to indicate whether the relocation is based on an Offset or change in the Beacon Slot number. Relocation SlotID type will only be used when operating in Coordinated Mode. The new value for Relocation SlotID may be smaller or larger than the current Beacon Slot ID.
- A 1-bit Leaving Group Flag (LGF) indicates whether the CCo is leaving the group.
- A 17-bit unsigned Relocation Offset (RLO) field that indicates the offset of the new Beacon location from the current Beacon location. Figure 5.4 shows an example of Beacon relocation. This field indicates the time in multiples of 0.32 μs. Relocation SlotID is set to indicate to the value of Beacon Slot ID when the Offset becomes effective.
- Relocation SlotID (RLSlotID) is a 3-bit field that indicates the new Beacon Slot ID in a group of networks. Relocation Offset will be set to zero when Relocation Type is Relocation SlotID.

5.2.2.8.11 AC Line Sync Countdown BENTRY is used to indicate the number of Beacon Periods in which the AC Line Cycle Synchronization of the Current CCo is going to change. When this BENTRY is present, the Countdown field will decrement by one in each Beacon Period until the countdown is complete. This BENTRY contains the following fields:

- A 6-bit Countdown field that indicates the number of Beacon Periods in which the AC Line Cycle Synchronization of the Current CCo is going to change.
- A 2-bit Reason Code field that indicates the reason for the AC Line Sync Countdown. A value of **0b00** indicates that the AVLN is Shut Down or the CCo

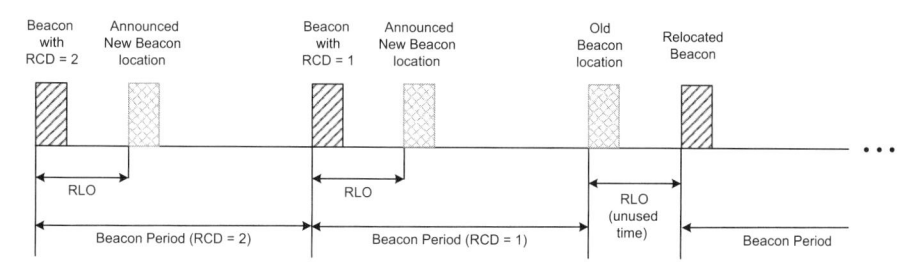

FIGURE 5.4 Example of Beacon Relocation.

is leaving group. A value of **0b01** indicates that the AC Line Cycle Synchronization handover to a CCo in a smaller Beacon Slot.

5.2.2.8.12 Change NumSlots BENTRY is used to change or update NumSlots for a group of networks. This BENTRY will be present in Discover and Proxy Beacons when present in the Central Beacon. This BENTRY contains the following fields:

- A 6-bit NumSlot Change Countdown (NSCCD) field that indicates the number of Beacon Periods after which the new NumSlot will become effective. When this BENTRY is present, NSCCD will decrement by one in each Beacon Period until the countdown is complete.
- A 3-bit New NumSlot Value (NewNumSlot) field that is the value of the NumSlot field in the Beacon MPDU payload that will become effective for all networks in a group of networks when the countdown is complete.

5.2.2.8.13 Change HM BENTRY is used to change or update the Hybrid Mode (HM) field for a group of networks. This BENTRY will be present in Discover and Proxy Beacons when present in the Central Beacon. This BENTRY contains the following fields:

- A 6-bit Hybrid Mode Change Countdown (HMCCD) field that indicates the number of Beacon Periods after which the new value for HM will become effective. When this BENTRY is present, HMCCD is decremented by one in each Beacon Period until the countdown is complete.
- A 2-bit New Hybrid Mode value (NewHM) field that is used to update the HM field in the Beacon MPDU payload for all networks in a group of networks when the countdown is complete.

5.2.2.8.14 Change SNID BENTRY is used to change or update the SNID (Short Network Identifier) field for an AVLN. This BENTRY will be present in Discover and Proxy Beacons when present in the Central Beacon. This BENTRY contains the following fields:

- A 4-bit SNID Change Countdown (SCCD) field that indicates the number of Beacon Periods after which the new SNID value will become effective. When this BENTRY is present, the SCCD is decremented by one in each Beacon Period until the countdown is complete.
- A 4-bit New SNID value (NewSNID) used to update the SNID field in the Beacon MPDU, the SOF, and elsewhere for all STAs in an AVLN when the countdown is complete.

5.2.2.8.15 Vendor-Specific BENTRY enables vendor specific extensions to the Beacon Management Information. The ability to transmit and receive vendor-specific BENTRIES is optional.

The first three octets of the Vendor-Specific BENTRY will contain the IEEE-assigned Organizationally Unique Identifier (OUI). The remaining fields in this BENTRY are defined by the vendor.

5.2.2.9 Beacon Payload Check Sequence (BPCS) The Beacon Payload Check Sequence (BPCS) is a 32-bit field that is used to check the integrity of the Beacon payload. BPCS is a 32-bit CRC computed on the Beacon payload excluding the BPCS field. CRC-32 is be computed using the following standard generator polynomial of degree 32,

$$G(x) = x^{32} + x^{26} + x^{23} + x^{22} + x^{16} + x^{12} + x^{11} + x^{10} + x^8 + x^7 + x^5 + x^4 + x^2 + x + 1$$

5.3 START-OF-FRAME (SOF)

The Start-of-Frame MPDU provides the primary means for stations to exchange data and management information. A SOF MPDU includes a 25-bit HomePlug 1.0 Frame Control (only present in Hybrid Mode), 128-bit SOF Frame Control followed by one or more PHY Blocks. The MAC Data Plane described in Chapter 6 converts the data and management messages into PHY Blocks. Each PHY Block maps on to a single FEC block at the physical layer. This enables MAC Layer to efficiently retransmit corrupt FEC blocks. The SOF payload for unicast transmissions is typically modulated using channel adapted Tone Maps to maximize the performance. Section 5.3.1 describes the contents of SOF Frame Control. The format of the SOF payload is presented in Section 5.3.2 (Figure 5.5).

5.3.1 Start-of-Frame (SOF) Frame Control

Start-of-Frame Frame Control fields can be divided into groups based on the functionality they support as shown in Table 5.13. The interpretation of these fields and their functionality is described in the following sections. SOF MPDUs includes variable length unicast or broadcast payload.

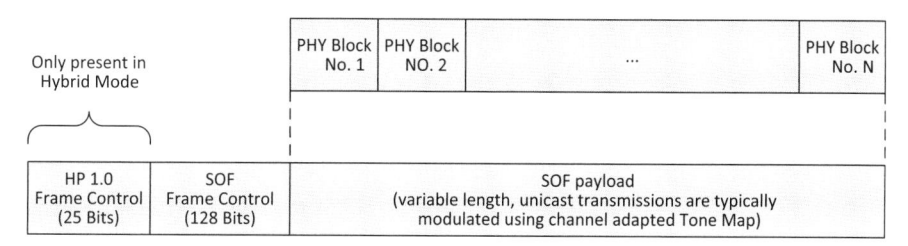

FIGURE 5.5 Start-of-Frame MPDU format.

TABLE 5.13 Start-of-Frame Frame Control Fields Grouped based on Functionality

Functionality	Fields	Field Size (Bits)
Addressing	Source Terminal Equipment Identifier (STEI)	8
	Destination Terminal Equipment Identifier (DTEI)	8
	Multicast Flag (MCF)	1
	Multinetwork Broadcast Flag (MNBF)	1
Queue	Link Identifier (LID)	8
	Management MAC Frame Stream Command (MFSCmdData)	3
	Data MAC Frame Stream Command (MFSCmdMgmt)	3
Bursting	MPDU Count (MPDUCnt)	2
	Bidirectional Burst Flag (BBF)	1
	Max Reverse Transmission Frame Length (MRTFL)	4
	Bit Map SACK info (BM-SACKI)	4
	Data MFS Response (MFSRspData)	2
	Management MFS Response (MFSRspMgmt)	2
Payload Demodulation	Number of Symbols (NumSym)	2
	PHY Block Size (PBSz)	1
	HomePlug AV Tone Map Index (TMI_AV)	5
	HomePlug AV Frame Length (FL_AV)	12
TDMA Allocations	Contention Free Session (CFS)	1
	Pending PHY Blocks (PPB)	8
	Bit-loading Estimate (BLE)	8
SACK Retransmission	Burst Count (BurstCnt)	2
	Request SACK Retransmission (RSR)	1
Encryption Key	Encryption Key Select (EKS)	4
Detection Status	HomePlug 1.0.1 Detected Flag (HP10DF)	1
	HomePlug 1.1 Detect Flag (HP11DF)	1
	Beacon Detect Flag (BDF)	1
Participation in Multiple Network	Different CP PHY Clock Flag (DCPPCF)	1
Convergence Layer	Convergence Layer SAP Type (CLST)	1

5.3.1.1 Addressing-Related Fields contain information about the source and destination of the SOF MPDU payload.

- Source Terminal Equipment Identifier (STEI) is an 8-bit field that is used to indicate the Terminal Equipment Identifier assigned to the station transmitting the SOF delimiter.
- Destination Terminal Equipment Identifier (DTEI) is an 8-bit field used to indicate to the TEI assigned to the HomePlug AV station that is the intended destination of the MPDU.

START-OF-FRAME (SOF)

- Multicast Flag (MCF) is a 1-bit field that is set to 0b1 to indicate that the SOF payload contains broadcast information and all stations in the AVLN should process the payload irrespective of the setting of DTEI field. The MCF field is used for partial acknowledgment of multicast/broadcast data (refer to Section 6.13.2).
- Multinetwork Broadcast Flag (MNBF) is a 1-bit field that is set to 0b1 to indicate that the SOF payload needs to be received by stations in all AVLNs irrespective of SNID contained in the SOF. This enables inter-network communications (refer to Section 6.9).

5.3.1.2 Queue-Related Fields Queues are maintained within each station to provide differentiated services based on MSDU priority. Queue-related fields provide information on queues to which the contents of the MPDU payload.

- Link Identifier (LID) is an 8-bit field that indicates the link (or queue) associated with the MPDU payload.
- Data MAC Frame Stream Command (MFSCmdData) is a 3-bit field that is used for buffer management and flow control of data queues (refer to Section 6.7).
- Management MAC Frame Stream Command (MFSCmdMgmt) is used for buffer management and flow control of management queues.

The SOF payload can include both data and management information. Hence, MAC Frame Stream Command for both data and management is included in each SOF.

5.3.1.3 Bursting-Related Fields Bursting enables AV stations to transmit multiple SOF and RSOF MPDUs without relinquishing control of the medium (refer to Sections 6.11 and 6.12). Bursting-related fields in the SOF indicate whether the SOF is part of a Burst and if the receiver is provided control of the medium after the end of transmission.

- MPDU Count (MPDUCnt) is a 2-bit field that indicates the number of MPDUs to expect in the current burst transmission. Up to four MPDUs can be supported in a Burst. A value of **0b00** indicates that this is the last MPDU in the MPDU Burst.
- Bidirectional Burst Flag (BBF) is set to 0b1 to indicate that a Bidirectional Burst may continue after this MPDU. In this case, the receiver can transmit a RSOF instead of a SACK.
- Maximum Reverse Transmission Frame Length (MRTFL) indicates the maximum frame length a receiver may use in a Reverse SOF. This field is valid only when BBF is set to 0b1.
- Data MAC Frame Stream Response (MFSRspData) is a 2-bit field that contains the response from the receiver's Data MAC Frame Stream to the corresponding command (MFSCmdData) from the transmitter (i.e., transmitted as part of RSOF payload).

- Management MAC Frame Stream Response (MFSRspMgmt) is a 2-bit field that contains the response from the receiver's Management MAC Frame Stream to the corresponding command (MFSCmdData) from the transmitter (i.e., transmitted as part of RSOF payload).
- Bit Mapped SACK Information (BM-SACKI) field in the SOF acknowledges the PBs transmitted in the previous Reverse SOF. LSB indicates the reception status of the first PB and so on.

5.3.1.4 Payload Demodulation-Related Fields contain information necessary for the receiver to demodulate and receive contents of the SOF payload.

- PHY Block Size (PBSz) is a 1-bit field that indicates the MPDU payload block size. A value 0b0 indicates that the SOF payload contains one or more 520 Octet PHY Blocks. A value of 0b1 indicates that the MPDU payload contains a single 136 Octet PHY Block.
- Number of Symbols (NumSym) is a 2-bit field that indicates the number of OFDM Symbols used for transmitting the MPDU payload. This field is used to indicate whether the SOF payload contains zero, one, two, or greater than two OFDM symbols. The Receiver uses the Number of Symbols field to determine the RIFS_AV used in this transmission. Note that FL_AV includes the duration of SOF payload OFDM symbols as well as the Response Interframe Space. Further, RIFS_AV is different for zero, one, two, and greater than two OFDM symbol payloads. Support for variable RIFS_AV (ranging from 30 to 160 μs) is intended to provide flexibility for implementers while maintaining interoperability.
- Tone Map Index (TMI_AV) is a 5-bit field that indicates the Tone Map to be used by the receiver in demodulating the MPDU payload. TMI_AV can indicate a ROBO Tone Map (i.e., Std-ROBO, Mini-ROBO, HS-ROBO) or a receiver specified tone map.
- Frame Length (FL_AV) is a 12-bit field that indicates the duration of MPDU payload and the interframe space (BIFS if MPDUCnt $>$ 0 or RIFS_AV if MPDUCnt $=$ 0) that follows the MPDU payload. FL_AV is measured in multiples of 1.28 μs. Figure 5.6 shows the measurement of the FL_AV parameter.

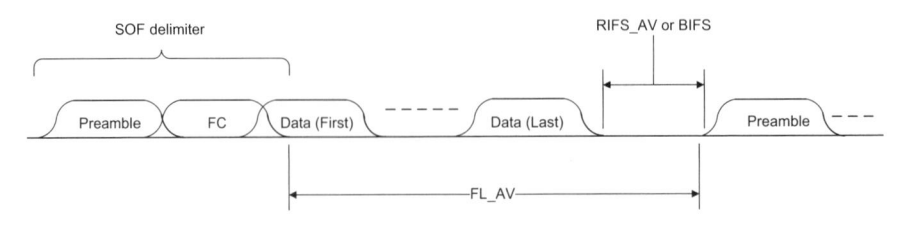

FIGURE 5.6 Measurement of FL_AV.

START-OF-FRAME (SOF)

RIFS_AV is a variable duration of time, based on the Tone Map used for modulating the MPDU payload and the number of Orthogonal Frequency Division Multiplexing (OFDM) symbols used for carrying the MPDU payload. The station(s) that is the intended destination of the MPDU payload can use its knowledge of RIFS_AV to determine the duration of MPDU payload. All other stations use FL_AV to determine the location of the response delimiter.

The maximum duration of Frame Length (MaxFL_AV) that can be supported by AV station is 5242.88 μs. However, it is optional for stations to support Frame Lengths greater than 2501.12 μs. Channel Estimation Indication is used by the receiver to indicate the maximum value of FL_AV that it is capable of receiving. Further, during CSMA allocations Frame Lengths greater than 2501.12 μs are not allowed to limit the cost of collisions.

When operating in the presence of HomePlug 1.0.1 stations, AV stations restrict their Frame Length durations to enable coexistence with HomePlug 1.0.1 (i.e., maintain proper virtual carrier sense, refer to Section 11.4).

5.3.1.5 TDMA Allocation-Related Fields
enable the CCo to dynamically adjust the duration of TDMA allocations.

- Pending PHY Blocks (PPB) is an 8-bit field that indicates the total number of PHY blocks pending transmission at the completion of the current MPDU. This information is included in all MPDUs transmitted during a TDMA allocation and is used by the CCo to determine if the TDMA allocation size needs to be adjusted to maintain the negotiated QoS.
- Bit-loading Estimate (BLE) is an 8-bit field that represents the number of user data bits (i.e., data bit prior to Forward Error Correction (FEC) encoding) that can be carried on the channel per microsecond. It takes into account the overhead due to the cyclic prefix and FEC, but does not include overhead associated with the PPDU format (e.g., delimiter) and IFS. This is used by the CCo to increase or decrease the TDMA allocation as channel conditions change.
- Contention Free Session (CFS) flag is a 1-bit field that is set to 0b1 on SOFs transmitted during TDMA (or Contention Free) allocations. This field enables other stations to determine that the SOF is part of a Contention Free transmission.

5.3.1.6 SACK Retransmission-Related Fields
enable transmitter to request the receiver to retransmit SACK for prior transmission (refer to Section 6.13.1).

- Request SACK Retransmission (RSR) is a 1-bit field that is set to 0b1 to indicate that the SOF is transmitted to request receiver to retransmit SACK information. RSR can only be used for Global Links (i.e., TDMA streams).
- Burst Count (BurstCnt) is a 2-bit field that is incremented after each MPDU burst is transmitted on the media for a particular Global Link when RSR = 0b0.

The Burst Count value for a Global Link is not affected by transmissions for other Global Links. The BurstCnt field is used by the station receiving the SOF MPDU to resolve ambiguities in the request SACK retransmission protocol.

5.3.1.7 Encryption-Related Fields The content of the SOF payload can be encrypted to maintain security. Encryption Key Select (EKS) is a 4-bit field that indicates the index of the AES-128 bit Encryption Key used for encrypting segments in the SOF payload. The mapping between EKS and Encryption Key is provided by the CCo for authenticated stations within the AVLN.

EKS value of 0xF is used to indicate that the SOF payload is unencrypted. Unencrypted transmissions are typically used for inter-network communication and during Authentication/Authorization. Encryption is mandatory for all communications between stations that are authenticated within the AVLN (i.e., all transmissions within an AVLN are always encrypted).

5.3.1.8 Detection Status-Related Fields are used by stations to indicate detection status of HomePlug 1.0.1 transmissions, HomePlug 1.1 transmissions and CCo Beacons.

- HomePlug 1.0.1 Detect Flag (HP10DF) is a 1-bit field that is set to 0b1 when the Station detected a HomePlug 1.0.1 transmission. This information is used by the CCo determine if the network should operate in AV-only mode or Hybrid Mode.
- HomePlug 1.1 Detect Flag (HP11DF) is a 1-bit field that is set to 0b1 when the station detected a HomePlug 1.1 transmission. HomePlug 1.1 stations are basically HomePlug 1.0.1 stations with extensions to better coexist with HomePlug AV. This information is used by the CCo determine if the network needs to operate in Hybrid Mode or AV-only mode. Currently, there are no HomePlug 1.1 capable products. It is likely that this feature will never be used in practice.
- Beacon Detect Flag (BDF) is a 1-bit field that indicates whether the station heard the Central Beacon transmission from the CCo of the AVLN to which it is associated during the current period.

5.3.1.9 Participation in Multiple Networks-Related Fields AV stations can be part of multiple networks at the same time. A station participating in multiple networks needs to be able to receive SOF MPDUs from any of the networks during Contention Period. This creates a challenge as the station needs to know which networks Network Time Base (NTB) is to be applied before receiving the SOF MPDU.

Different CP PHY Clock Flag (DCPPCF) is a 1-bit field that is set to 0b1 when the station uses a different PHY Receive Clock Correction during the Contention Period (CP) than the Short Network Identifier (SNID) indicated in the SOF. Otherwise, it will be set to 0b0. This enables the receiver of to know that it needs to precede any unicast transmission to this station with an RTS/CTS exchange.

START-OF-FRAME (SOF)

Stations participating in multiple networks use the SNID in the RTS to determine the correct NTB to be applied to receive the subsequent SOF MPDU (refer to Section 7.5.1.2).

5.3.1.10 Convergence Layer SAP Type (CLST) is a 1-bit field that indicates the Convergence Layer SAP for which the current MPDU payload is designated. AV stations always set this field to 0b0 to indicate that they are using Ethernet SAP.

5.3.2 SOF Payload

SOF MPDU payload have one of two formats:

- One or more 520-octet PBs.
- A single 136-octet PB.

HomePlug AV supports PB sizes of 520 octets and 136 octets as shown in Figure 5.7. Each PB is mapped onto a single FEC block at the physical layer. Each PB contains a PHY Block Header (PBH), PHY Block Body (PBB), and PHY Block Check Sequence (PBCS). PBH and PBCS are four octets each. When the PB size is 520 octets, the PB Body is 512 octets long. When the PB size is 136 octets, the PBB is 128 octets long.

At high data rates, multiple 520-octet PBs can fit in a single OFDM Symbol. In such cases, HomePlug AV transmissions are restricted to have only one PB ending in

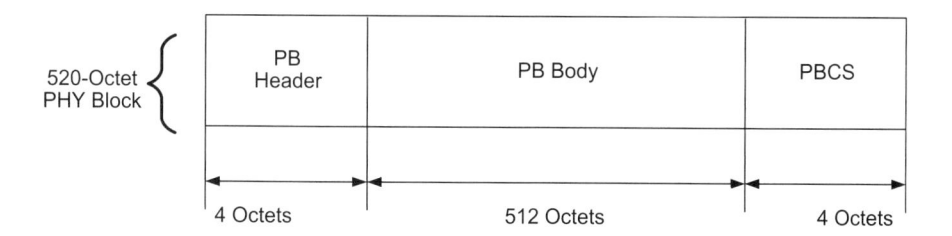

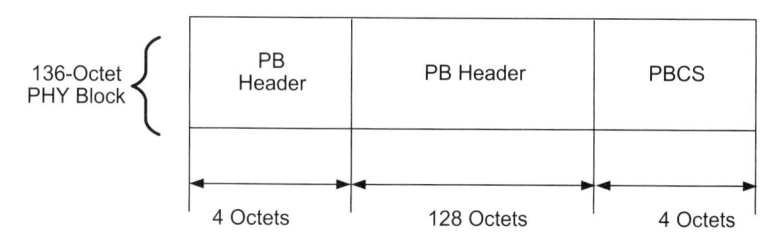

FIGURE 5.7 PHY Block formats.

TABLE 5.14 PHY Block Header Fields

Field	Field Size (Bits)
Segment Sequence Number (SSN)	16
MAC Frame Boundary Offset (MFBO)	9
Valid PHY Block Flag (VPBF)	1
Management Message Queue Flag (MMQF)	1
MAC Frame Boundary Flag (MFBF)	1
Oldest Pending Segment Flag (OPSF)	1

the last OFDM symbol of the MPDU. This restriction is intended to allow sufficient processing time for the last PHY block within an MPDU and generate a SACK based on the negotiated RIFS_AV.

Table 5.14 shows the fields in the 4-octet PB Header field. The format of the PHY Block Header is independent of the PB size being used. The interpretation of fields in the PHY Block is as follows:

- Segment Sequence Number (SSN) is a 16-bit field that is initialized to 0 and is incremented by 1 for each new segment in a transmit queue. The SSN is used by the receiver to properly reorder the received segments for reassembly.
- MAC Frame Boundary Offset (MFBO) is a 9-bit field that carries the offset in octets of the MAC Frame boundary (i.e., the first octet of the first new MAC Frame) within the PHY Block Body to enable the receiver to extract MAC Frames from the received segments.. This field is valid only when the MAC Frame Boundary Flag is set to 0b1. A value of 0x000 indicates the first octet and so on.
- Valid PHY Block Flag (VPBF) is set to 0b1 to indicate that the current PHY block contains valid data. When VPBF is set to 0b0, it indicates that the PB is empty. Empty PBs do not carry valid data and are discarded by the receiver. Empty PB(s) may be added to the end of an MPDU to ensure that only one PB is ending in the last OFDM Symbol of the long PPDU.
- Management Message Queue Flag (MMQF) is a 1-bit field that is set to 0b0 on PHY Blocks carrying data and is set to 0b1 on PHY Blocks carrying Management messages. Since Long MPDU payload can include segments from both data and management queues, this information is necessary to determine the queue to which the PB belongs during the reassembly process.
- MAC Frame Boundary Flag (MFBF) is a 1-bit flag that is set to 0b1 to indicate that a MAC Frame boundary is present in the current PBB. Note that when MAC Frame is large (e.g., 1524 octets), some of the segments will not contain a MAC Frame boundary.
- Oldest Pending Segment Flag (OPSF) is a 1-bit field that is set to 0b1 to indicate that the segment contained in the PBB is the oldest pending segment in the corresponding MAC Frame Stream at the transmitter. This is used to

synchronize the transmitter and receiver with respect to the Minimum Segment Sequence Number.
- PHY Block Body (PBB) carries the encrypted segment as the payload. The PBB field is either 512 or 128 octets, depending on the PB size. Note that a segment may have to be padded before encryption to ensure that it fits exactly into the PBB.
- PB Check Sequence (PBCS) contains a 32-bit CRC and is computed over the PB Header and the encrypted PB Body. CRC-32 is be computed using the following standard generator polynomial of degree 32.

$$G(x) = x^{32} + x^{26} + x^{23} + x^{22} + x^{16} + x^{12} + x^{11} + x^{10} + x^8 + x^7 + x^5 + x^4 + x^2 + x + 1$$

5.4 SELECTIVE ACKNOWLEDGMENT (SACK)

Selective Acknowledgments (SACKs) are used to indicate the reception status of the PHY Blocks in SOF and RSOF. SACK MPDU contains 25-bit HomePlug 1.0 Frame Control (only present in Hybrid Mode) and a 128-bit SACK Frame Control as shown in Figure 5.8. The SACK MPDU does not include any MPDU payload.

The SACK Frame Control fields can be divided into groups based on the functionality they support as shown in Table 5.15. The interpretation of these fields and their functionality is described in the following sections.

5.4.1 Addressing-Related Field

Addressing-related fields contain information about the destination of the SACK.

- Destination Terminal Equipment Identifier (DTEI) is an 8-bit field used to indicate to the TEI assigned to the HomePlug AV station that is the intended destination of the SACK.

Source TEI is not explicitly communicated in the SACK. Since SACKs are expected at a fixed time after the end of transmission, the transmitter can implicitly

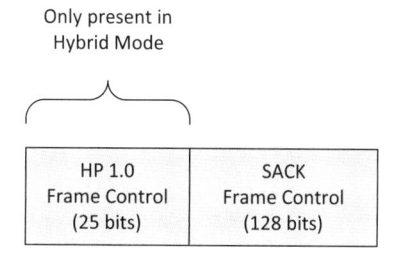

FIGURE 5.8 SACK MPDU format.

TABLE 5.15 SACK Frame Control Fields Grouped based on Functionality

Functionality	Fields	Field Size (Bits)
Addressing	Destination Terminal Equipment Identifier (DTEI)	8
Queue	Data MAC Frame Stream Response (MFSRspData)	2
	Management MAC Frame Stream Response (MFSRspMgmt)	2
	Receive Window Size (RxWSz)	4
Bursting	Request Reverse Transmission Flag (RRTF)	1
	Request Reverse Transmission Length (RRTL)	4
TDMA Allocations	Contention Free Session (CFS)	1
Detection Status	Beacon Detect Flag (BDF)	1
SACK Data	SACK Data (SACKD)	Var
Version	SACK Version Number (SVN)	1

determine the source station that transmitted the SACK. Elimination of STEI from the SACK was primarily done to allow more space for SACK Data.

5.4.2 Queue-Related Field

Queue-related fields in the SACK enable queue management and flow control.

- Data MAC Frame Stream Response (MFSRspData) is a 2-bit field that contains the response from the receiver's Data MAC Frame Stream to the corresponding command (MFSCmdData) from the transmitter (refer to Section 5.3.1.2).
- Management MAC Frame Stream Response (MFSRspMgmt) is a 2-bit field that contains the response from the receiver's Management MAC Frame Stream to the corresponding command (MFSCmdData) from the transmitter.
- Receive Window Size (RxWSz) is a 4-bit field that indicates the reassembly buffer available for the corresponding MAC Frame Stream at the receiver. The Receive Window Size field is only present in the SACK for Priority Links.

The Link Identifier (LID) of the queue is not explicitly communicated in the SACK. Since SACKs are expected at a fixed time after the end of transmission, the transmitter can implicitly determine the LID to which the queue-related information applies. Elimination of LID from the SACK was primarily done to allow more space for SACK Data.

5.4.3 Bursting-Related Field

Bursting-related fields in the SACK are used to request a reverse transmission opportunity using the bidirectional bursting mechanism (refer to Section 6.12).

SELECTIVE ACKNOWLEDGMENT (SACK) **87**

- Request Reverse Transmission Flag (RRTF) is a 1-bit field that is set to 0b1 to indicate the receiver is requesting the transmitter to initiate a bidirectional burst.
- Request Reverse Transmission Length (RRTL) is a 4-bit field that specifies the minimum required duration for a Reverse Transmission.

5.4.4 TDMA Allocation-Related Fields

SACK includes a 1-bit Contention Free Session (CFS) field that is set to indicate that the SACK is part of a Contention Free transmission.

5.4.5 Detection Status-Related Field

Beacon Detect Flag (BDF) is a 1-bit field that indicates whether the station heard the Central Beacon transmission from the CCo of the AVLN to which it is associated during the current period.

5.4.6 Version-Related Fields

SACK Version number is indicated using the SVN field and is used to enable extensions to the SACK Frame Control contents in future versions of the AV standard. AV 1.1 stations set this field to 0b0.

5.4.7 SACK Data

SACK Data includes acknowledgment information for the PHY Blocks in MPDU Burst to which the SACK applies. SACK Data can support up to four MPDUs in the Burst.

SACK Data includes a SACK Type (SACKT) and SACK Information (SACKI) for four MPDUs. SACK Type is a 2-bit field that indicates whether

- SACK information contains Bit Map SACK information (i.e., mixture of good and bad PBs indicated using 1-bit per PB).
- SACK information contains compressed Bit Map SACK information. In this case, a compression scheme defined in HomePlug AV is applied on the bit map SACK information to reduce the number of bits required for transmitting the SACK information.
- MPDU was not received (i.e., either no Preamble was detected or a corrupt Frame Control was detected). In this case, SACK information is not present.
- Uniform SACK information, in this case SACK Data can indicate
   o All PBs in the corresponding MPDU were received with error.
   o All PBs in the corresponding MPDU were received properly.
   o Tone Map used for the MPDU is invalid.

o PHY Block decryption error, that is, receiver does not have the Network Encryption Key indicated by the Encryption Key Select in the SOF Frame Control to decrypt the PHY Blocks in the MPDU payload.

It should be noted that roughly 72 bits are available for transmission of SACK Data. At high PHY data rate the number of PHY Blocks transmitted in a Burst can be significantly larger than 72. Further, due to noise, the PB error rate is typically in the range of 1–10%. Hence, compressed version of bit map is quite commonly used in SACK Data.

5.5 REQUEST TO SEND (RTS)/CLEAR TO SEND (CTS)

Request to Send (RTS) and Clear to Send (CTS) are primarily intended to handle hidden station within the network. In HomePlug AV, RTS/CTS can also be used for other functions such participation in multiple network (refer to Section 7.5.1.2) and multinetwork broadcasting (refer to Section 6.9.1).

RTS and CTS MPDUs share the same MPDU (or delimiter) type and have the same format. The RTS Flag in the RTS/CTS Frame Control is used to distinguish between RTS and CTS. RTS/CTS contains 25-bit HomePlug 1.0 Frame Control (only present in Hybrid Mode) and a 128-bit RTS/CTS Frame Control as shown in Figure 5.9. RTS/CTS MPDU does not include any MPDU payload.

RTS/CTS frame control fields can be divided into groups based on the functionality they support as shown in Table 5.16. The interpretation of these fields and their functionality is described in the following sections.

5.5.1 Addressing-Related Fields

Addressing-related fields contain information about the source and destination of the RTS/CTS MPDU.

- Source Terminal Equipment Identifier (STEI) is an 8-bit field that is used to indicate the Terminal Equipment Identifier assigned to the station transmitting the RTS/CTS delimiter.

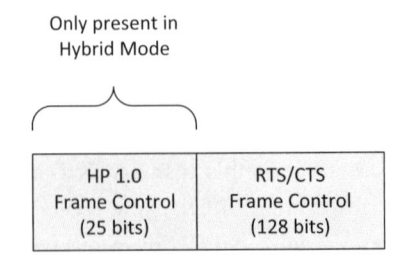

FIGURE 5.9 RTS/CTS MPDU format.

TABLE 5.16 RTS/CTS Frame Control Fields Grouped based on Functionality

Functionality	Fields	Field Size (Bits)
Addressing	Source Terminal Equipment Identifier (STEI)	8
	Destination Terminal Equipment Identifier (DTEI)	8
	Multicast Flag (MCF)	1
	Multinetwork Broadcast Flag (MNBF)	1
Queue	Link Identifier (LID)	8
TDMA Allocations	Contention Free Session (CFS)	1
Detection Status	HomePlug 1.0.1 Detect Flag (HP10DF)	1
	HomePlug 1.1 Detect Flag (HP11DF)	1
	Beacon Detect Flag (BDF)	1
RTS Flag	RTS Flag (RTSF)	1
Immediate Grant	Immediate Grant Flag (IGF)	1
Duration	Duration (DUR)	14

- Destination Terminal Equipment Identifier (DTEI) is an 8-bit field used to indicate to the TEI assigned to the HomePlug AV station that is the intended destination of the RTS/CTS MPDU.
- Multicast Flag (MCF) is a 1-bit field that is set 0b1 to indicate that the subsequent SOF MPDU payload will contain broadcast information and all stations in the AVLN should process the payload irrespective of the setting of DTEI field. MCF field is used for partial acknowledgment of multicast/broadcast data (refer to Section 6.13.2).
- Multinetwork Broadcast Flag (MNBF) is a 1-bit field that is set to 0b1 to indicate that the subsequent SOF MPDU payload needs to be received by stations in all networks irrespective of SNID contained in the SOF. This enables internetwork communications (refer to Section 6.9).

5.5.2 Queue-Related Fields

Queues are maintained within each station to provide differentiated services based on MSDU priority. RTS/CTS include an 8-bit Link Identifier (LID) that indicates the link (or queue) of the subsequent SOF MPDU payload.

5.5.3 TDMA Allocation-Related Fields

RTS/CTS include a 1-bit Contention Free Session (CFS) field that is set to indicate that they are transmitted as part of a Contention Free transmission.

5.5.4 Detection Status Fields

Detection status fields are used by stations to indicate their of detection status of HomePlug 1.0.1, HomePlug 1.1, and CCo Beacons.

- HomePlug 1.0.1 Detect Flag (HP10DF) is a 1-bit field that is set to 0b1 when the Station detected a HomePlug 1.0.1 transmission. This information is used by the CCo determine if the network should operate in AV-only mode or Hybrid Mode.
- HomePlug 1.1 Detect Flag (HP11DF) is a 1-bit field that is set to 0b1 when the station detected a HomePlug 1.1 transmission. HomePlug 1.1 stations are basically HomePlug 1.0.1 stations with extensions to better coexist with HomePlug AV. This information is used by the CCo determine if the network needs to operate in Hybrid Mode or AV-only mode. Beacon Detect Flag (BDF) is a 1-bit field that indicates whether the station heard the Central Beacon transmission from the CCo of the AVLN to which it is associated during the current period.

5.5.5 Immediate Grant-Related Fields

Immediate Grant Flag (IGF) is a 1-bit field that is set to 0b1 in the RTS to indicate that the receiver is provided with a reverse transmission for the duration indicated in the DUR field. This mechanism is used by the CCo to provide transmission opportunities to stations when operating in the presence of HomePlug 1.0 stations. In practice, Immediate Grant is very rarely used.

5.5.6 Virtual Carrier Sense (VCS)-Related Fields

Duration (DUR) is a 14-bit field that indicates the duration of medium reservation in multiples of 1.28 μs. RTS and CTS are separated by RTS-to-CTS Gap (RCG). The gap between CTS and the subsequent transmission is CTS-to-MPDU Gap (CMG, 120 μs) when IGF is set to 0b0. The duration field is measured from the end of the corresponding RTS or CTS until the end of the SACK as shown in Figure 5.10.

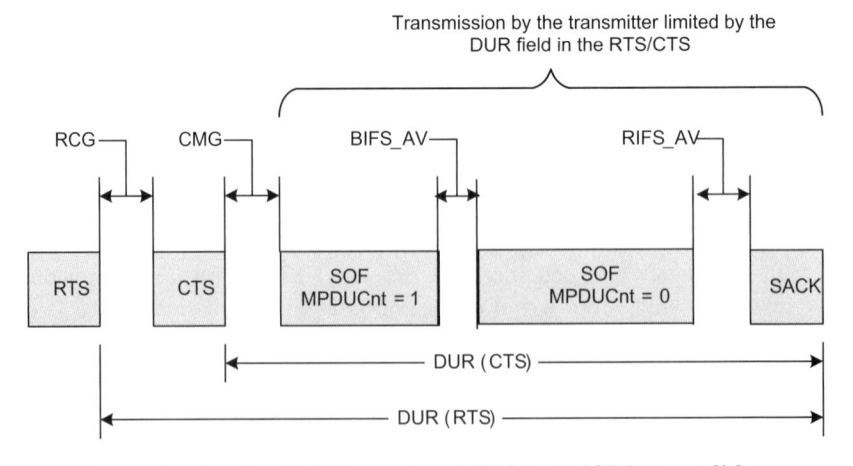

FIGURE 5.10 Duration field in RTS/CTS when IGF is set to 0b0.

SOUND

5.5.7 RTS Flag

RTS Flag (RTSF) is a 1-bit field that indicates whether the corresponding MPDU is an RTS or CTS. A value of 0b1 indicates that the MPDU is a RTS. A value of 0b0 indicates that the MPDU is CTS.

5.6 SOUND

The Sound MPDU is used during channel estimation. Sound MPDUs that include a MPDU payload are referred to as Long Sound MPDU. Long Sound MPDUs are transmitted by a station to enable the receiving station to estimation the channel. Stations acknowledge the reception of Long Sound MPDU using a Sound ACK. A Sound ACK does not include any MPDU payload. Long Sound MPDUs and Sound ACKs share the same MPDU (or delimiter) Type and are distinguished based on the "Sound ACK Flag" field in the Sound Frame Control.

Long Sound MPDUs consist of a 25-bit HomePlug 1.0 Frame Control (only present in Hybrid Mode), 128-bit Sound Frame Control followed by either a 136-octet Sound payload transmitted using Mini-ROBO or a 520 Octet Sound payload transmitted using Std-ROBO. Long Sound payloads consist of a zero pad followed by 4-octet Sound Payload Check Sequence (SPCS) as shows in Figure 5.11. Since the contents are Sound payload are fixed, the receiver can use this information to accurately estimate the channel.

Sound ACK consists of a 25-bit HomePlug 1.0 Frame Control (only present in Hybrid Mode) followed by 128-bit Sound Frame Control.

The Sound Frame Control format is described in Section 5.6.1. The format of the Sound payload is described in Section 5.6.2. Sound ACK does not include any MPDU payload.

5.6.1 Sound Frame Control

Sound Frame Control fields can be divided into groups based on the functionality they support as shown in Table 5.17.

5.6.1.1 Addressing Addressing-related fields contain information about the source and destination of the Sound MPDU.

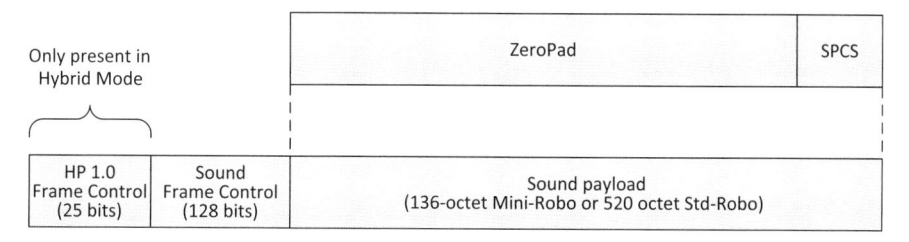

FIGURE 5.11 Long Sound MPDU format.

TABLE 5.17 Sound Frame Control Fields Grouped based on Functionality

Functionality	Fields	Field Size (Bits)
Addressing	Source Terminal Equipment Identifier (STEI)	8
	Destination Terminal Equipment Identifier (DTEI)	8
Queue	Link Identifier (LID)	8
Bursting	MPDU Count (MPDUCnt)	2
Payload Demodulation	PHY Block Size(PBSz)	1
	HomePlug AV Frame Length (FL_AV)	12
TDMA Allocations	Contention Free Session (CFS)	1
	Pending PHY Blocks (PPB)	1
Detection Status	Beacon Detect Flag (BDF)	1
Sound SACK	Sound ACK Flag (SAF)	1
Sound Complete	Sound Complete Flag (SCF)	1
Sound Reason Code	Sound Reason Code (SRC)	8
Max Tone Maps Requested	Max Tone Maps Requested (REQ_TM)	3

- Source Terminal Equipment Identifier (STEI) is an 8-bit field that is used to indicate the Terminal Equipment Identifier assigned to the station transmitting the Sound MPDU.
- Destination Terminal Equipment Identifier (DTEI) is an 8-bit field used to indicate to the TEI assigned to the HomePlug AV station that is the intended destination of the Sound MPDU.

5.6.1.2 Queue Link Identifier (LID) is an 8-bit field that indicates the link (or queue) associated with the Sound MPDU. It should be noted that channel adaptation between a pair of stations is same for all Links between them. LID in Sound MPDU is primarily intended for use by Global Links as this will enable the STA to indicate the PPBs associated with the Global Link to the CCo. The use of Sound MPDUs by Global Links will also notify the CCo that channel conditions have changed and the TDMA allocations need to be adjusted to provide the negotiated QoS.

5.6.1.3 Bursting Bursting enables AV stations to transmit multiple Sound MPDUs without relinquishing the control of the medium. Sound Frame Control includes a 2-bit MPDUCnt field that indicates the number of Sound MPDUs to expect in the current burst transmission. Up to four MPDUs can be supported in a Burst. A value of **0b00** indicates that this is the last MPDU in the MPDU Burst. Bidirectional Bursting is not allowed with Sound MPDUs.

5.6.1.4 Payload Demodulation Payload Demodulation-related fields contain necessary information for receiver to demodulate and receive contents of the Sound payload. Sound MPDU payload is always modulated using either Mini-ROBO or Std-ROBO modulation.

SOUND

- PHY Block Size (PBSz) is a 1-bit field that indicates the MPDU payload block size. A value 0b0 indicates that the Sound payload contains one 520 octet PHY Blocks modulated using Std-ROBO. A value of 0b1 indicates that the Sound payload contains a single 136 octet PHY Block modulated using Mini-ROBO.
- Frame Length (FL_AV) is a 12-bit field that indicates the duration of MPDU payload and the interframe space (BIFS if MPDUCnt > 0 or RIFS_AV if MPDUCnt = 0) that follows the MPDU payload. The interpretation of FL_AV is similar to that of the corresponding field in SOF.

5.6.1.5 TDMA Allocations TDMA allocation-related fields enable CCo to dynamically adjust the duration of TDMA allocations.

- Pending PHY Blocks (PPB) is an 8-bit field that indicates the total number of PHY blocks pending transmission at the completion of the current MPDU. This information is included in all MPDUs transmitted during TDMA allocation and is used by CCo to determine if the TDMA allocation size needs to be adjusted to maintain the negotiated QoS.
- Contention Free Session (CFS) flag is a 1-bit field that is set to 0b1 on SOFs transmitted during TDMA (or Contention Free) allocations. This field enables other stations to determine that the SOF is part of a Contention Free transmission.

5.6.1.6 Detection Status-Related Field Sound frame control includes a 1-bit Beacon Detect Flag that indicates whether the station heard the Central Beacon transmission from the CCo of the AVLN to which it is associated during the current period.

5.6.1.7 Sound ACK Sound ACK Flag (SAF) is a 1-bit field that indicates whether the MPDU is a Long Sound MPDU or a Sound ACK MPDU. The Long Sound MPDU contains a payload that is used by the receiving station to estimate the channel characteristics. Sound ACK MPDU indicates the reception status and completion of the sounding process. The Interframe Spacing between the Sound MPDU with MPDUCnt of **0b00** and corresponding Sound ACK is the RIFS_AV_default (140 µs).

5.6.1.8 Sound Complete Flag Sound Complete Flag is a 1-bit field that is used by the station transmitting the Sound ACK to indicate whether it received sufficient number of Sound MPDUs to complete channel estimation.

5.6.1.9 Sound Reason Code Sound Reason Code is an 8-bit field that indicates the reason for transmitting the Sound MPDU. The various sound reason codes are as follows:

- Sound MPDU is transmitted to obtain the Tone Map corresponding to a TMI_AV. This can be used when the receiver specified a Tone Map index

but the transmitter does not recognize the Tone Map. The five LSBs of this field contain the TMI_AV. In response to this reason code, the receiver may resend the Tone Map Data (TMD) corresponding to TMI_AV or it may resend all valid Tone Maps upon the reception of this Sound Reason Code.
- Sound MPDU is transmitted to indicate a Tone Map error condition detected at the transmitter. The Receiver will resend all valid Tone Maps.
- Sound MPDU is transmitted as part of Initial Channel Estimation.
- Sound MPDU is transmitted in an interval where the receiver has indicated that no AC Line Cycle adapted Tone Maps are available.
- Sound MPDU is transmitted in an interval specified as unusable by the receiver.

5.6.1.10 Max Tone Maps Max Tone Maps Requested (REQ_TM) is a 3-bit field that indicates the maximum number of TMs that the transmitting STA can support for the receiving station. The receiving STA should not generate more than this number of distinct TMs during the channel estimation procedures. ROBO Mode Tone Maps are not counted toward this limit. This field is valid only when the Sound ACK Flag (SAF) is zero.

5.6.2 Format of Sound MPDU Payload

The Sound MPDU payload carries 136 or 520 octets of payload modulation using Mini-ROBO or ROBO Modulation, respectively. Sound MPDUs are used during the channel estimation process, with the transmitter specifying whether the Sound MPDU has 136 or 520 octets of payload. The format of the Sound MPDU payload is shown in Table 5.18. The Sound MPDU payload is not encrypted.

- Zero Pad (ZPAD) field consists of 132 or 516 octets of zeros.
- Sound Payload Check Sequence (SPCS) is used to check the integrity of the Sound payload. SPCS is a 32-bit CRC computed on the Sound payload. CRC-32 is be computed using the following standard generator polynomial of degree 32

$$G(x) = x^{32} + x^{26} + x^{23} + x^{22} + x^{16} + x^{12} + x^{11} + x^{10} + x^8 + x^7 + x^5 + x^4 + x^2 + x + 1$$

TABLE 5.18 Sound Payload Fields

Field	Field Size (Octets)
Zero Pad (ZPAD)	132 or 516
Sound Payload Check Sequence (SPCS)	4

5.7 REVERSE START-OF-FRAME (RSOF)

The Reverse SOF (RSOF) is a long MPDU used to carry both SACK information and payload (i.e., piggy backs data with acknowledgment information). RSOF is used for reverse transmissions during bidirectional bursting.

A RSOF MPDU includes a 25-bit HomePlug 1.0 Frame Control (only present in Hybrid Mode), 128-bit SOF Frame Control followed by one or more PHY Blocks as shown in Figure 5.12. The RSOF payload is typically modulated using channel adapted Tone Maps to maximize the performance. Section 5.7.1 describes the contents of SOF Frame Control. The Format of the Reverse SOF payload is the same as the SOF payload as described in Section 5.3.2.

5.7.1 Reverse SOF (RSOF) Frame Control

RSOF Frame Control fields can be divided into groups based on the functionality they support as shown in Table 5.19. The interpretation of these fields and their functionality is described in the following sections.

5.7.1.1 Addressing-Related Field Addressing-related fields contain information about the destination of the RSOF.

- Destination Terminal Equipment Identifier (DTEI) is an 8-bit field used to indicate to the TEI assigned to the HomePlug AV station that is the intended destination of the RSOF MPDU.

Source TEI is not explicitly communicated in the RSOF. Since RSOF is expected at a fixed time after the end of transmission, the transmitter can implicitly determine the Source station that transmitted with RSOF. Elimination of STEI from the RSOF was primarily done to allow more space for SACK Data.

5.7.1.2 Queue-Related Field Queue-related fields enable queue management and flow control.

- Data MAC Frame Stream Response (MFSRspData) is a 2-bit field that contains the response from the receiver's Data MAC Frame Stream to the corresponding command (MFSCmdData) from the transmitter (refer to Section 5.3.1.2).

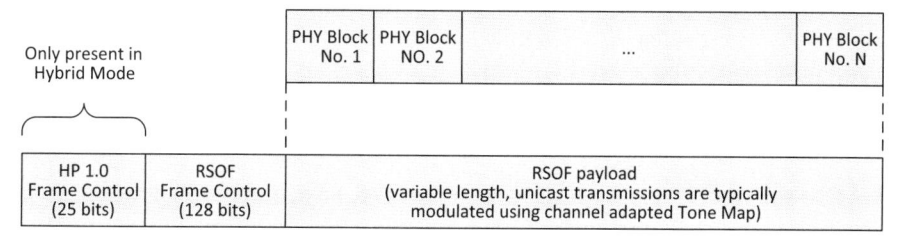

FIGURE 5.12 Reverse SOF MPDU format.

TABLE 5.19 Reverse SOF Frame Control Fields Grouped based on Functionality

Functionality	Fields	Field Size (Bits)
Addressing	Destination Terminal Equipment Identifier (DTEI)	8
Queue	Receive Window Size (RxWSz)	4
	Data MAC Frame Stream Response (MFSRspData)	2
	Management MAC Frame Stream Response (MFSRspMgmt)	2
Bursting	Request Reverse Transmission Flag (RRTF)	1
	Request Reverse Transmission Length (RRTL)	4
	Data MAC Frame Stream Command (MFSCmdData)	3
	Management MAC Frame Stream Command (MFSCmdMgmt)	3
TDMA Allocations	Contention Free Session (CFS)	1
Detection Status	Beacon Detect Flag (BDF)	1
Version	SACK Version Number (SVN)	1
Selective Data	SACK Data (SACKD)	Var
Payload Demodulation	Number of Symbols (NumSym)	2
	PHY Block Size (PBSz)	1
	HomePlug AV Tone Map Index (TMI_AV)	5
	Reverse SOF Frame Length (RSOF_FL_AV)	10

- Management MAC Frame Stream Response (MFSRspMgmt) is a 2-bit field that contains the response from the receiver's Management MAC Frame stream to the corresponding command (MFSCmdData) from the transmitter.
- Receive Window Size (RxWSz) is a 4-bit field that indicates the reassembly buffer available for the corresponding MAC Frame Stream at the receiver. The Receive Window Size field is only present in the SACK Data of Priority Links.
- Data MAC Frame Stream Command (MFSCmdData) is a 3-bit field that is used for buffer management and flow control of data queues.
- Management MAC Frame Stream Command (MFSCmdMgmt) is used for buffer management and flow control of management queues.

Link Identifier (LID) of the queue is not explicitly communicated in the RSOF. Since RSOF is expected at a fixed time after the end of transmission, the transmitter can implicitly determine the LID to which the queue-related information applies. Elimination of LID from the RSOF was primarily done to allow more space for SACK Data.

5.7.1.3 Bursting-Related Field Bursting-related fields in the RSOF are used to request a reverse transmission opportunity using the bidirectional bursting mechanism (refer to Section 6.12).

- Request Reverse Transmission Flag (RRTF) is a 1-bit field that is set to 0b1 to indicate the receiver is requesting the transmitter to continue the bidirectional burst.
- Request Reverse Transmission Length (RRTL) is a 4-bit field that specifies the minimum required duration for a Reverse Transmission. This field includes the payload duration as well as the subsequent RIFS_AV that is being requested for the reverse transmission.
- Data MAC Frame Stream Command (MFSCmdData) is a 3-bit field that is used for buffer management and flow control of data queues for RSOF payload.
- Management MAC Frame Stream Command (MFSCmdMgmt) is used for buffer management and flow control of management queues for RSOF payload.

5.7.1.4 TDMA Allocation-Related fields RSOF includes a 1-bit Contention Free Session (CFS) field that is set to indicate whether the RSOF is part of a Contention Free transmission.

5.7.1.5 Detection Status-Related Field Beacon Detect Flag (BDF) is a 1-bit field that indicates whether the station heard the Central Beacon transmission from the CCo of the AVLN to which it is associated during the current period.

5.7.1.6 Version-Related Fields SACK Version number is indicated using the SVN field and is used to enable extensions to the RSOF contents in future versions of the AV standard. AV stations set this field to 0b0.

5.7.1.7 Selective Acknowledgment-Related Field Selective Acknowledgments in AV are generated after the last MPDU in the MPDU Burst (i.e., MPDUCnt = 0). SACK contains acknowledgment information for each of the MPDUs in the MPDU Burst.

The contents of this field are the same as the corresponding fields in the SACK.

5.7.1.8 Payload Demodulation-Related Fields Payload demodulation-related fields contain necessary information for the receiver to demodulate and receive contents of the RSOF payload.

- PHY Block Size (PBSz) is a 1-bit field that indicates the MPDU payload block size. A value 0b0 indicates that the SOF payload contains one or more 520 octet PHY Blocks. A value of 0b1 indicates that the MPDU payload contains a single 136 octet PHY Block.
- Number of Symbols (NumSym) is a 2-bit field that indicates the number of OFDM Symbols used for transmitting the MPDU payload. This field is used to indicate whether the RSOF payload contains zero, one, two, or greater than two OFDM symbols. The Receiver uses the Number of Symbols field to determine the RIFS_AV used in this transmission. Note that FL_AV includes the duration of SOF payload OFDM symbols as well as the Response Interframe Space.

Further, RIFS_AV is different for zero, one, two, and greater than two OFDM symbol payloads.
- Tone Map Index (TMI_AV) is a 5-bit field that indicates the Tone Map to be used by the receiver in demodulating the MPDU payload. TMI_AV can indicate a ROBO Tone Map (Std-ROBO, Mini-ROBO, HS-ROBO) or a receiver specified tone map.
- RSOF Frame Length (RSOF_FL_AV) is a 10-bit field that indicates the duration of RSOF MPDU payload and the interframe space (BIFS if MPDUCnt > 0 or RIFS_AV if MPDUCnt $= 0$) that follows the MPDU payload.

5.8 SUMMARY

This chapter provided a description of the structure and operation of the HomePlug AV MAC protocol Data Unit, and explained the formats used for key delimiters (such as SOF and RTS/CTS) as well as the Beacon, Sound, and Data MPDUs.

In Chapter 6 we turn our attention to the HomePlug Data Plane which processes the MAC Service Data Units, and provides mechanisms for queue management and the handling of multiple networks.

6

MAC DATA PLANE

6.1 INTRODUCTION

The HomePlug AV MAC Data Plane provides reliable and efficient transportation services to MAC Service Data Units (MSDUs) generated by the higher layer. The format of the MSDU frame is based on standard Ethernet frame format as shown in Figure 6.1, with the following fields:

- *Original Destination Address:* 6-Octet MAC Address of the destination station(s) of the MSDU. The format of this field is same as described in the IEEE standard 802-2001. The prefix "original" is used to emphasize that the Original Destination Address (ODA) can be a bridged station (e.g., an Ethernet station that is bridged through a powerline station).
- *Original Source Address:* 6-Octet MAC Address of the station that is the source of the MSDU. The format of this field is same as described in the IEEE standard 802-2001. The prefix "original" is used to emphasize that the Original Source Address (OSA) can be a bridged station.
- *Optional VLAN Tag:* 4-Octet VLAN Tag field that is formatted as described in IEEE 802.1Q.
- *Ethertype/Length:* 2-Octet Ethernet II Type/Length field.
- *Data:* 46–1500 Octets of data.

HomePlug AV and IEEE 1901: A Handbook for PLC Designers and Users, First Edition.
Haniph A. Latchman, Srinivas Katar, Larry Yonge, and Sherman Gavette.
© 2013 by The Institute of Electrical and Electronics Engineers, Inc. Published 2013 by John Wiley & Sons, Inc.

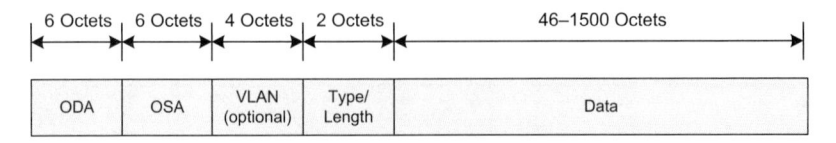

FIGURE 6.1 MSDU Format.

The MAC Data Plane also provides transportation service for MAC management messages. Ethernet-based formatting is also used for management messages. A unique IEEE-assigned Ethertype of **0x88e1** is used to distinguish AV management messages from other Ethernet frames. Further details of AV management messages are provided in Chapter 8.

The MAC Data Plane at the transmitter converts MSDUs and management messages into MAC Protocol Data Unit (MPDUs). The sequence of steps involved in the process is as follows:

- *MAC Frame Generation:* This step involves encapsulating MSDU and management messages with additional MAC layer information.
- *MAC Frame Streams:* This step involves grouping of MAC Frames into queues based on destination and quality of service (QoS) requirements.

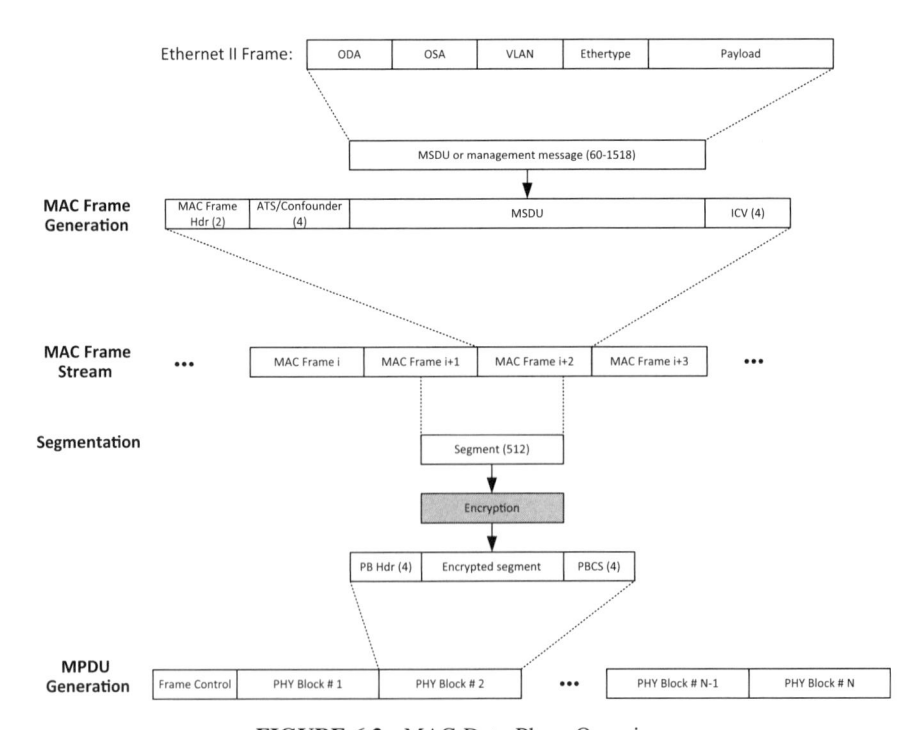

FIGURE 6.2 MAC Data Plane Overview.

- *Segmentation:* This step involves conversion of MAC Frame Streams into 512-octet logical segments.
- *MPDU Generation:* This step involves converting segments into PHY Blocks and inserting them into the MPDU payload.

An overview of the AV MAC Data Plane is shown in Figure 6.2. The length of various fields is shown within parenthesis. Detailed descriptions of these steps are presented in the following sections.

6.2 MAC FRAME GENERATION

The MAC processes MSDU and management messages to generate a MAC Frame. A MAC Frame is composed of a MAC Frame Header, optional Arrival Time Stamp (ATS) or random confounder, MSDU, or management message, and an Integrity Check Value (ICV) as shown in Figure 6.3. Note that a MAC Frame can contain either an MSDU or management message, but not both.

The fields in the MAC Frame are as follows:

- **MAC Frame Header** is a 2-octet field that consists of a 2-bit MAC Frame Type and 14-bit MAC Frame Length.
  - MAC Frame Type (MFT) is a 2-bit field that indicates the type of information contained in the MAC Frame, with the following four possibilities:
    - **0b00**: Indicates the presence of a bit pad in the MAC Frame Stream. This is used when MAC Frame Stream has to be padded to generate a segment (refer Section 6.4).
    - **0b01**: Indicates the presence of the MSDU without an associated ATS.
    - **0b10**: Indicates the presence of the MSDU along with an associated ATS.
    - **0b11**: Indicates the presence of a MAC Management Entry with associated confounder.
  - MAC Frame Length (MFL) is a 14-bit field that specifies the MAC Frame Length in octets, excluding the 2-octet MAC Frame Header field and the 4-octet ICV, but including the ATS or confounder (if either is present). A value of **0x0000** indicates a length of 1 octet, and so on.

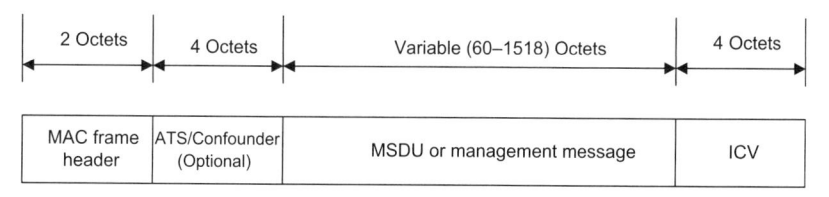

2 Octets	4 Octets	Variable (60–1518) Octets	4 Octets
MAC frame header	ATS/Confounder (Optional)	MSDU or management message	ICV

FIGURE 6.3 MAC Frame Format.

- **Arrival Time Stamp** is a 4-octet field that contains the Network Time Base at the time when the MSDU arrived from the Higher Layer Entity. ATS in the MAC Frame is used by the receiver to provide fixed latency (or controlled jitter) within the powerline network (refer to Section 7.5.1.1). The presence of ATS is negotiated during connection setup, based on the QoS requirement specified in the Connection Specification (CSPEC). ATS is not present in MAC Frames carrying management messages.
- **Confounder** consists of a 4-octet pseudorandom value. This is always present when the MAC Frame contains a management message and is used to render identical messages as different cipher texts. This enhances the security by defeating recognized cipher text attacks.
- **MSDU or Management Message** is the payload of the MAC Frame. The contents of the MSDU or management message are not modified by the MAC Data Plane.
- **Integrity Check Value (ICV)** is a CRC-32 computed over a MAC Frame. The ICV does not cover the MAC Frame Header, ATS (if present), or confounder (if present). CRC-32 is computed using the following standard generator polynomial of degree 32:

$$G(x) = x^{32} + x^{26} + x^{23} + x^{22} + x^{16} + x^{12} + x^{11} \\ + x^{10} + x^8 + x^7 + x^5 + x^4 + x^2 + x + 1$$

6.3 MAC FRAME STREAMS

MAC Frames that belong to the same stream (or queue) are concatenated together into a MAC Frame Stream. Within the transmitter, the stream associated with a MAC Frame depends on the following:

- **Destination Terminal Equipment Identifier** (DTEI) associated with the MAC Frame. DTEI of a MAC is determined based on the ODA of the MAC Frame. The ODA might not be the MAC addresses of HomePlug AV stations within the AVLN. Such cases will occur when the destination of the MAC Frame is bridged across the AVLN. HomePlug AV MAC uses the bridging function (refer to Section 11.1) to map the ODA with the powerline destination MAC addresses (i.e., DA). Within an AVLN, each powerline station is provided with an unique Terminal Equipment Identifier (TEI) by the CCo. The mapping from each station's MAC address with their corresponding TEIs is provided by the CCo to all stations in the AVLN. This information is used to determine the DTEI associated with the MAC Frame from the DA.
- **Link Identifier** (LID) associated with the MAC Frame (refer to Section 9.4.1).
- Whether the MAC Frame contains MSDU (i.e., data) or management messages.

MAC FRAME STREAMS

The AV MAC segregates MAC Frames carrying MSDUs (i.e., data) based on the {DTEI, LID} with which they are associated and concatenates them to form a MAC Frame Stream. Since the MAC Frame Streams carry regular data (as opposed to management messages), these are also referred to as "data streams."

The HomePlug AV MAC segregates MAC Frames carrying management messages based on the DTEI with which they are associated and concatenates them to form a MAC Frame Stream. These streams are referred to as "Management Streams."

Each stream belongs to one of the following categories:

- *Data streams for established connections* MAC Frames that belong to each established connection are provided with a separate MAC Frame Stream at the transmitter. This enables provisioning of QoS guarantees. The LLID or GLID uniquely identifies a link at the transmitter. The LLID is used along with the STEI to uniquely identify a stream within an AVLN.
- *Data stream for connectionless MSDUs* Connectionless MAC Frames that belong to the same {DTEI, PLID} are concatenated to form a separate MAC Frame Stream at the transmitter. In HomePlug AV, a DTEI of **0xFF** indicates a multicast/broadcast transmission. As a result, all connectionless multicast and broadcast frames that belong to the same Priority Link Identifier (PLID) (i.e., priority) are concatenated into a single MAC Frame Stream at the transmitter.
- *Management Streams* MAC Frames that carry management messages and belong to the same DTEI are concatenated to form a separate MAC Frame Stream at the transmitter.

The MAC Frame Stream is the basic entity that is subjected to the MAC Segmentation and Reassembly process.

6.3.1 Priority of Management Streams

Management messages intended for the same DTEI are concatenated regardless of the PLID (i.e., priority) associated with them. This choice is intended to reduce the number of management MAC Frame Streams that must be managed by the stations. Priority promotion of management streams is used to ensure that high-priority management messages are delivered in a timely manner. When transmitting segments of a management stream over the medium, each management stream is treated as having a priority (or PLID) equal to the highest-priority management message (or its associated segment) pending in the stream.

For example, consider a management stream that contains the three MAC Frames {MF1, MF2, MF3}. In this example, MF1 is the oldest MAC Frame and MF3 is the most recent. Furthermore, consider the PLIDs {MF1, MF2, MF3} to be {0, 3, 0} respectively. Since the highest priority of the management message in this stream is 3, the PLID of the stream for this instance is treated as 3. Once all the segments carrying MF2 are transmitted to the destination (and assuming that segments from MF3 are still pending), the priority of the MAC Frame Stream is reduced to a PLID of 0.

6.4 SEGMENTATION

Each MAC Frame Stream is segmented into 512-octet segments for transportation as part of an MPDU payload. Each segment maps onto a single forward error correction (FEC) Block of a PPDU at the physical layer. Since PHY errors occur on an FEC-Block basis, segmentation ensures that only corrupted data is retransmitted [17].

A segment can be generated from a MAC Frame Stream whenever there are enough octets to form a new segment. A MAC Frame Stream is treated as an octet stream for segmentation purposes. Thus, a segment can contain a fraction of a MAC Frame and/or multiple MAC Frames, depending on their sizes. For each segment, the MAC tracks the offset of the first MAC Frame Boundary within the segment. This information is transmitted along with the segment (in the PHY Block Header) and is used by the receiver to demarcate the MAC Frames from the received segments. Each segment is also associated with a Segment Sequence Number (SSN). The SSN is initialized to zero for the first segment in a MAC Frame Stream and incremented by one when a new segment is generated. SSN associated with a segment is indicated to the receiver using the PHY Block Header. SSNs enable reception of out-of-order segments and duplicate detection at the receiver.

The end of the MAC Frame Stream might not contain enough octets to fill a segment completely at the time when the last MAC Frame should be sent. In such cases, the MAC Frame Stream can be padded, so a segment can be formed. Padding adds overhead and, therefore, should be avoided whenever possible. It is recommended that padding of MAC Frame Streams to generate a segment be done just before the segment can be transmitted on the medium. The first two bits of the first (or only) octet in the pad is set to **0b00** to indicate the presence of an octet pad in the remainder of the segment. The receiver can then identify the octet pads and start processing the next segment.

Once a segment is formed as described earlier, it is treated as a single entity targeted for reliable delivery services by the MAC. Each segment is encrypted and then inserted into a PHY Block Body (PBB). A PB comprises the data bits of an FEC Block at the PHY layer. All PBs have a PB Header, PBB, and PHY Block Check Sequence (PBCS). The PB Header field carries the sequence number and MAC Frame Boundary offset associated with the segment. The PB Header also indicates whether the segment belongs to a data stream or a management stream. The PBCS field is used to check the integrity of the PHY Block at the receiver.

6.5 LONG MPDU GENERATION

A long MPDU consists of Frame Control information followed by one or more PBs as shown in Figure 6.4. Each long MPDU payload can carry segments from a data stream and/or a management stream. The following rules are used when segments from management streams are combined with segments from a data stream within a single MPDU:

LONG MPDU GENERATION

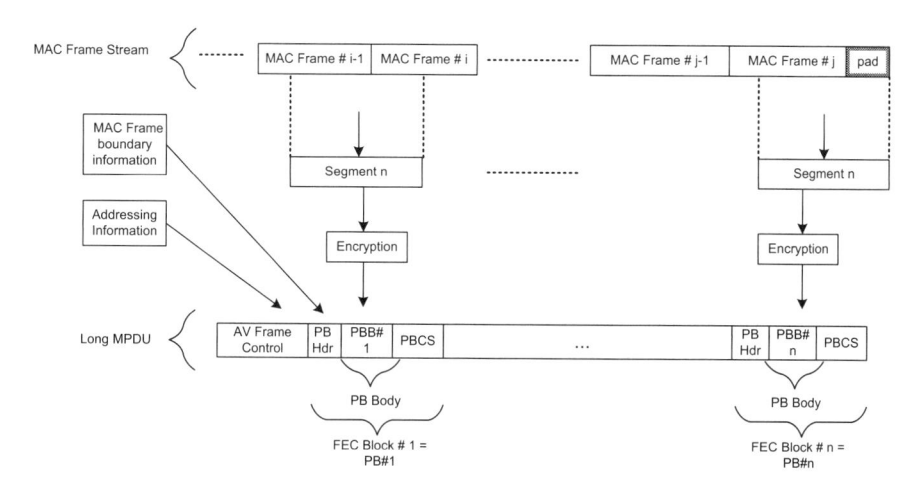

FIGURE 6.4 MAC Segmentation and MPDU Generation.

- Connectionless data streams can only be combined with management streams that are associated with the same {DTEI} and have the same PLID.
- Global connection-based data streams (i.e., LID indicates a Global Link Identifier) can only be combined with management streams that are associated with same DTEI and have a PLID of 3.
- Local connection-based data streams (i.e., LID indicates a Local Link Identifier) can only be combined with management streams that are associated with the same DTEI and have the same channel access priority.

MAC traffic may often require delivery of short messages. This is particularly true of management messages. When a segment containing 128 octets or less of MAC Frame Stream data is the only pending segment in a MAC Frame Stream, it may be transmitted as a padded segment in a 136-octet PB. The first octet of the segment pad indicates that the remainder of the segment is padded to the next 512-octet boundary, eliminating the need to send the rest of the padding in PBs. All receivers are required to be capable of receiving a 136-octet PB containing a shortened 512-octet segment.

The presence of an octet pad to the next 512-octet boundary is implicit for a 128-octet segment. For example, if a station needs to send a 20-octet management message to another station and has no other management message segments pending to that station, it should transmit it using an MPDU with a single 136-octet PB containing a 512-octet segment shorted to 128 octets.

An MPDU will only contain PBs of the same size. This means that short management messages that are combined with data streams using 520-octet PBs must also use 520-octet PBs for transmission.

Note: Only one 136-octet PB is allowed per MPDU.

Once a long MPDU is generated, it is handed over to the physical layer for delivery.

6.6 REASSEMBLY

The information contained in the AV Frame Control and PB Header is used by the receiver to determine uniquely the stream to which a segment belongs. The relevant fields in the Frame Control are {STEI, DTEI, MCF, and LID}. The relevant field in the PB Header is MMQF. When the Multicast Flag (MCF) field indicates the presence of a broadcast/multicast payload, the DTEI is assumed to be **0xFF** for reassembly purpose (and the actual value DTEI present in the Frame Control is ignored, refer to Section 6.13.2).

Segments belonging to the same stream are reordered and duplicate segments are discarded by the receiver using the segment sequence number (SSN) in the PB Header. MAC Frame boundary information in the PB Header and the MAC Frame Length in the MAC Frame Header are used to extract MAC Frames from the MAC Frame Stream. Once a MAC Frame is extracted, the receiver checks its integrity using ICV and successfully received MSDUs and management messages are delivered to their intended destination.

6.7 BUFFER MANAGEMENT AND FLOW CONTROL

Buffer management and flow control mechanisms in HomePlug AV plays an important role in preventing packet loss due to buffer overflow at the receiver. These mechanisms also enable HomePlug devices to

- provision buffer resources based on QoS requirements,
- dynamically adjust buffer allocation based on demand (e.g., based on number of active MAC Frame Streams),
- Effectively handle low data rate host interfaces (e.g., AV receiver bridging traffic via a low data rate wireless networks),
- Operate reliably even in the presence of limited buffers (e.g., low-cost implementations with on-chip memory).

For each active MAC Frame Stream, the receiver assigns a dedicated buffer. This is referred to as the Receive Window Size (RxWSz). The RxWSz for various MAC Frame Streams is obtained as follows:

1. For MAC Frame Streams of established connections, RxWSz is negotiated during connection setup and remain fixed until the connection is terminated.
2. For connectionless MAC Frame Streams carrying unicast data, RxWSz is initialized to 16 and is dynamically updated based on the RxWSz advertised in the SACK/RSOF delimiter.
3. For MAC Frame Streams (both unicast and multicast/broadcast) carrying management messages, RxWSz is set to 8. Compared with connectionless data streams, no negotiation of RxWSz is necessary for management streams.

BUFFER MANAGEMENT AND FLOW CONTROL

TABLE 6.1 MAC Frame Stream Command Interpretation

MFSCmd	Interpretation
0b000	INIT
0b001	IN_SYNC
0b010	RE_SYNC
0b011	RELEASE
0b100	NOP (No operation)
Others	Reserved

4. Broadcast and multicast data transmissions that are connectionless always use an RxWSz of 16.

MFSCmd and MFSRsp in the frame control are used by the transmitter and receiver, respectively, for flow control. Tables 6.1 and 6.2 show the interpretation of MFSCmd and MFSRsp. Since HomePlug AV allows transmission of segments belonging to one data and one management stream in an MPDU, separate instances of these commands and responses are present in the frame control (i.e., MFSCmdData, MFSCmdMgmt, MFSRspData and MFSRspMgmt fields in the frame control). For connectionless unicast data MAC Frame Streams, the receiver uses RxWSz field in the SACK/RSOF to indicate changes to its Receive Window Size. The behavior of the transmitter and receiver is described in the following sections.

6.7.1 Transmit Buffer Management

For each MAC Frame Stream, the transmitter tracks the Minimum Transmit Segment Sequence Number (MinTxSSN) and Maximum Transmit Segment Sequence Number (MaxTxSSN). MinTxSSN and MaxTxSSN are used to perform flow control at the transmitter. MinTxSSN is the SSN of the oldest pending segment in the MAC Frame Stream. If there are no pending segments, MinTxSSN indicates the SSN that will be assigned to the next newly generated segment. MaxTxSSN is the SSN of the most recent segment that can be transmitted in the MAC Frame Stream before the segment with SSN of MinTxSSN is acknowledged successfully. The MaxTxSSN is

TABLE 6.2 MAC Frame Stream Response Interpretation

MFSRsp	Interpretation
0b00	ACK
0b01	NACK
0b10	FAIL
0b11	HOLD

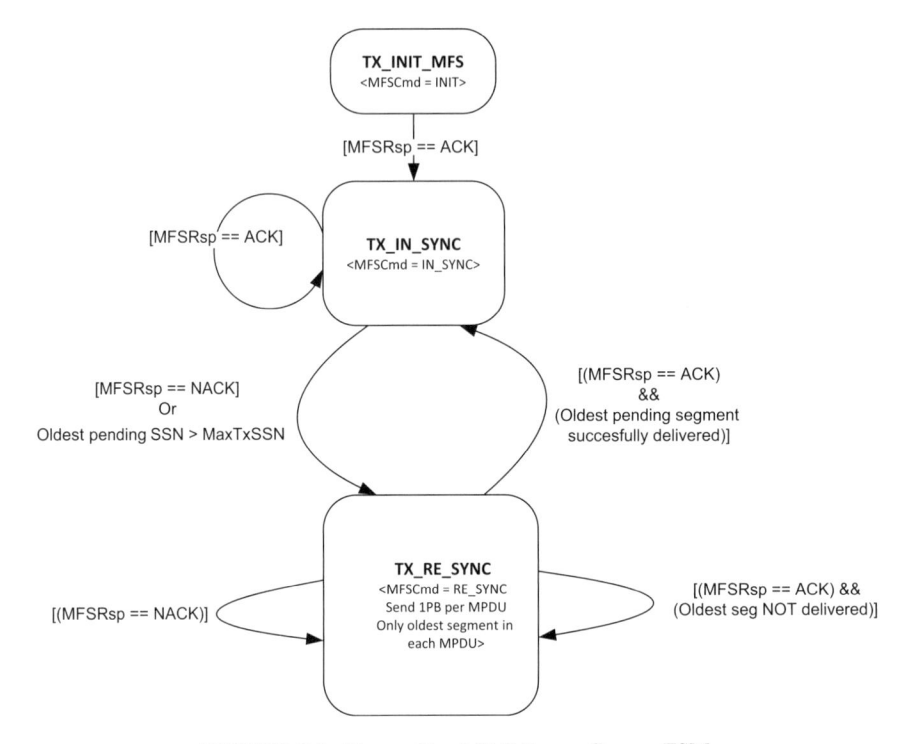

FIGURE 6.5 Transmitter MAC Frame Stream FSM.

the sum of RxWSz and the MinTxSSN minus one (i.e., MaxTxSSN = RxWSz + MinTxSSN − 1). MinTxSSN is set to **0x0000** when the MAC Frame Stream is initially formed. MinTxSSN is subsequently updated when the oldest pending segment in the MAC Frame Stream is successfully delivered.

The Finite-State-Machine (FSM) used by the transmitter to maintain flow control on unicast transmissions is shown in Figure 6.5. The transmitter sets the MFSCmd based on the MAC Frame Stream's FSM state. The various states of the transmitter FSM and the corresponding behavior is as follows:

- **TX_INIT_MFS**: When the transmitter initiates a new MAC Frame Stream, it enters the TX_INIT_MFS state and sets the MinTXSSN to 0. In this state, the transmitter sets the MFSCmd = INIT to indicate to the receiver that the segments correspond to a new MAC Frame Stream. The transmitter continues to operate in this state until it receives MFSRsp = ACK. Reception of MFSRsp = ACK indicates that the receiver has successfully established the receive MFS and causes the transmitter to enter TX_IN_SYNC state.
- **TX_IN_SYNC**: The Transmitter operates in TX_IN_SYNC state when it is properly synchronized to the receiver with regards to the MinTxSSN and MaxTxSSN (i.e., transmitter will not cause buffer overflow at the receiver).

In this state, the transmitter sets MFSCmd = IN_SYNC and continues to operate in this state until one of the following two conditions occur:
- MFSRsp = NACK is received. This indicates that the receiver intends to reduce the RxWSz. This causes the transmitter to transition to TX_RE_SYNC state.
- In some cases, all the segments in the [MinTxSSN, MaxTxSSN−1] range may expire before the transmitter can deliver them to the receiver. In such cases, the oldest pending segment at the transmitter is beyond the receiver's buffer range. This condition also causes the transmitter to transition to TX_RE_SYNC state.

- **TX_RE_SYNC**: The transmitter operates in TX_RE_SYNC state while it is trying to get back in synchronism with the receiver. In this state, the transmitter sets MFSCmd = RE_SYNC and only includes the Oldest Pending Segment in the MPDU. The transmitter continues to be in this state until the receiver sends MFSRsp = ACK and the Oldest Pending Segment is successfully delivered. Note that successful delivery of the Oldest Pending Segment is necessary since the SSN is communicated in the PB header. Once the Oldest Pending Segment is successfully delivered to the receiver, the transmitter transitions to IN_SYNC state.

The transmitter can cause the receiver to release a MAC Frame Stream by setting MFSCmd = Release. The transmitter releases the MAC Frame stream when it gets MFSRsp = FAIL from the receiver. The receiver can cause the transmitter to delay its transmissions by sending MFSRsp = HOLD. MFSRsp = HOLD does not affect the state of the transmitter's FSM. It merely causes the transmitter to delay transmitting segments from the corresponding MAC Frame Stream for one Beacon Period. For Broadcast/Multicast MAC Frame Streams, the transmitter always sets the MFSCmd = NOP.

6.7.2 Receive Buffer Management

For each reassembly stream, the receiver tracks the Minimum Receive Segment Sequence Number (MinRxSSN) and the Maximum Receive Segment Sequence Number (MaxRxSSN). MinRxSSN indicates the SSN of the oldest expected segment. MaxRxSSN is the SSN of the most recent segment that can be received. The receive window size (RxWSz) of the reassembly buffer is the difference between the MaxRxSSN and the MinRxSSN plus one (i.e., MaxRxSSN = RxWSz + MinRxSSN − 1).

The receiver processes each received segment belonging to a valid MAC Frame Stream as follows:

- Reception of a segment with an SSN equal to MinRxSSN cause the receiver to process that segment and any remaining contiguous set of segments. The MinRxSSN is updated to the oldest expected SSN.

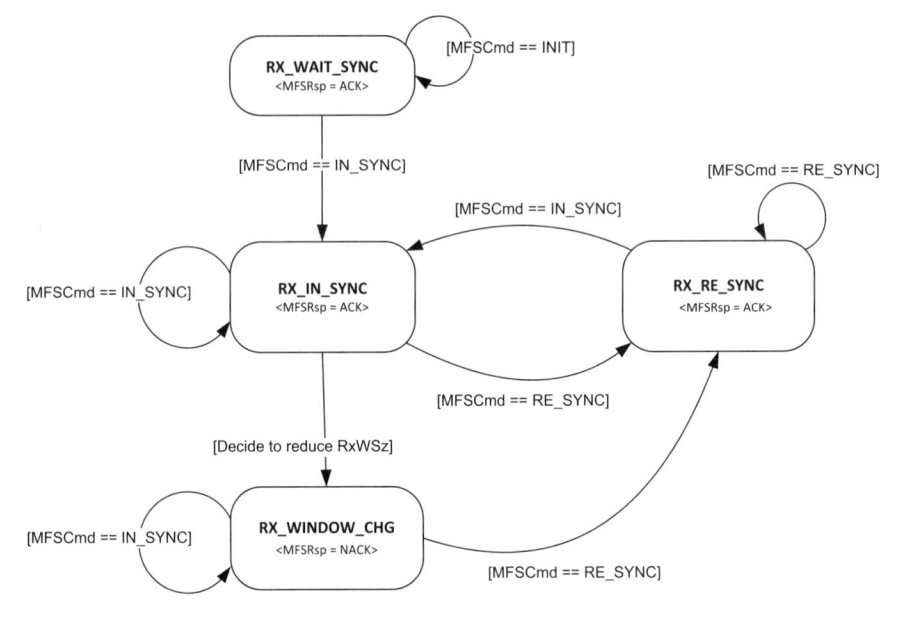

FIGURE 6.6 Receiver MAC Frame Stream FSM.

- Reception of a segment with an SSN in the range (MinRxSSN+1, MaxRxSSN) inclusive, will cause the receiver to treat the segment as being out-of-order. The receiver stores out-of-order segments.
- If a segment is received with an SSN less than MinRxSSN, the segment is discarded.
- For broadcast/multicast MFS, reception of a segment with an SSN greater than the MaxRxSSN and the oldest pending segment flag (OPSF) set to **0b0** causes the MaxRxSSN to be updated to the SSN of the received segment. The received segment is stored for subsequent reassembly. The MinRxSSN is updated appropriately.
- Reception of an MPDU with a single segment with an SSN greater than MaxRxSSN and an OPSF set to **0b1** cause the receiver to set the MinRxSSN to the SSN of the received segment plus one. All pending segments with sequence numbers less than the new MinRxSSN are dropped.

The Finite-State-Machine (FSM) used by the receiver to coordinate flow control on unicast transmissions is shown in Figure 6.6. The receiver sets the MFSRsp based on the MAC Frame Stream's FSM state. The various states of the receiver FSM and the corresponding behavior is as follows:

- **RX_WAIT_SYNC:** The receiver enters RX_WAIT_SYNC state when it initiates a new MAC Frame Stream based on MFSCmd = INIT from the transmitter. In this state, the receiver sends MFSRsp = ACK. Reception of

MFSCmd=IN_SYNC from the transmitter causes the receiver to transition to RX_IN_SYNC.
- **RX_IN_SYNC:** The receiver operates in RX_IN_SYNC state when it is properly synchronized with the transmitter with regard to the MinRxSSN and MaxRxSSN. The receiver sets MFSCmd to ACK and continues to operate in this state until one of the following two conditions occur:
  - MFSCmd = RE_SYNC is received. This is an indication from the transmitter that it is out of synchronism and causes the receiver to enter RX_RE_SYNC state.
  - In some cases, the receiver might decide to reduce the RxWSz associated with a MAC Frame Stream. Such conditions can occur, for example, when the receiver reassigns some of the memory associated with this MAC Frame Stream to other MAC Frame Streams. In such cases, the receiver enters RX_WINDOW_CHG state.
- **RX_WINDOW_CHG:** The receiver enters RX_WINDOW_CHG state to force the transmitter to resynchronize. In this state, the receiver sets MFSRsp to NACK and continues to operate in this state until it receives MFSCmd = RE_SYNC. Reception of a MFSCmd = RE_SYNC causes the receiver to enter the RX_RE_SYNC state.
- **RX_RE_SYNC:** The receiver stays in the RX_RE_SYNC state until the transmitter confirms that it has successfully resynched by transmitting MFSCmd = N_SYNC. Reception of MFSCmd = IN_SYNC causes the receiver to transition to the RX_IN_SYNC state.

In some cases, the receiver might not have sufficient buffer resources to continue receiving more Segments from the transmitter. This can occur when the STAs has a slow H1 interface or when the station has limited processing capabilities. Under such conditions, upon the reception of MPDUs with one or more Segments, the receiver can send a SACK with MFSRsp set to HOLD. This will cause the transmitter to refrain from sending more Segments of the MFS for one Beacon Period. The receiver can continue to send MFSRsp = HOLD until it has sufficient buffer resources.

For broadcast MAC Frame streams, the transmitter sets MFSCmd = NOP and the receiver responds with MFSRsp = ACK.

The receiver delivers all received MSDUs in a MAC Frame Stream in the order they arrived at the corresponding transmitter. This will require the receiver to store out-of-order segments until all the expected previous segments are received. The receiver stores segments as they are correctly received, combining the oldest contiguous run of segments to form the MACFrame Stream. The oldest undelivered MAC Frame is determined by using the MAC Frame Header length information and the PB Header MAC Frame Boundary Offset information. When all of its constituent segments have been received, the MAC Frame is formed and its ICV checked. If the ICV is correct, the MAC Frame is scheduled for delivery, either immediately or at the time specified by the ATS and the CSPEC for the stream.

When all of the MAC Frames associated with a segment have been delivered, the segment may be discarded.

6.8 COMMUNICATION BETWEEN ASSOCIATED BUT UNAUTHENTICATED STAs

Communication between associated but unauthenticated STAs refers to communication between a pair of STAs that are associated with the same AVLN and at least one of these STAs is not authenticated. Association and authentication status are known to associated STAs from the TEI Map information provided by the CCo. Communications with associated but unauthenticated STA has the following restrictions:

- Data MSDUs are not allowed. In general, HomePlug AV prohibits exchange of MSDUs in clear text.
- Exchange of management messages is limited to unicast message that facilitate authentication of the unauthenticated stations. Segments are transmitted in clear text.

6.9 COMMUNICATION BETWEEN STAs NOT ASSOCIATED WITH THE SAME AVLN

There are several instances where STA(s) that are not associated with the same AVLN or are not associated with any AVLN need to communicate with each other. For example, neighboring network CCos might exchange management messages for neighbor network coordination or a STA might need to send association messages (**CC_ASSOC.REQ/CNF**) to the CCo, and so on.

The following restrictions are placed when communicating with STAs that are not associated with the same AVLN:

- Data MSDUs are not allowed. In general, HomePlug AV prohibits exchange of MSDUs in clear text.
- Exchange of management messages is limited to messages that facilitate association/authentication and messages necessary for neighbor network coordination. Segments are transmitted in clear text.

Transmitters treats each management message that needs to be transmitted to STAs that are not associated with its AVLN (if any) or to STAs that are not associated with any AVLN, as belonging to a new MAC Frame Stream. Furthermore, the maximum length of each management message transmitted will be limited to 502 octets. The payload of the Long MPDU is limited to one segment and each segment contains only one MAC management message.

Transmissions to a STA that is associated with a different AVLN may be unicast (i.e., unicast DTEI) in instances where the {SNID, DTEI} and Network Time Base (NTB) of the destination are known. In this case, a PHY clock correction based on the NTB of the destination is used. Furthermore, the STEI is set to **0x00** (i.e., new station without a TEI) in the SOF.

Transmissions to an unassociated STA or a STA whose association status (i.e., the AVLN to which it is associated) is not known, uses a Multi-Network Broadcast (MNBC) transmission mechanism. Multicast/broadcast transmissions to STA(s) that might not be associated with the transmitter's AVLN (if any) also uses an MNBC transmission mechanism.

Receivers identify transmissions from unassociated STAs based on either the STEI set to **0x00** or the MNBF set to **0b1**. Each received unassociated STA transmission is assembled using a new receive MAC Frame Stream. Duplicate rejection cannot be achieved for management messages received from unassociated STAs. Processing of such management messages should be designed to gracefully handle duplicate management messages.

6.9.1 Multinetwork Broadcast (MNBC)

A Multinetwork Broadcast (MNBC) transmission mechanism is used when the management messages contained in the MPDU need to be received by all stations in the vicinity irrespective of the network association. MNBC transmissions have the MNBF flag in the Frame Control set to **0b1**. This indicates to the receiver that the MPDU is a broadcast to all stations regardless of the SNID or network association.

MNBC transmissions use an RTS/CTS exchange before the Long MPDU carrying the management message is transmitted. RTS/CTS notifies the receiving stations to apply the correct PHY Receive Clock Correction for the Long MPDU that follows based on the network identified by the SNID in the RTS. If a broadcast/multicast proxy station is not available, a CTS is not required to transmit the MPDU following the RTS.

Figure 6.7 shows an example of the Multinetwork Broadcast transmission mechanism. In this example, STA1 is tracking the PHY Clock of SNID1 and transmits an MNBC transmission. The MNBF is set to **0b1** and the SNID is set to SNID1 in RTS, CTS, and SOF. The Long MPDU is transmitted using the PHY Clock Correction

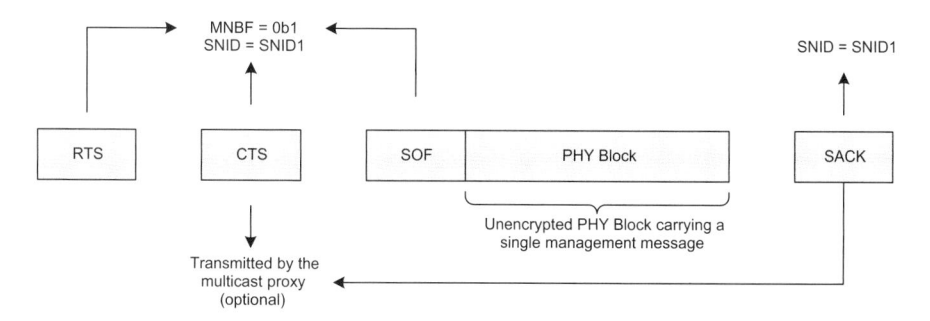

FIGURE 6.7 Illustration of Multinetwork Broadcast Transmission.

based on SNID1. Consider a receiving STA, STA2, which is tracking the PHY Clock of SNID2. Subsequent to reception of the RTS, it will determine that this is an MNBC based on the MNBF. The receiver then configures itself to receive the following Long MPDU using the PHY Clock correction of SNID1. This will enhance the reliability of the reception of the Long MPDU payload. Subsequent to the reception of the Long MPDU, STA2 will revert to using the PHY Clock correction of SNID2.

6.10 DATA ENCRYPTION

Encryption in HomePlug AV STAs is performed independently on each segment. HomePlug AV STAs uses 128-bit AES-based encryption in Cipher Block Chaining (CBC) mode, which is known to be cryptographically superior to non-CBC encryption in the sense that it eliminates certain vulnerabilities in the non-CBC encryption such as used in Wi-Fi Systems. The Initialization Vector is generated from the following:

- 12 Octets of the Frame Control Variant fields,
- 3 Octets from the PB Header,
- 1 Octet PB Count that contains the relative location of the PB in an MPDU. The first PB transmitted in the MPDU has a PB count of **0x00**, the second PB transmitted in the MPDU has a PB count of **0x01**, and so on.

Since the Initialization Vector depends on the MPDU in which the segment is transmitted, segments need to be encrypted separately for each retransmission. HomePlug AV implementations use hardware support for encryption and treat encryption as part of the physical layer processing flow (i.e., encryption of segments is done in real-time just before the Turbo FEC Encoding). This architecture enables HomePlug AV implementation to eliminate latency incurred for encryption.

Typical encryption schemes used by communication systems (e.g., IEEE 802.11) uses additional bits to communicate the Initialization Vector for each encrypted block. Furthermore, encryption schemes typically require padding so that the encrypted block is an integral multiple of the key size (e.g., 16-byte multiples for AES-128). HomePlug AV is designed to eliminate these overheads. In particular, the generation of Initialization Vector from Frame Control and other known information eliminates the need for sending a separate Initialization Vector. Furthermore, since encryption is done on 128-byte or 512-byte PHY Block Body (both of which are multiple of 16-bytes), no padding is necessary. These design choices further improve the efficiency of HomePlug AV.

6.11 MPDU BURSTING

MPDU bursting is the process in which a station transmits multiple Long MPDUs in a Burst (without relinquishing the medium) before soliciting a response. An MPDU

BIDIRECTIONAL BURSTING

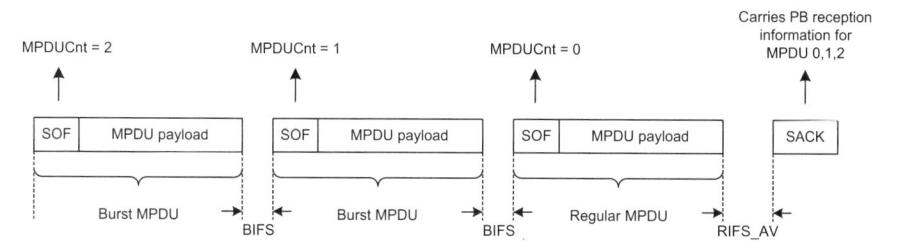

```
MPDUCnt = 2              MPDUCnt = 1              MPDUCnt = 0         Carries PB reception
                                                                      information for
                                                                      MPDU 0,1,2
   ↑                        ↑                        ↑                        ↑
┌─────┬──────────────┐  ┌─────┬──────────────┐  ┌─────┬──────────────┐  ┌──────┐
│ SOF │ MPDU payload │  │ SOF │ MPDU payload │  │ SOF │ MPDU payload │  │ SACK │
└─────┴──────────────┘  └─────┴──────────────┘  └─────┴──────────────┘  └──────┘
      Burst MPDU      →│←      Burst MPDU      →│←     Regular MPDU    →│←
                       BIFS                     BIFS                    RIFS_AV
```

FIGURE 6.8 Example of MPDU Bursting.

Burst can include either multiple Long SOF MPDUs or multiple Long Sound MPDUs. When a Burst of SOF MPDUs is transmitted, the SACK transmitted at the end of the MPDU Burst contains the reception status of all the PBs in the Burst. When a Burst of Sound MPDUs is transmitted, the Sound ACK that is transmitted at the end of the MPDU Burst will indicate the reception status of the Sound MPDUs.

The MPDUCnt field in the SOF and the Sound Frame Control is used to indicate the number of additional MPDUs the transmitter intends to send. For example, MPDUCnt = 1 indicates that the transmitter intends to send one more MPDU in this MPDU Burst. MPDUCnt = 0 indicates that the corresponding MPDU is the last MPDU in the Burst. A transmitter can send up to four MPDUs in a single MPDU Burst. The receiving station stores the acknowledgement information for each MPDU in the Burst and sends it to the transmitter after receiving the last MPDU in the Burst (i.e., MPDU with MPDUCnt = 0). During CSMA, all stations in the network defer from contending for the medium until the Burst transmission is completed.

Figure 6.8 shows an example of MPDU bursting with three SOF MPDUs in a Burst. The first, second, and third (or last) MPDUs are indicated by MPDUCnt values of 2, 1, and 0, respectively. The SACK contains reception status of PBs in all three MPDUs. The gap between consecutive MPDUs in a Burst is referred to as Burst Inter Frame Space (BIFS) and is fixed at 20 μs.

It should be noted that the payload of each MPDU is modulated using a single tone map. Since it is common to have different tone maps at different locations of the AC line cycle, MPDU Bursting enables AV stations to transmit MPDUs across multiple tone map regions in a single channel access. This enables HomePlug AV stations to operate efficiently.

Support for MPDU Bursting is mandatory for the receiver and optional at the transmitter. During CSMA, the maximum duration of MPDU Burst is limited to 5 ms to reduce the cost of collision.

6.12 BIDIRECTIONAL BURSTING

HomePlug AV stations may optionally support bidirectional bursting. This procedure allows a transmitting station to allocate part of a burst to a receiving station, so the receiving station can send data to the transmitting station over their "reverse channel."

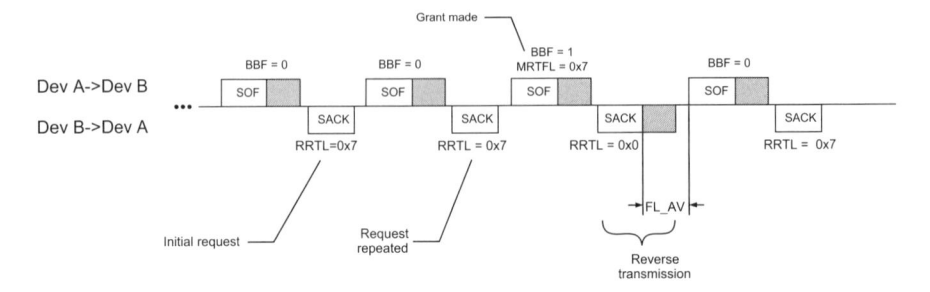

FIGURE 6.9 Bidirectional Burst Mechanism.

Bidirectional bursting enables HomePlug AV to significantly improve MAC efficiency as well as to reduce latency for supporting bidirectional traffic. For example, the TCP protocol requires TCP acknowledgments to be received in a timely manner to maintain the flow. The use of Bidirectional bursting for TCP facilitates the timely delivery of TCP acknowledgments and also improve MAC efficiency (i.e., there is no contention for transmitting the TCP acknowledgement, and there is a lower overhead due to concatenation of SACK information with TCP acknowledgments in the Reverse Start of Frame (RSOF)).

The receiving station initiates bidirectional bursting using the Request Reverse Transmission Flag (RRTF) and the Request Reverse Transmission Length (RRTL) fields in the Frame Control of the SACK. The RRTL field specifies the minimum required frame length for a Reverse SOF MPDU in the reverse direction.

On receiving the request, the original transmitter decides whether the request will be honored and its duration. The original transmitting station grants the reverse transmission request by setting the **Bidirectional Burst Flag** (BBF) and the Max Reverse Transmission Frame Length (MRTFL) field in the SOF. The MRTFL field includes payload and subsequent RIFS_AV, but not the Reverse Start of Frame (RSOF) delimiter.

Figure 6.9 shows the bidirectional burst mechanism. When Device B (Dev B) determines that it requires reverse direction transmission, it sets the RRTF and RRTL fields in the SACK or RSOF. This is set until Dev A responds with a grant for reverse transmission (by setting the BBF flag set to **0b1** and indicating the maximum duration of reverse transmission in MRTFL) or until there is no longer a need to request a transmission in the reverse direction.

Figure 6.10 shows the various interframe spaces during a bidirectional burst. An interframe spacing of RIFS_AV always follow a Reverse SOF. The FL_AV field in the Reverse SOF will include the Reverse SOF payload and the subsequent RIFS_AV.

6.12.1 Bidirectional Bursting During CSMA

Figure 6.11 shows the usage of bidirectional bursts during CSMA. When Dev A gains access to the channel, it can set the BBF field in the SOF to indicate that the

BIDIRECTIONAL BURSTING

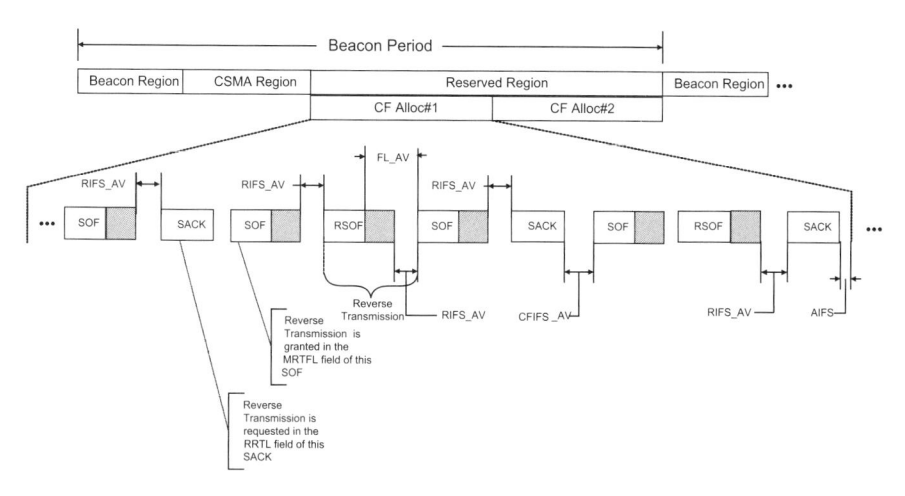

FIGURE 6.10 Interframe Spacing during Bidirectional Burst.

channel will not be relinquished after the first transmission. Dev B, by sending a Reverse SOF indicates to the other devices that they cannot access the channel following this transmission. Dev A can immediately follow the receipt of Reverse SOF with the transmission of another MPDU. If Dev B continues to request a bidirectional burst (as indicated by the RRTF and RRTL fields) and Dev A has no data to send, it may continue the burst by sending a short SOF with the BBF field set to one. Similarly, if Dev A has granted time for a reverse transmission and Dev B

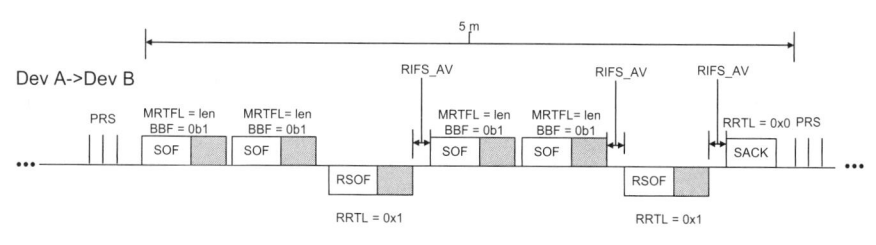

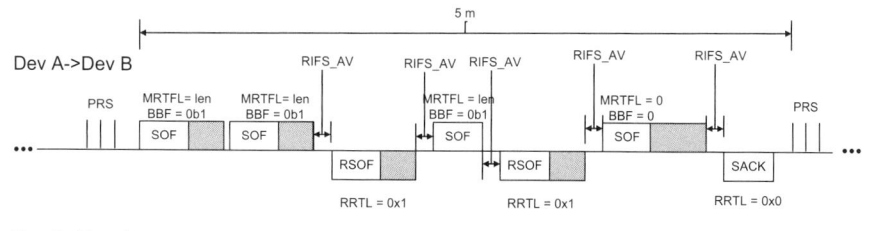

FIGURE 6.11 Bidirectional Bursts during CSMA.

does not have any data to transmit, Dev B should continue the burst by sending a short RSOF. The sequence of bidirectional bursts will be terminated with either Dev A or Dev B transmitting a SACK. Dev A may also instruct Dev B to terminate the burst by setting the BBF field to zero in the SOF. If either device suspects a collision (all received PBs bad), it should terminate the sequence of bidirectional burst with a SACK. Likewise, if no frame control is received, the sending device should assume a collision has occurred.

During a bidirectional burst, listening stations (i.e. stations not participating in the bidirectional burst) will defer to the two stations participating in the bidirectional burst until the end of the burst. On receiving a SOF with MPDUCnt set to **0b00** and BBF set to **0b0**, the third-party stations would infer that the bidirectional burst is ending and they would start priority contention at the end of the expected SACK transmission. If they receive an SOF with MPDUCnt set to **0b00** and BBF set to **0b1**, they would start looking for a reverse transmission. If they receive a RSOF, they would continue to look for a SOF. If they receive a SACK at any time, they will start priority contention immediately.

During CSMA, the total duration of bidirectional burst is limited to 5 ms.

6.12.2 Connections and Links During Bidirectional Bursts

The Reverse SOF FC (the reverse direction) does not contain an LID field, instead the LID is implied/derived from the LID in the preceding Long SOF FC (the forward direction).

For connectionless priority links (i.e., LIDs 0 to 3), the PLID of the data stream in the payload of the Reverse SOF will be the same as the PLID contained in the corresponding SOF. For connections with a single forward link or a single reverse link, only management message at any priority can be transmitted in the payload of the Reverse SOF. For bidirectional connections, the LID of the data stream in the reverse direction should be the LID of the other link in the connection. Optionally, management messages are also allowed in the payload of the Reverse SOF.

6.12.3 Encryption of RSOF Payload

Reverse SOF FC does not contain an Encryption Key Select (EKS) field. The EKS of the encryption key used in encrypting the Reverse SOF payload will be the same as the EKS contained in the preceding SOF FC.

6.13 AUTOMATIC REPEAT REQUEST (ARQ)

HomePlug AV uses Selective ACKnowldgement (SACK) Automatic Repeat reQuest (ARQ)) at the segment level. Segments are sent as PHY Block Body and each PB is contained in its own FEC Block. An MPDU contains a variable number of FEC blocks, depending on the data rate and payload duration, and up to four MPDUs may be sent as a burst with a single SACK in response.

The SACK format provides support for up to four SACK Type (SACKT) fields, one per MPDU in a burst within the SACK delimiter. The SACKT and SACKI fields indicate whether

- all the PBs in the corresponding MPDU were received correctly,
- all the PBs in the corresponding MPDU were received with errors,
- the corresponding MPDU was not detected (i.e., either no Preamble was detected or a corrupt Frame Control was detected),
- a mixture of good and bad PBs were found in the corresponding MPDU. In this case, the reception status of the PBs can be indicated using 1 bit per PB, 1 bit for a pair of PBs, or as a compressed version.

6.13.1 Request SACK Retransmission

The Request SACK Retransmission mechanism enables the transmitter to request the receiver to retransmit the SACK information associated with a previously transmitted MPDU Burst. This enables the transmitter to avoid retransmission of all the segments in the MPDU Burst. The Request SACK Retransmission function is optional for the transmitter and mandatory for the receiver. The Request SACK Retransmission function is restricted to Global (TDMA) Links.

When a SACK MPDU is transmitted in response to an SOF with RSR = **0b0** (i.e., when SACK retransmission is NOT requested), the receiver stores the SACKT, SACKI fields of the SACK MPDU and the GLID, BurstCnt, and STEI fields of the SOF. This data is updated by the receiver on receipt of another SOF MPDU with the same GLID, STEI, and RSR = **0b0**. The transmitter requests retransmission of a missing SACK by sending an SOF with RSR = **0b1**. The BurstCnt in the SOF is set to the BurstCnt of the previously transmitted Burst for which SACK retransmission is requested.

On receipt of an SOF with RSR = **0b1**, if the receiver has SACKT and SACKI data corresponding to the GLID, STEI, and BurstCnt fields of the SOF, the receiver responds with a SACK containing these SACKT and SACKI fields. Otherwise, the receiver responds with all SACKT and SACKI fields set to indicate that SACK information is not available (i.e., SACKT = **0b11** and SACKI = **0b11**).

6.13.2 Broadcast/Multicast and Partial Acknowledgment

Multicast/broadcast transmissions cannot make use of the standard ARQ mechanism, because there can be more than one destination that would acknowledge the transmission. The MAC improves the information available to the transmitter through a "partial ARQ" scheme in which one station in the group serves as a proxy to provide the response. Using this mechanism, the DTEI of the multicast or broadcast transmission is set to the TEI of the proxy station. The presence of multicast payload in the MPDU is indicated by setting the Multicast Flag (MCF) in the Frame Control. All stations that receive the transmissions with MCF set to **0b1** will reassemble the corresponding MPDU irrespective of the DTEI. The transmitter uses the SACK information sent by the proxy to determine whether the transmission is successful.

When no proxy station is selected, the DTEI of the multicast MPDU will be set to 0xFF. In this case, the transmitter assumes that all transmissions are successful.

6.14 SUMMARY

In this chapter, the generation of MAC Frames and MAC Frame Streams were discussed, together with the efficient segmentation and reassembly processes used in HomePlug AV. The chapter also examined the issues of buffer management and flow control, MPDU bursting, bidirectional bursting, and selective acknowledgements as used in HomePlug AV.

Having now presented the basic features of the HomePlug AV PHY (Chapter 4) and MAC (Chapters 5 and 6), Chapter 7 proceeds with a discussion of the selection, role, and operation of the Central Coordinator (CCo) in HomePlug AV.

7

CENTRAL COORDINATOR

7.1 INTRODUCTION

HomePlug AV uses a centralized network architecture with Central Coordinator (CCo) providing several network management functions such as Association, Authentication, Time Division Multiple Access (TDMA) Admission Control, QoS guarantees, etc. One of the key design elements of HomePlug AV is to hide the intricacies of the CCo functionality from the user. In particular,

- The user does not have to distinguish between a normal station and a CCo. All HomePlug AV devices are required to be able to operate as a CCo.
- The user does not have to appoint a specific station as a CCo. CCo selection is automatically made by HomePlug devices and the most suitable station becomes the CCo.
- The user does not have to be involved in handling CCo failures or when CCo functionality is handed over from one station to another. HomePlug AV is designed to provide seamless network operation across CCo changes and CCo failures.

These features enable an effective plug-and-play experience for the user, while maintaining the high-level network controllability that can be achieved by using a centralized architecture.

HomePlug AV and IEEE 1901: A Handbook for PLC Designers and Users, First Edition.
Haniph A. Latchman, Srinivas Katar, Larry Yonge, and Sherman Gavette.
© 2013 by The Institute of Electrical and Electronics Engineers, Inc. Published 2013 by John Wiley & Sons, Inc.

Sections 7.2–7.4 provide details on the CCo selection and CCo failure recovery and transfer/handover mechanisms, respectively. The details of various network management functions provided by the CCo are described in Section 7.5.

7.2 CCo SELECTION

In HomePlug AV networks, the CCo can be automatically selected and does not require the user to have any knowledge about the CCo function or its operation. This functionality is referred to as "CCo Auto-Selection." Typically, the first station (STA) to instantiate a network automatically assumes the role of a CCo for this Home AV Logical Network (AVLN). As the network evolves with more STAs joining or leaving the AVLN, another STA may be more suitable to fulfill the role of CCo. The current CCo applies the Auto-Selection procedure on an ongoing basis to identify the best STA within the AVLN to perform the function. If a more suitable STA is identified by the selection process, the current CCo hands over the function to the more suitable STA. All AV STAs support the Auto-Selection function. Alternately, the STA operating as the CCo may also be appointed by the user. The "User-Appointed CCo" feature is intended for use by advanced users in managed networks (e.g., service provider deployment).

7.2.1 CCo Selection for a New AVLN

When an Unassociated STA determines that a new AVLN needs to be formed based on the Management Message Entries (MMEs) it received, the STAs determine whether it should become the CCo for the new AVLN based on the CCo Capability field and the Original Source Address (OSA) contained in these MMEs.

If the CCo Capability of the Unassociated STA is greater than that of the other STAs detected, or if the STA's MAC address is greater than the other STAs' when the CCo Capability is equal to the greatest capability detected, the STA must become the CCo, possibly in Coordinated Mode if neighboring networks are detected, and begin transmitting the Central Beacon. For comparing MAC addresses, the Individual/Group (I/G) bit of the 48-bit MAC address must be treated as the least-significant bit in the least-significant octet.

An exception is an Unassociated STA that is in SC-Add state will always become the CCo (refer Section 10.3.4.3).

7.2.2 Auto-Selection of CCo

The Auto-Selection function is used by the current CCo to determine the most suitable station in the AVLN to assume the role of CCo using the following criterion (in order of precedence) for CCo selection:

- *User-Appointed CCo:* A station that is appointed by the user as a CCo is always selected by the Auto-Selection function.

- *STA Capability:* In the absence of a User-Appointed CCo, STA capability is the highest criterion for ranking STAs. Various STA capabilities are described in Section 5.2.2.5. An STA with Level-1 CCo Capability is ranked higher than an STA with Level-0 Capability and so on.
- *Number of Discovered STAs:* The number of discovered stations (refer to Section 7.5.2) by an STA is the next most important ranking criterion. The number of discovered stations indicates the number of stations in the AVLN with which the station can directly communicate (i.e., without a repeater). Thus, this criterion enables the selection of the STA that has the best visibility in the network to become the CCo.
- *Number of Discovered Networks:* The number of networks discovered by an STA is the next most important ranking criterion. The STA in the network that discovers the largest number of neighbor networks has the highest potential for coordinating with these neighbor networks, and hence may be selected to become the CCo. (refer Section 7.5.2.)

7.2.2.1 CCo Capability All HomePlug AV stations are required to support CCo functionality. The CCo Capability of each HomePlug AV station is classified into four categories:

- *Level-0 CCo:* A Level-0 CCo uses CSMA-Only mode of operation and does not support contention-free allocations.
- *Level-1 CCo:* A Level-1 CCo supports CSMA-Only mode as well as Uncoordinated Mode. Level-1 CCos typically use Uncoordinated Mode when there are no neighbor networks and fall back to CSMA-Only mode when neighboring networks are detected. When operating in Uncoordinated Mode, Level-1 CCos can provide contention-free allocation and bandwidth management functions.
- *Level-2 CCo:* A Level-2 CCo supports CSMA-Only mode, Uncoordinated Mode as well as Coordinated Mode. Coordinated Mode enables Level-2 CCos to provide contention-free allocations and bandwidth management functions in the presence of neighboring networks.
- *Level-3 CCo:* Level-3 CCo is a future-generation CCo.

An AVLN with a Level-x CCo is also referred to as Level-x AVLN. It is mandatory for all HomePlug AV stations to support Level-0 or Level-1 or Level-2 CCo capabilities. Furthermore, it is mandatory for all HomePlug AV stations to operate as an STA in a Level-0, Level-1, or Level-2 AVLN. The CCo Capability of each STA is provided to the CCo in the CC_ASSOC.REQ message at the time of association. Every STA also indicates its CCo Capability in its Discover Beacon transmission. The CCo Capability of the CCo of an AVLN is indicated in all Central, Proxy, and Discover Beacons.

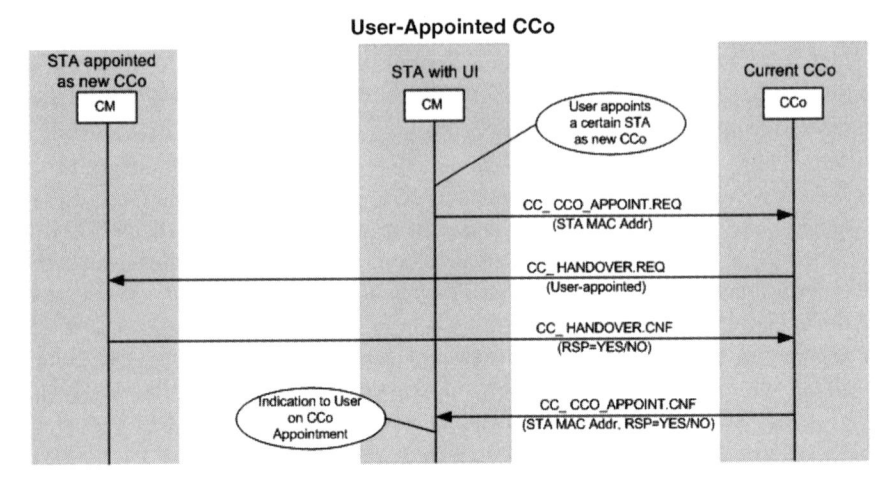

FIGURE 7.1 User-Appointed CCo.

7.2.3 User-Appointed CCo

An user can appoint a specific station in the AVLN to always assume the role of CCo by using the User-Appointed CCo functionality. The following procedure describes the User-Appointed CCo process. Figure 7.1 shows this function.

- The user enters the MAC address of the STA that should be assigned the role of CCo. The user enters this MAC address into a User Interface (UI) made available by an STA that is already associated and authenticated with the network.
- The UI STA communicates with the existing CCo via a CC_CCO_APPOINT. REQ message, with the Request Type indicating a request to appoint an STA with the MAC address contained in the CC_CCO_APPOINT.REQ message as a User-Appointed CCo.
- The current CCo responds by querying the appointed STA with a CC_HANDOVER.REQ message, requesting the STA to assume the role of CCo. The message indicates that the handover is due to user appointment.
- The STA responds with a CC_HANDOVER.CNF message.
- The current CCo passes on this response to the UI STA through the CC_CCO_APPOINT.CNF message.
- The current CCo then carries out the remaining steps of the handover function (refer Section 7.4).

A User-Appointed CCo does not perform the Auto-Selection of the CCo function. A User-Appointed CCo can be unappointed as a User-Appointed CCo by transmitting a CC_CCO_APPOINT.REQ message with a Request Type indicating unappointment of the existing CCo as a User-Appointed CCo. Upon unappointment as a User-Appointed CCo, the CCo continues to act as a CCo and starts performing Auto-Selection of the CCo function.

7.3 BACKUP CCo AND CCo FAILURE RECOVERY

Backup CCo and CCo failure recovery procedures enable HomePlug AV networks to recover from CCo failure.

7.3.1 Backup CCo

The function of the Backup CCo is to assume the role of CCo in the event of a CCo failure. The CCo designates an STA as the Backup CCo by sending a **CC_BACKUP_ APPOINT.REQ** message. The STA that is identified responds with a **CC_BACKUP_ APPOINT.CNF** message. The CCo keeps the Backup CCo up-to-date with the list of associated and authenticated stations in the AVLN by sending **CC_HANDOVER_ INFO.IND**. The Backup CCo sends a **CC_HANDOVER_INFO.RSP** message to indicate successful reception of **CC_HANDOVER_INFO.IND**. The current CCo may also send **CC_LINK_INFO.IND** message(s) to the Backup CCo to transfer the Connection Specification (CSPEC) and Bit-Load Estimate (BLE) information about all Global Links in the AVLN. The Backup CCo acknowledges the proper reception of this message using **CC_LINK_INFO.RSP**.

7.3.2 CCo Failure Recovery

It is possible for the existing CCo to drop out of the network without warning either because of equipment failure or because the user unknowingly unplugged the STA that was serving as the CCo.

If a Backup CCo is appointed in the network and if the Backup CCo does not receive any Central Beacons from the CCo for ≥ 2 (actual value used is implementation dependent) Beacon Periods, and during the same time period, the Beacon Detect Flag in the Frame Controls of all transmissions indicate Beacons that are not detected, then the Backup CCo assumes the role of the new CCo.

All STAs in the AVLN reinitiate the power-on network procedure (refer Section 10.2) when a CCo failure occurs and the Backup CCo (if any) fails to take over as the new CCo. An STA should wait for at least 10 Beacon Periods before reinitiating the power-on network procedure.

7.4 TRANSFER/HANDOVER OF CCo FUNCTIONS

The transfer of the CCo function from the current CCo to another STA (or the new CCo) in the network is shown in Figure 7.2. The handover may be initiated when the user has appointed a new CCo or a new CCo is selected by the Auto-Selection process.

Every CCo supports "hard handover": every CCo-capable STA is required to take over the role of the current CCo when requested and start transmitting Beacons for the network when the handover countdown expires.

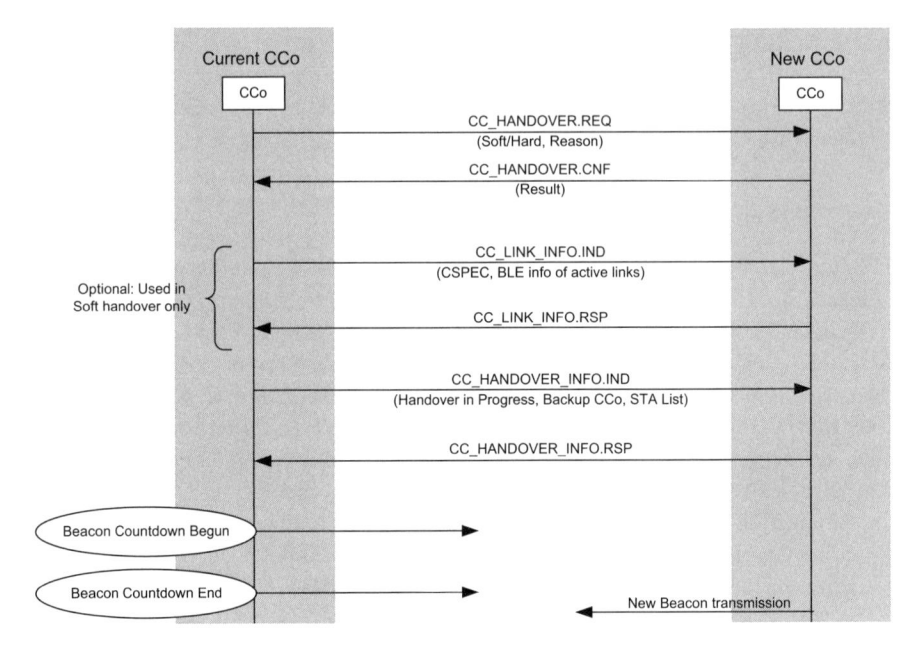

FIGURE 7.2 Transfer of CCo function.

The current CCo and the new CCo may optionally exchange CSPEC and BLE information about active connections in the network during the handover process so that the new CCo may be able to maintain uninterrupted service at the agreed upon QoS level for these connections. This optional process is called "soft handover."

The following steps describe the handover process:

- The current CCo sends a CC_HANDOVER.REQ message to the STA, requesting it to assume the role of CCo. The message indicates whether a soft or hard handover is requested.
- The STA responds with a CC_HANDOVER.CNF message.
- The current CCo sets the Handover-In-Progress (HOIP) bit in the Beacon to indicate that handover is in progress.
- If it is a soft handover, the current CCo sends a **CC_LINK_INFO.IND** message to the new CCo to transfer the CSPEC and BLE information about all Global Links in the AVLN. The new CCo acknowledges the proper reception of this message using **CC_LINK_INFO.RSP**. If it is a hard handover, **CC_LINK_INFO.IND/RSP** messages are not exchanged.
- The CCo initiates a transfer of relevant network information to the new CCo via the CC_HANDOVER_INFO.IND message. The message includes the list of associated and authenticated STAs in the network. The new CCo sends a **CC_HANDOVER_INFO.RSP** message to indicate successful reception of **CC_HANDOVER_INFO.IND**.

- The current CCo begins a handover countdown using the CCo Handover BENTRY.
- When the countdown expires, the current CCo stops transmitting the Beacon and the new CCo will take over the Beacon transmission.

7.5 CCo NETWORK MANAGEMENT FUNCTIONS

The following sections provide details on various network management functions of the CCo. The CCo of a network is responsible for managing the AVLN. These functions include

- assignment of Terminal and Network Identifiers (refer Chapter 10),
- managing Network Security (refer Chapter 10),
- bandwidth management for TDMA links within the AVLN (refer Chapter 9),
- coordination with neighbor networks (refer Chapter 12),
- Network Time Base synchronization (refer Section 7.5.1),
- Discover Process to facilitate maintenance of network topology information (refer Section 7.5.2).

7.5.1 Network Time Base Synchronization

AV stations use ± 25 parts-per-million (ppm) local clocks for signal processing. Network Time Base synchronization is used as a part of network-wide mechanism to reduce the clock error between stations in the network to within a fraction of a ppm. The CCo maintains a 32-bit timer, called the Network Time Base (NTB), clocked by a 25-MHz clock derived from the CCo's station clock (STA_Clk). The NTB is transmitted by the CCo in the Beacon and each STA in the AVLN receiving the Beacon synchronizes to the NTB. The CCo's STA_Clk and the Network Time Base are used to

- correct the PhyClk used for the processing of transmit and receive signals at all non-CCo STAs in the AVLN. This enables stations to support higher order modulation (e.g., 1024 QAM) without the need for more accurate (and more expensive) local clocks,
- announce the future Beacon position and the schedule within the Beacon Period. This enables stations to precisely determine the start and end of TDMA allocations,
- derive the ATS and perform the jitter control function (refer Section 7.5.1.1).

The CCo embeds a 32-bit Beacon Time Stamp (BTS) into the Beacon Frame Control. The BTS is the value of the NTB at the time instant at which the first nonzero sample of the Beacon PPDU is transmitted. All stations in the AVLN maintain a local

32-bit timer, called the NTB_STA, which must be synchronized in frequency and absolute value to the NTB of the Central Coordinator. Synchronization is normally achieved through the reception of the Central Beacon. If the Central Beacon cannot be heard reliably, the station may synchronize to a Proxy Beacon. If neither the Central Beacon nor a Proxy Beacon can be heard reliably, the Discover Beacon of another station may be used for this function until a Proxy Beacon is established.

One approach for achieving synchronization is to compute the frequency error between the CCo's STA_Clk and the STA's STA_Clk, and also compute the offset between the corresponding time bases (i.e. the values in the 32-bit timers). When a Beacon is detected, the receiver stores the 32-bit value of its local timer at the time of reception of the beginning of the AV Preamble signal of Beacon n, denoted as **LTmr$_n$**. The received Beacon contains the Beacon Time Stamp, denoted as **BTS$_n$**. Propagation delay between the transmitter and receiver can be ignored; however, if it is known, it could be compensated for in the computation of the offset.

The following formulas can be used to estimate the clock frequency and timer offset errors:

```
FreqError₁ = (BTS₁ - BTS₀) / (LTmr₁ - LTmr₀) - 1
Offset₁ = BTS₁ - LTmr₁
FreqErrorₙ = FreqErrorₙ₋₁ + wf ((BTSₙ - BTSₙ₋₁) / (LTmrₙ - LTmrₙ₋₁) -
1 - FreqErrorₙ₋₁) , n ≥ 2
Offsetₙ = Offsetₙ₋₁ + FreqErrorₙ (LTmrₙ - LTmrₙ₋₁) + wo ((BTSₙ -
LTmrₙ ) - (Offsetₙ₋₁ - FreqErrorₙ (BTSₙ - LTmrₙ₋₁))) , n ≥ 2
```

Note: w_o and w_f are weighting constants of the form $(1/2)^k$, where **k** is a positive integer. Larger values of **k** provide better filtering of the uncertainty between each **BTS$_n$** and **LTmr$_n$** pair caused by factors such as preamble detection jitter. Larger values of **k** result in a longer period of time to achieve convergence to the correct estimate of frequency error. When tracking a Beacon, it is recommended that a small value of **k** be used for the first samples and increased to the ultimate value.

Once *FreqError$_n$*, and *Offset$_n$* are known to be accurate, the relationship between the Network Time Base estimate (**NTB_STA$_i$**) and the local timer at instance *i* is

```
NTB_STAᵢ = LTmrᵢ + Offsetₙ + FreqErrorₙ (LTmrᵢ - LTmrₙ)
LTmrᵢ = (NTB_STAᵢ + FreqErrorₙ LTmrₙ - Offsetₙ) /
(1 + FreqErrorₙ)
```

where **Offset$_n$**, **FreqError$_n$**, and **LTmr$_n$** are the values computed from the most recently received Beacon.

Proxy stations inserts a BTS in the Proxy Beacon to enable hidden stations to track the NTB. All stations insert a BTS in Discover Beacon to enable new stations (that may not hear the Central Beacon) joining the network to track the NTB.

CCo NETWORK MANAGEMENT FUNCTIONS   129

7.5.1.1 Arrival Time Stamp for MSDU Jitter and Delay Control Each STA uses its NTB_STA in generating the Arrival Time Stamp (ATS). The ATS is the value of the NTB or NTB_STA when the first octet of the MSDU arrives at the Convergence Layer of a station and is used to provide jitter control and smoothing functions across the HomePlug AVLN.

The terms "Smoothing," and "Jitter Control" all refer to the smooth delivery of packets to the destination (rendering) application. Smoothing is necessary for many AV applications to provide an acceptable user experience. When smoothing is performed, each packet will be delivered to the rendering application at a constant delay interval after the packet was generated at the source application.

The Higher Level Entity (HLE) may request the Convergence Layer to provide ATS to HLE or perform smoothing for MSDUs belonging to established connections using CSPEC (ATS to HLE in CINFO and Smoothing). If the transmitting station detects either an ATS request or a smoothing request, it prepends a 32-bit ATS field before each MAC Service Data Unit (MSDU) that carries user traffic. The ATS is the value of the Network Time Base of the transmitting station when the MSDU is received by the Convergence Layer.

At the receiving station, the presence or absence of ATS field(s) in a MAC Frame is indicated by the MFT field in the MAC Frame Header. If ATS fields are present and if smoothing has been requested, the receiving station delivers MSDUs to the application based on the ATS value of the received MSDU in a way that will enable the stream to meet its latency and jitter requirements negotiated during the connection setup. If ATS fields are present and if an Arrival Time Stamp has been requested, the receiving station delivers the ATS to HLE.

7.5.1.2 PHY Clock Correction When Participating in More Than One Network
AV stations can optionally support the ability to be a part of multiple networks at the same time. A station participating in multiple networks needs to be able to receive Start-of-Frame (SOF) MPDUs from any of the networks during Contention Period. This creates a challenge as the station needs to know which networks NTB to be applied before receiving the SOF MPDU. HomePlug AV overcomes this problem by requiring stations to use RTS/CTS when the destination station(s) are known to be using an NTB that is different than the local AVLN's NTB. The SNID in the RTS is used by stations participating in multiple networks to apply the correct NTB for receiving the subsequent SOF.

A station that is using a different NTB, notifies the CCo using CC_DCPPC.IND message. The CCo maintains a list of all stations in its network. If the list contains one or more stations, the CCo sets RTS Broadcast Flag (RTSBF) in the Beacon. This causes all stations in the AVLN to send broadcast transmission with RTS/CTS, thus enabling the stations participating in multiple networks to properly receive the broadcast MPDU.

A station that is using a different NTB also sets the Different CP PHY Clock Flag (DCPPCF) in the SOF. This enables the receiver to know that it needs to precede any unicast transmission to this station with an RTS/CTS exchange, thus enabling the stations participating in multiple networks to properly receive the unicast MPDU.

7.5.2 Discover Process

The Discover Process is a periodic, low-overhead background process that is ongoing within the network where each associated and authenticated STA takes turn in transmitting a Discover Beacon as instructed by the CCo.

The purposes of the Discover Process are

- to allow the CCo and STAs to determine the identity and capability of other STAs in the network. Each STA creates and updates its Discovered STA List as an output of this function,
- to allow the CCo to discover networks that it cannot detect directly. Each STA creates and updates a Discovered Network List as an output of this function,
- the Discovered STA Lists and Discovered Network Lists of the CCo and of all STAs are used by the CCo to create the Topology Table which is used by the CCo in the CCo selection process,
- to allow hidden STAs (HSTAs), that is, STAs that cannot directly detect the Central Beacons transmitted by the CCo, but can detect Discover Beacons transmitted from certain STAs, to communicate with the CCo and to associate and authenticate with the network. In this case, the STA transmitting the Discover Beacon acts as a Proxy to relay MMEs.

The Discover Process consists of the following steps:

- At least once in every Discover Period, the CCo will schedule an opportunity to transmit a Discover Beacon for every associated and authenticated STA in the network (including HSTAs) and the CCo itself. In Uncoordinated and Co-ordinated Modes, a special contention-free allocation is provided for transmission of Discover Beacon. In CSMA-Only mode, the STA sends a Discover Beacon using CSMA/CA channel access at CAP2. The Discover BENTRY is used to identify the STA that transmits the Discover Beacon.
- The STA identified by the Discover BENTRY broadcasts a Discover Beacon when scheduled by the CCo.
- Every STA that receives a Discover Beacon and is able to correctly decode the Beacon updates its own Discovered STA List with the MAC address of the STA transmitting the Discover Beacon. The receiving STA also records the CCo Capability of the STA transmitting the Discover Beacon. STAs may also update their Discovered STA List every time it decodes a transmission from another STA successfully, in addition to the updates based on the receipt of Discover Beacons.
- Periodically, the CCo queries all STAs associated with the CCo to obtain their individual Discovered STA Lists and Discovered Network Lists. The CCo sends a **CC_DISCOVER_LIST.REQ** message to an STA to query for its Discovered STA List and Discovered Network List. The STA responds with the information in a **CC_DISCOVER_LIST.CNF** message. STAs in the AVLN

CCo NETWORK MANAGEMENT FUNCTIONS   131

will also send the **CC_DISCOVER_LIST.IND** message to the CCo in an unsolicited manner when they detect a new neighboring AVLN. The CCo constructs and updates its Topology Table using this information.

7.5.2.1.1 Discover Beacons   A Discover Beacon is a special type of Beacon that is transmitted by an associated and authenticated STA in the network during the Discover Process. It contains information including the Network Identifier (NID) of the network, the Terminal Equipment Identifier (TEI), the MAC address, the number of discovered STAs and networks, and the CCo Capability of the transmitting STA. The CCo schedules each and every associated and authenticated STA in the network and the CCo itself to transmit a Discover Beacon at least once in every 10 s.

The Discover Beacon contains a copy of the Regions BENTRY and Schedule BENTRIES of the Central Beacon. Schedule BENTRIES provide the locations of the persistent and nonpersistent CSMA allocations as well as all the CF allocations in the Beacon Period. HSTAs that can detect the Discover Beacon can use this information to exchange association and authentication messages with the CCo, using the STA transmitting the Discover Beacon as a relay. Each STA updates its Discovered STA List and Discovered Network List when it detects a Discover Beacon from another STA.

The CCo uses the Discovered Info BENTRY contained in the Discover Beacon for maintaining it network Topology Table. If the BENTRY indicates that the content of the Discovered STA List or Discovered Network List has been updated recently, the CCo may choose to query the STA transmitting the Discover Beacon for the latest Discovered STA List and Discovered Network List.

7.5.2.1.2 Discovered STA List and Discovered Network List   Each STA records the identity and attributes of every STA (from its own network and from different networks), whose Discover Beacons it can decode correctly. This information is maintained in the Discovered STA List. The Discovered STA List contains the MAC address of the STA that was heard, a flag to indicate whether the discovered STA is associated with the same or a different network, and the Short Network Identifier (SNID) of the network with which the discovered STA is associated. The CCo, PCo, and Backup CCo Capability and their corresponding status of the transmitting STA are also to be recorded. The signal level and average BLE may optionally be recorded. This optional information may be used by the CCo to determine whether or not to coordinate with another CCo or Group. The Discovered STA List should be updated every time the STA receives a Discover Beacon from another STA.

Each STA also maintains a Discovered Network List. This list is updated when the STA receives and decodes a Central, Proxy, or Discover Beacon with an NID that is different from the NID of its own network. Each entry of the Discovered Network List contains the NID, SNID, network mode, Hybrid Mode Flag, number of Beacon Slots, and start time of the Beacon Region of that network relative to the start of the Beacon Period.

An aging mechanism is used to remove stale entries from the Discovered STA List and Discovered Network List. An entry from the Discovered STA List is removed if a

Discover Beacon or other transmission from this STA has not been detected for 3–5 min. An entry from the Discovered Network List will be removed if a Central, Proxy, or Discover Beacon or other transmission from this network has not been detected for 3–5 min.

7.5.2.1.3 Topology Table The CCo maintains a Topology Table, which is a composite of the Discovered STA Lists and the Discovered Network Lists of all the STAs and HSTAs associated and authenticated with the CCo, together with the CCo's own Discovered STA List and Discovered Network List. The Topology Table contains the MAC addresses of all STAs and the Network Identifiers of all networks discovered by every STA and HSTA associated and authenticated with the CCo.

The CCo uses its Topology Table to make decisions such as identifying HSTAs, identifying suitable PCos and establishing PxNs, and determining which STA can best fulfill the role of PCo. The CCo may use the Topology Table to try to avoid interfering with a neighbor network.

7.6 SUMMARY

This chapter considered the functions and operation of the Central Coordinator (CCo) in HomePlug AV, including the CCo selection, backup and failure recovery, as well as the CCo discovery process. Since the CCo is central to the operation of HomePlug AV, these CCo functions and operations are referenced as needed throughout this book.

In Chapter 8, HomePlug AV channel access mechanisms are discussed with reference to the Beacon structure for Hybrid CSMA/TDMA operation. This chapter considers CSMA-Only as well as Coordinated and Uncoordinated Modes in Home Plug AV. The details are also provided in the CSMA channel access mechanism which are essential for those used in Home Plug AV 1.0.1.

8

CHANNEL ACCESS

8.1 INTRODUCTION

HomePlug AV provides a hybrid Carrier Sense Multiple Access/Time Division Multiple Access (CSMA/TDMA) channel access mechanism that uses a Beacon-based periodic schema. This section provides details on the dependence of the Beacon Period on the underlying AC line cycle frequency and the operation of the channel access methodology within each Beacon Period. The information contained in the Beacon and the relevant timing of allocations within the Beacon period are critical to the overall operation of the network. In addition, channel adaptation and Tone Map generation rely on AC line cycle synchronization to adapt optimally to synchronous powerline noise. Hence, it is important to understand how the Beacon period is synchronized in each local network by the Central Coordinator (CCo).

As discussed earlier, a HomePlug AV network consists of a set of AV stations connected to the AC powerline and logically separated by a privacy mechanism based on 128-bit Advanced Encryption Standard (AES) Network Encryption Key (NEK). A set of stations that use the same NEK form an AV Logical Network (AVLN) and each AVLN is managed by a Central Coordinator that performs network management functions such as authentication, association, admission control, and scheduling. Figure 8.1 shows a sample of HomePlug AV network architecture.

The CCo manages and transmits the Beacon Frame to accomplish its various functions within each Beacon Period. The Beacon Period structure consists of a Beacon Region followed by CSMA and TDMA Regions (Figure 8.2). The Beacon

---

HomePlug AV and IEEE 1901: A Handbook for PLC Designers and Users, First Edition.
Haniph A. Latchman, Srinivas Katar, Larry Yonge, and Sherman Gavette.
© 2013 by The Institute of Electrical and Electronics Engineers, Inc. Published 2013 by John Wiley & Sons, Inc.

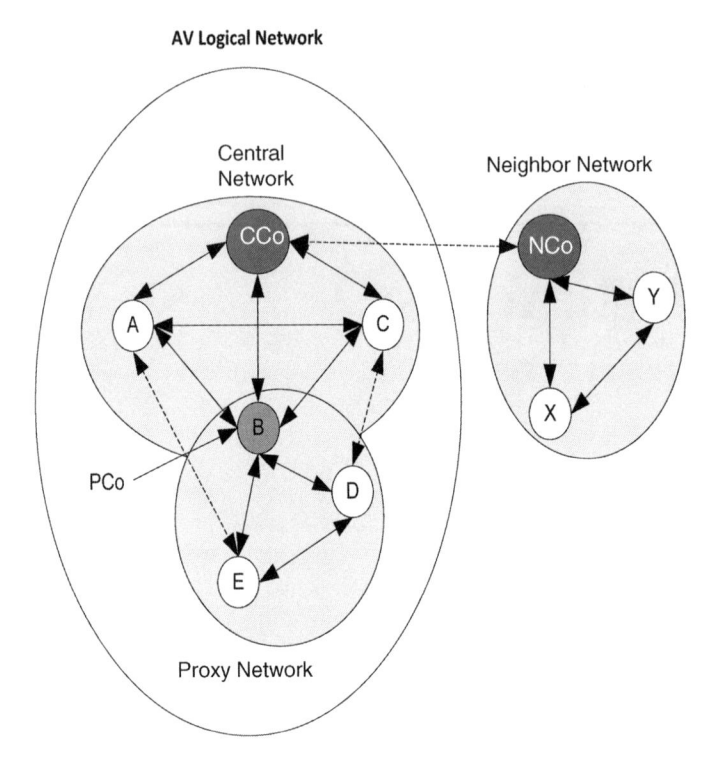

FIGURE 8.1 HomePlug AV network architecture.

Region contains the Beacon transmitted by the CCo and each Beacon consists of a Preamble, Frame Control, and a 136-byte payload. Information describing the allocations within the Beacon Period is broadcast in the Beacon payload using "mini-ROBO," a very robust modulation method.

TDMA allocations are provided by the CCo for streams requiring Quality of Service (QoS) whereas CSMA allocations in the Beacon Period are used by connectionless traffic and by connections that do not have strict QoS requirements. The CSMA channel access mechanism used by HomePlug AV is the same as the one used by HomePlug 1.0 [18,19] and is briefly described in Section 1.4.

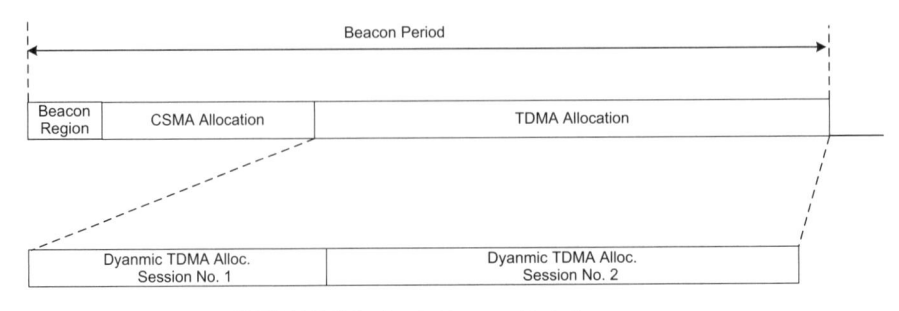

FIGURE 8.2 Basic Beacon Period structure.

The CCo exchanges messages (request, response, and confirm) with neighbor CCos (i.e., NCos) in its Interfering Network List (INL) to request new Reserved Regions for each network. The CCo first sends a request to all NCos that specifies the time intervals that the CCo wants to use as its new Reserved Regions. Each NCo will send a response back to the CCo. If all the responses indicate that the request is accepted, then the CCo will send a positive confirm message to all NCos, update its Beacon Period structure to include the new Reserved Regions, and start using them.

8.2 BEACON PERIOD AND AC LINE CYCLE SYNCHRONIZATION

To improve performance and QoS stability in the presence of common powerline noise, the Beacon Period is set to be twice the AC line cycle period (50 or 60 Hz) and is synchronized with the underlying AC line cycle. There are variations in the phase and frequency of the AC line cycle from the power generation and distribution system, hence the Beacon Period can vary as the AC line cycle period varies. The CCo is responsible for performing an effective synchronization function and the HomePlug AV specification recommends an approach to provide a consistent behavior and performance among different manufacturers.

8.2.1 Line Cycle Synchronization

The CCo uses a phase lock mechanism to synchronize the Beacon transmission to the AC line cycle and in this way also provides synchronization for all stations in the network. This will ensure that the Tone Map remains valid, even when the underlying AC line cycle frequency and phase characteristics change. Line-cycle synchronization is achieved by having the CCo track a particular point in the AC line cycle using a Digital Phase Locked Loop (DPLL) or similar system. Using a filter or digital lock loop at the CCo is essential to eliminate noise events or jitter in the measurement of the line-cycle phase. The CCo uses its local tracking history to also predict future locations of the Beacons that it announces to all stations in the Beacon schedule.

To ensure that stations with persistent or on-going allocations can transmit even when a Beacon is not detected, the CCo provides information about the location of future Beacons within the Beacon Frame Control (refer to Section 5.2.1.).

8.3 BEACON PERIOD STRUCTURE

The HomePlug AV Beacon Period is twice the underlying AC line cycle period. Thus, when operating in powerline environments with an AC line cycle frequency of 60 Hz, the Beacon Period will be 33.33 ms. Similarly, when operating in powerline environments with an AC line cycle frequency of 50 Hz, the Beacon Period will be 40 ms.

There are three types of Beacons: Central Beacon, Proxy Beacon, and Discover Beacon.

- Central Beacons are issued by the CCo, and contain coordination information for the STAs within its AVLN, and for CCos of neighboring AVLNs.
- Discover Beacons are used to exchange information about network topology and, in particular, to discover neighbor networks (refer to Section 7.5.2).
- Proxy Beacons are issued by Proxy CCos within an AVLN to relay the information contained in the Central Beacon to hidden stations (refer to Section 11.5).

When and how these various types of Beacon are sent (and expected by a STA) depends on the network mode of the AVLN, which is indicated in the Beacon Delimiter. The channel access information is communicated between CCos of neighboring AVLNs using the Regions BENTRY, from which a CCo determines when STAs within its AVLN may use Carrier Sense Multiple Access/Collision Avoidance (CSMA/CA) access or be assigned exclusive access through TDMA. The CCo then conveys this information to the STAs within its AVLN using Persistent and Nonpersistent Schedule BENTRIES. A STA uses this schedule information to determine when it is eligible to access the channel using CSMA/CA or TDMA.

The overall structure of a Beacon Period consists of periods for transmission of Beacons and periods for transmission of data by STAs in the AVLNs sharing the medium. The terms Contention Period (CP) and Contention Free Period (CFP) are used to indicate intervals of time where CSMA/CA and TDMA-based channel access mechanisms are, respectively, used for medium sharing. How this is done depends on the AVLN's network mode and the QoS needs of the streams in the AVLN. Regions BENTRIES specify what activity is possible for an AVLN for every interval of the Beacon Period, and are only relevant to coordination between neighbor networks (refer to Chapter 12); STAs within an AVLN do not interpret Regions BENTRIES. Schedule BENTRIES specify for a particular AVLN its CPs, CFPs, and by omission, periods during which no access is allowed (again, depending on the AVLN's network mode). The "big picture" intervals described in the Regions BENTRIES are called *regions*, while the AVLN-specific access intervals carried in Schedule BENTRIES are called *allocations*.

Figure 8.3 shows the Beacon Period synchronized to AC line cycle with Beacon, CSMA, and Reserved Regions, along with Persistent and NonPersistent Contention-Free and CSMA allocations, for the case of an AVLN that is operating in stand-alone or Uncoordinated mode and is supporting three TDMA Sessions.

A Beacon Region consists of one or more Central Beacons. A Beacon consists of a Preamble, a Frame Control, and a 132-octet Beacon payload. The CCo generates the Central Beacon and transmits it in the Beacon Region in stand-alone or "Uncoordinated" mode as well as in "Coordinated mode"— when there are interfering neighbor networks. The location of the Beacon Region is specified in the Regions BENTRY for neighbor networks, but STAs within an AVLN calculate its location and duration from other information carried in the Beacon. In CSMA-only mode, there is no Beacon Region.

Information describing the allocations within the Beacon Period is broadcast in the Beacon payload by using one or more Schedule Beacon Entries. This information

BEACON PERIOD STRUCTURE

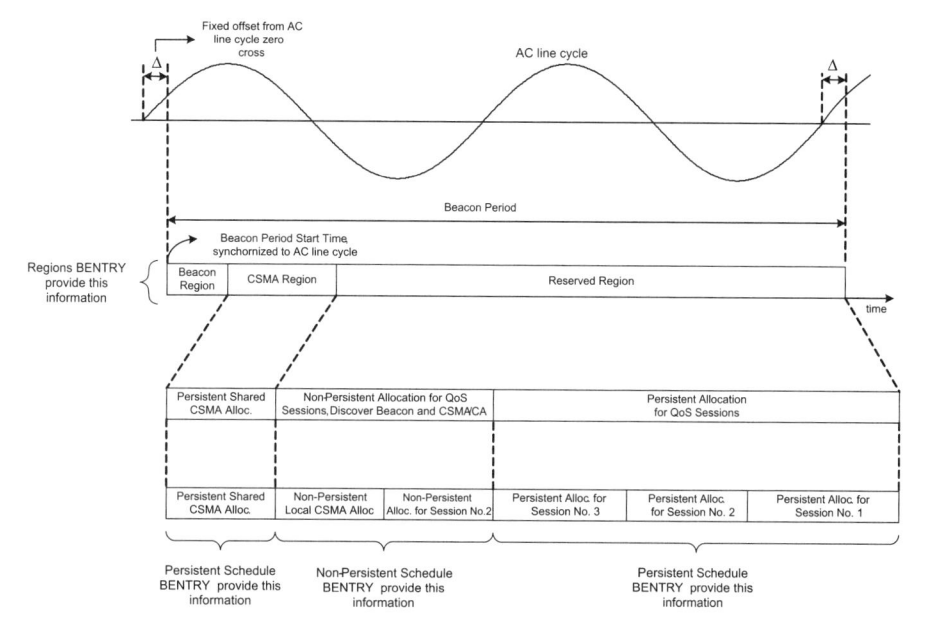

FIGURE 8.3 Example of Beacon Period structure in Uncoordinated Mode.

is used by STAs within the network to coordinate sharing of bandwidth. The Beacon carries two types of scheduling information:

- Nonpersistent Scheduling information.
- Persistent Scheduling information.

Persistent CFP Schedule information is carried in the Persistent Schedule BENTRY. Persistent Schedule information is valid for multiple periods and can contain the following types of allocations:

- Persistent CSMA/CA Allocations (shared or local).
- Persistent TDMA Allocations.

The Persistent Schedule BENTRY has two fields that are used to interpret the persistence of schedule information:

- Current Schedule Countdown.
- Preview Schedule Countdown.

If the schedule is not changing, the schedule information reflects the current schedule and the Preview Schedule Countdown (PSCD) is zero. In this case, the Current Schedule Countdown (CSCD) indicates the minimum number of Beacon Periods for which the current schedule may be assumed valid. The Current

Schedule Countdown value must not be smaller than the previous Current Schedule Countdown value minus one. In this way, stations that miss Beacons will know how long they may use the current schedule information they have. It is important to note that to transmit in a TDMA allocation, a STA must have knowledge of the Beacon Period Start Time (BPST) as well as the schedule information for that Beacon Period.

The Current Schedule Countdown value may indicate indefinite persistence which is needed when Beacons are sent using CSMA/CA and such indefinitely persistent allocations remain valid until superseded by newer schedule information in a Central Beacon sent at some later time. When the new schedule information is sent, it should have a suitably large Current Schedule Countdown value depending on the reliability of Beacon reception in the AVLN. Any STA that does not receive the new schedule information by the time it becomes effective may transmit according to the old, indefinitely persistent allocation information, so caution is advised in the use of indefinite persistence.

In order to change the schedule, a CCO transmits a Persistent Schedule BENTRY announcing the Preview Schedule. In this BENTRY, the PSCD is set to a nonzero value. This value indicates that the schedule information is a new schedule (not the current schedule) and when the new schedule will take effect. In this case, the Current Schedule Countdown previews the value that the new schedule will have for its Current Schedule Countdown during the first Beacon Period when it takes effect. The Current Schedule Countdown value in this case is a preview value and must not change from its initial value. In this way, stations that miss Beacons will know when they can use the new schedule information they have and for how long it will be valid. This approach allows a number of repetitions of the new schedule to ensure that all stations have the relevant information, even if some stations miss the Beacon during the Beacon Period when the new schedule takes effect.

Each Central Beacon should carry a Persistent Schedule BENTRY with the Preview Schedule when the schedule is changing. It may also carry a separate Persistent Schedule BENTRY with the Current Schedule within the same Beacon. If it does, the Current Schedule Countdown field in this BENTRY is decremented to indicate the start of the new schedule.

Only one Preview Schedule is allowed to be in existence at any time. If the CCo has announced a Preview Schedule, that Preview Schedule must become the Current Schedule before the CCo can announce a new Preview Schedule.

Figure 8.4 shows examples of schedule changes. Initially, Schedule A is in effect. In Beacon Period 2, the CCo determines that the schedule should change to Schedule B. So, beginning in Beacon Period 3, the CCo includes a BENTRY containing Schedule B with the values of PSCD and CSCD shown for the schedule. (The previewed schedules are shaded in the figure.) Although the CCo has the option of transmitting both Schedules A and B in separate BENTRIES, transmitting Schedule A is no longer necessary.

Once the CCo has announced Schedule B in Beacon Period 3, the earliest that Schedule B can be replaced by a new Schedule C is Beacon Period 9. That is because while the PSCD for Schedule B is nonzero, the CSCD is a preview of what the CSCD

BEACON PERIOD STRUCTURE   139

Schedule A												
PSCD	0	0	0	0	0							
CSCD	3	3	2	1	0							
Schedule B												
PSCD			3	2	1	0	0	0				
CSCD			2	2	2	2	1	0				
Schedule C												
PSCD					3	2	1	0	0	0	0	0
CSCD					1	1	1	1	1	1	1	...
					Current Schedule							
	A	A	A	A	A	B	B	B	C	C	C	C	C
	1	2	3	4	5	6	7	8	9	10	11	12	...

Beacon Period Number

FIGURE 8.4 Example of Beacon schedule persistence.

will be in the first Beacon Period that B is current (Beacon Period 6). Since the CCo chose a current schedule persistence preview value of 2, Schedule B must be the current schedule for three Beacon Periods.

Persistence of schedule information improves reliability, but decreases responsiveness to urgent needs. A STA that requires additional allocation might communicate its request to the CCo in the same Beacon Period (3) in which a new schedule (B) is announced. This requires the session to wait for the announced schedule (B) to take effect before a revised schedule (C) can be broadcast. The revised schedule (C) must then countdown before it becomes effective, so the session may be forced to wait several Beacon Periods before it can obtain the additional allocation it needs.

To allow rapid response to urgent allocation requests, some portion of the contention-free allocation schedule may be nonpersistent. This Nonpersistent allocation (provided using Nonpersistent Schedule BENTRY) provided for established CFP connections is termed Extra Allocation. It is necessary for a Station to receive the Beacon to utilize any Extra Allocation it is given in a Beacon Period, which makes Extra Allocations less reliable than the persistent allocations.

Nonpersistent CFP Schedule information is carried in the Nonpersistent Schedule BENTRY. Nonpersistent Schedule is valid only for the current period and can contain Extra Allocations, CSMA/CA Allocations or Discover Beacon Allocation.

The Discover Beacon Allocation provides Discover Beacon transmission opportunities for various stations in the network.

8.3.1 Beacon Period Structure in CSMA-Only Mode

In CSMA-only mode, the Beacon Period structure consists of CSMA Regions and Stay-out Regions and there is no Beacon Region, since the Beacon is transmitted in CSMA/CA mode. The locations of the CSMA Regions are inferred from the Schedule BENTRIES of the Beacon payload. Details on how Stay-out Regions are determined can be found in Section 12.3.

CSMA/CA will be used exclusively by all traffic in CSMA-only mode. There are no Global Links, and all stations can send traffic using CSMA/CA without regard to the start of the Beacon Period (i.e., a transmission can extend across the Beacon Period boundary since that region is not reserved for Central Beacon transmission in CSMA-only mode). Schedules specify the locations of CSMA/CA allocations. Persistent schedules may be made with indefinite persistence. This is needed due to the unreliability of Beacon reception when CSMA/CA is used. A STA that has not heard a Central Beacon or Proxy Beacon for several Beacon Periods uses its local clock to estimate the locations of the CSMA/CA allocations, and transmit in them, as long as the schedule information remains valid. An indefinitely persistent allocation remains valid until it is superseded by newer schedule information in a Central Beacon sent at some later time.

In CSMA-only mode, all Beacons are transmitted using CSMA/CA. The Central Beacon should be transmitted as near to the start of the Beacon Period as possible. Proxy Beacons (if any) should be transmitted as soon after the Central Beacon is received as possible, and both are sent at priority CAP3. Discover Beacons are transmitted as soon as possible after being designated by the CCo in the Central Beacon, and they should be sent at priority CAP2. Since CSMA/CA transmissions rely on carrier sensing, Beacon reliability can be expected to be significantly better than that of TDMA Beacons in environments where TDMA allocations cannot be guaranteed because of hidden nodes. Implementations may optionally use Request to Send/Clear to Send (RTS/CTS) to enhance Beacon reliability. It should be noted that a Beacon sent in CSMA-only mode does not have an explicit acknowledgment (i.e., SACK). Beacon transmission is followed by a Beacon-to-Beacon Interframe Space (B2BIFS, which is 90 μs), followed by Priority Resolution Slots.

In CSMA-only mode, all transmissions are sent using Hybrid Delimiters (refer to Section 5.1). This simplifies neighbor network detection and coordination. Optionally, STAs may use frame lengths not compatible with HomePlug 1.0 when there are no HomePlug 1.0 stations detected, as expressed in the value of the Hybrid Mode (HM) field of the Beacon (refer to Section 5.2.2.4). Since all transmissions use CSMA/CA, Contention Free Period Initiation (CFPI) allocations are never used in CSMA-only mode.

Figure 8.5 shows an example of the Beacon Period Structure in CSMA-only mode. Note that the Central Beacon is sent near the start of the Beacon Period, but had to first wait until another transmission ended, then contend for the channel using CSMA/CA with priority CAP3.

The Central Coordinator will maintain a logical Beacon Period Start Time by indicating one or more Beacon Transmission Offsets (BTOs) in the Frame Control of the Central Beacon. Maintenance of the logical BPST is essential for channel adaptation and neighbor network coordination in CSMA-only mode. Since CSMA collisions can result in Beacon loss, all stations in the AVLN should use local AC line cycle tracking to estimate the BPST when CCo Beacons cannot be detected. Together with indefinitely persistent CSMA/CA allocations, this allows a STA to transmit even when Beacons are not received reliably.

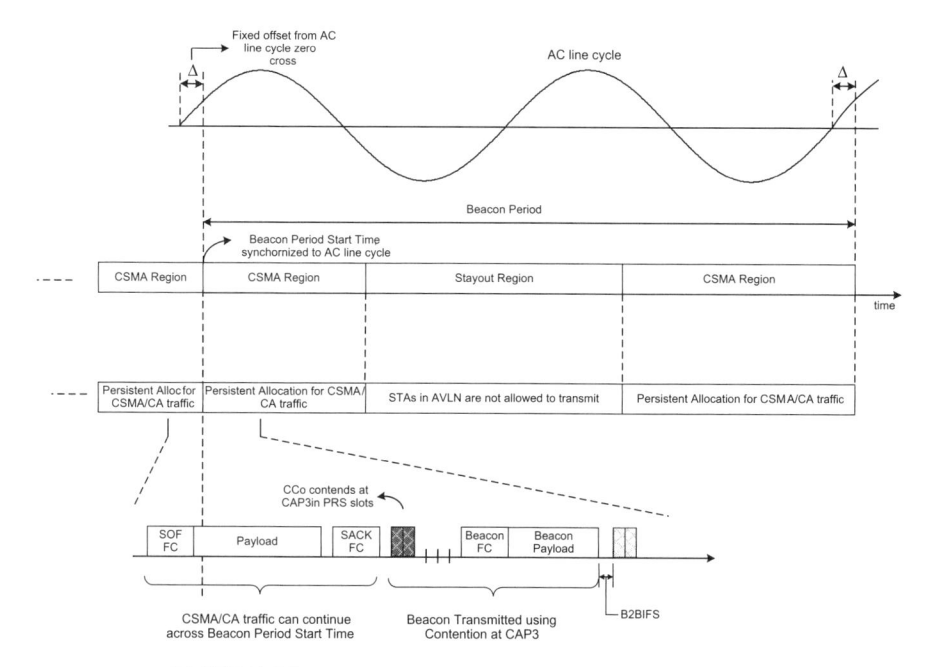

FIGURE 8.5 Beacon Period structure in CSMA-only mode.

8.3.2 Beacon Period Structure in Uncoordinated Mode

The Beacon Period structure in Uncoordinated Mode consists of a Beacon Region followed by a CSMA Region optionally followed by one or more Reserved Regions and CSMA Regions. The duration of the Reserved Regions and CSMA Regions are inferred from the Schedule BENTRIES of the Beacon payload.

A CSMA Region must immediately follow the Beacon Region, and it must have a minimum length of MinCSMARegion, defined in HomePlug AV to be 1500 µs. Additional CSMA Regions may also be present in the Beacon Period; the minimum length parameter does not apply to these additional CSMA regions. CSMA Regions use the HomePlug 1.0 CSMA/CA channel access mechanism described in Section 8.4.

If there are streams that require QoS, and the CCo is capable of QoS management, there may be persistent or nonpersistent allocations for specific streams in the Reserved Regions. Figure 8.6 shows the Beacon Period Structure in Uncoordinated mode.

The Central Beacon is always transmitted in Hybrid Mode using TDMA in the one and only Beacon slot in the Beacon Region, which is aligned with the start of the Beacon Period. All STAs refrain from transmitting during this interval. Proxy Beacons and Discover Beacons are also transmitted in Hybrid Mode using TDMA in intervals allocated for them in the Reserved Regions.

While all Beacons are always sent in Hybrid Mode, all other traffic is sent in a mode that depends on the Hybrid Mode field value in the Beacon and the region in which the transmission occurs (refer to Section 5.2.2.4).

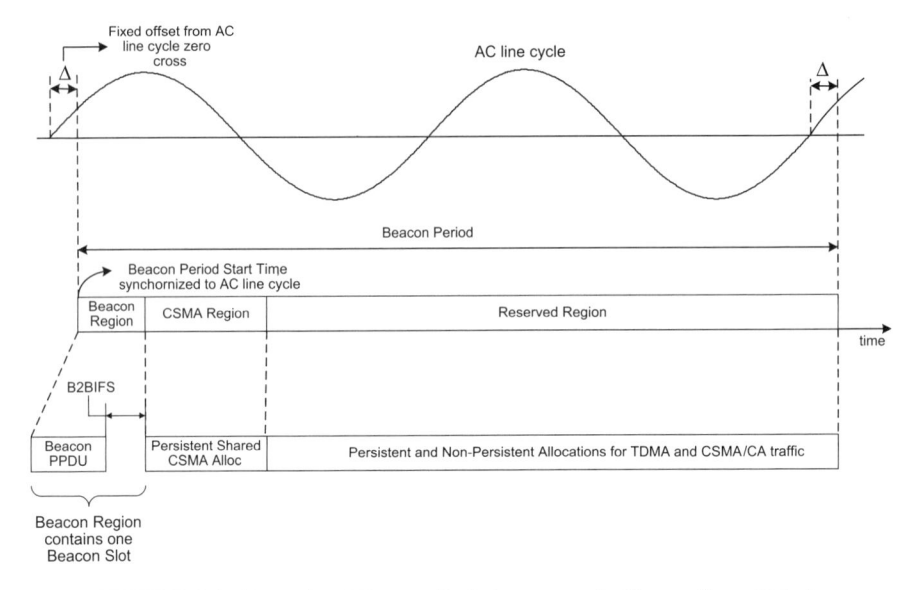

FIGURE 8.6 Example of Beacon Period structure in Uncoordinated Mode.

AC line cycle synchronization is performed by the CCo, and all STAs within the AVLN synchronize to the CCo using the Central and Proxy Beacons. Depending on the schedule persistence and the number of Beacon Transmission Offset values, a STA that misses one or more Beacons may still be able to transmit at the appropriate times (either TDMA assigned to one of its streams, or CSMA/CA), but once either the future Beacon location information from the BTOs or the persistence of the schedule information has been exhausted, a STA must not transmit as part of that AVLN until it receives another Beacon. Lacking schedule information, it cannot know when CSMA/CA regions occur, nor its TDMA allocations.

8.3.3 Beacon Period Structure in Coordinated Mode

Coordinated Mode is only used when one or more neighboring AVLNs are present. The Beacon Period structure in Coordinated Mode consists of a Beacon Region with one or more Beacon slots, followed by a CSMA Region, optionally followed by one or more Reserved Regions, CSMA Regions, Stay-out Regions, and Protected Regions. The duration of the Reserved Regions and CSMA Regions are inferred by STAs in the AVLN from the Schedule BENTRIES of the Beacon payload. A CCo computes these regions according to the methods described in Section 12.3.

Note: An AVLN Operating in Coordinated Mode might have only one Beacon slot in its Beacon Region.

A CSMA Region must immediately follow the Beacon Region, and it must have a minimum length of MinCSMA Region (1500 μs). Additional CSMA Regions may

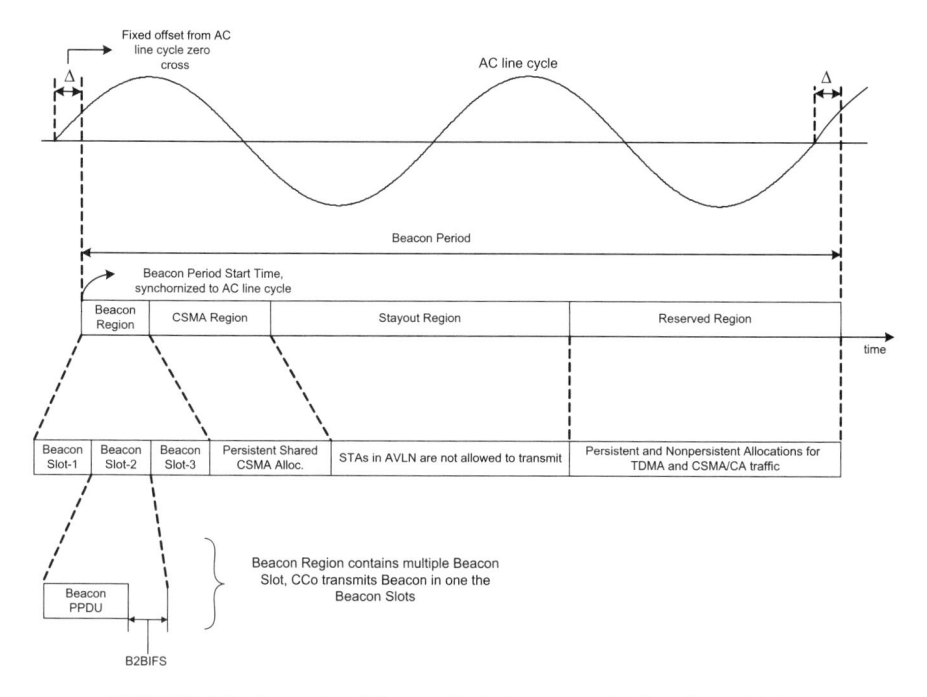

FIGURE 8.7 Example of Beacon Period structure in Coordinated Mode.

also be present in the Beacon Period; the minimum length parameter does not apply to these additional CSMA regions.

If there are streams that require QoS, and the CCo is capable of QoS management, then there may be allocations for specific streams in the Reserved Regions. These may be either persistent or nonpersistent.

Figure 8.7 shows the Beacon Period Structure in Coordinated Mode with a multislot Beacon Region, CSMA Regions, Reserved Regions, and Stay-out Regions.

The Central Beacon is always transmitted in Hybrid Mode using TDMA in one of the Beacon slots in the Beacon Region, which is aligned with the start of the Beacon Period. All STAs refrain from transmitting during this interval. Proxy Beacons and Discover Beacons are also transmitted in Hybrid Mode using TDMA in intervals allocated for them in the Reserved Regions.

While all Beacons are always sent in Hybrid Mode, all other traffic is sent in a mode that depends on the Hybrid Mode field value in the Beacon and the region in which the transmission occurs (refer to Section 5.2.2.4).

8.4 CSMA CHANNEL ACCESS

The CSMA channel access scheme used in HomePlug AV is essentially the same as used in HomePlug 1.0.1 [18,19]. Medium sharing is in CSMA allocations is accomplished by the CSMA/CA mechanism with priorities and a random backoff time. Prioritized access is achieved by a Priority Resolution Period in which stations signal the priority at which they intend to transmit, allowing only the highest priority available to continue in the contention process. The random backoff mechanism spreads the time over which stations attempt to transmit, thereby reducing the probability of collision, using a truncated binary exponential backoff mechanism similar to IEEE 802.11.

8.4.1 Carrier Sense Mechanism

Carrier sense is a fundamental part of the distributed access procedure. Physical Carrier Sense (PCS) is provided by the PHY upon detection of a Priority Resolution symbol, or upon detection of the Preamble. In the latter case, PCS stays high long enough for frame control to be detected and Virtual Carrier Sense (VCS) to be asserted by the MAC. A Virtual Carrier Sense mechanism is provided by the MAC by tracking the expected duration of channel occupancy. Virtual Carrier Sense is set by the content of the frame control received or upon collision. In these cases, Virtual Carrier Sense tracks the expected duration of the busy state of the medium. Virtual Carrier Sense is also set upon priority pre-emption. In this case, Virtual Carrier Sense tracks the expected duration of the Contention State. The medium is also considered busy when the station is transmitting.

Figure 8.8 depicts where each of these states occurs for the case in which a MAC Protocol Data Unit (MPDU) is transmitted or detected in the Contention State. Figure 8.9 depicts where each of these states occurs for the case in which a station gets pre-empted in the Priority Resolution Period and detects no MPDU transmission in the Contention State. Figure 8.10 depicts where each of these states occurs for the case in which MPDU frame control errors or collision lead to a busy state of the medium and no delimiter is detected for an EIFS_X period.

8.4.1.1 MAC-Level Acknowledgments MAC-level acknowledgments occur immediately following the Long MPDU (in case of a MPDU Burst, the last MPDU in the Burst) to which they are responding without relinquishing the channel to new transmissions. The HomePlug MAC also supports a partial ARQ scheme in

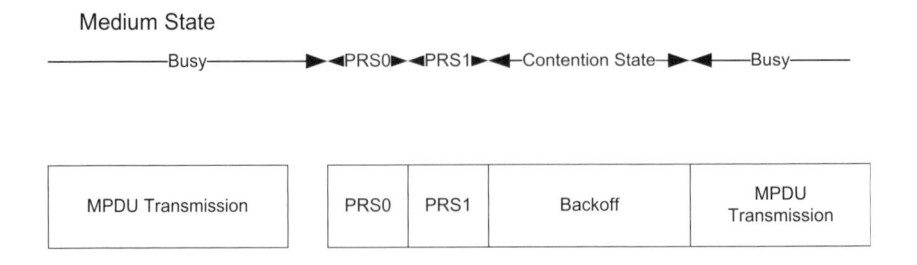

FIGURE 8.8 Medium States when a MPDU is transmitted or detected in Contention State.

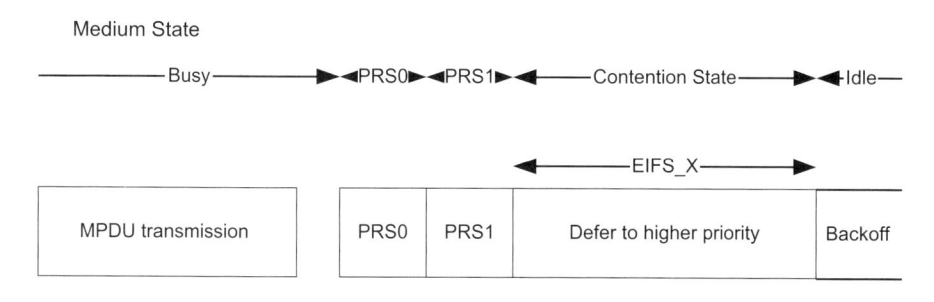

CSMA CHANNEL ACCESS

FIGURE 8.9 Medium States when a station gets pre-empted in the Priority Resolution Period and detects no MPDU transmission in Contention State.

which one member of the logical network acknowledges a multicast transmission as a proxy. Partial ARQ does not guarantee delivery to multicast groups but does provide an indication that the message was received by at least one station.

During CSMA allocations, collisions can result when multiple stations choose the same backoff value. AV stations infer a collision based on the reception status of responses transmitted by the receiver.

8.4.1.2 Setting of Virtual Carrier Sense (VCS) Timer A VCS timer is maintained by all HomePlug AV stations to improve reliability of channel access during CSMA. The setting of the VCS timer is shown in Table 8.1.

The value of EIFS_X depends on whether the transmissions in the corresponding CSMA interval are in Hybrid Mode or AV-only mode.

- When operating in AV-only mode, EIFS_X will be equal to EIFS_AV (2920.64 μs).
- When operating in Hybrid Mode, EIFS_X will be equal to EIFS (1695 μs).

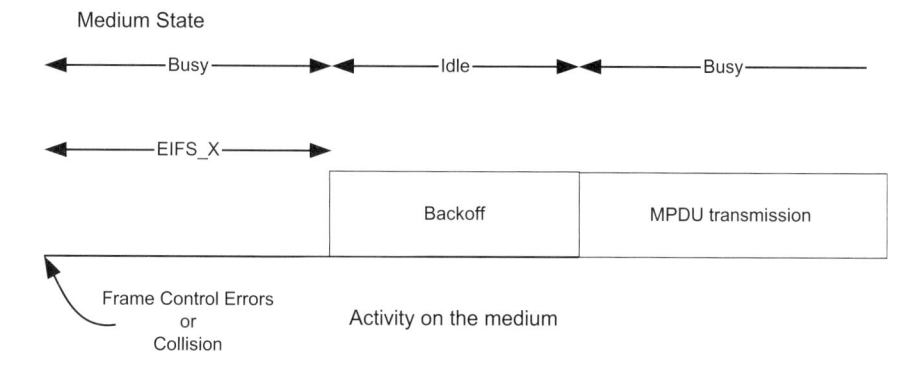

FIGURE 8.10 Medium States when MPDU frame control errors or collision lead to a busy state and no delimiter is detected for an EIFS_X period.

TABLE 8.1 Setting the VCS Timer

Event Type	New VCS Timer Value	Medium State When VCS Timer Expires
Collision	EIFS_X	Idle
Frame Control with bad FCCS is received	EIFS_X	Idle
Start-Of-Frame delimiter with MPDUCnt set to 0b00 is received	FL_AV*1.28 μs) + DelimiterTime + CIFS_AV	PRS0
Sound delimiter with MPDUCnt set to 0b00 is received	(FL_AV*1.28 μs) + DelimiterTime + CIFS_AV	PRS0
Start-Of-Frame delimiter with MPDUCnt set to 0b01, 0b10, or 0b11 is received	(FL_AV*1.28 μs)	Search for next MPDU in the Burst
Sound delimiter with MPDUCnt set to 0b01, 0b10, or 0b11 is received	(FL_AV*1.28 μs)	Search for next MPDU in the Burst
SACK delimiter is received	CIFS_AV	PRS0
RTS delimiter is received	RCG + DelimiterTime + CMG + DelimiterTime	Idle
CTS delimiter is received	(DUR*1.28 μs) + CIFS_AV	PRS0

8.4.1.3 RTS/CTS The Request to Send and Clear to Send delimiters can be used by HomePlug AV stations during CSMA allocations to handle hidden nodes.

8.4.2 Contention Procedure

To reduce the probability of collision between contending nodes, a random backoff algorithm is used to disperse the times at which stations with queued MPDUs attempt transmission. This section describes the behavior of the random backoff time.

The random backoff algorithm uses several counters to facilitate its operation. These are:

- BPC, the Backoff Procedure Event Counter.
- BC, the Backoff Counter.
- DC, the Deferral Counter.
- CW, the Contention Window.

When a station first attempts to deliver a MPDU, it implements the following procedure.

(1) The station first initializes BPC, BC, and DC by setting each to zero.

(2) The station then uses the carrier sense mechanism described in Section 1.4.1 to determine the Medium State as determined by the VCS and PCS. The station then performs one of the following steps:
- When the Medium State is idle, the station may transmit the MPDU without contention resolution and processing proceeds to step 5.
- When the Medium State is busy, the station proceeds to step 6. If the station does not transmit immediately, processing proceeds to step 4.
- When the Medium State is PRS0, PRS1, or Contention State, the station resolves priority contention according to the procedures of Section 1.4.2.1 and then proceeds to step 3.

(3) If priority contention was lost, the station sets VCS according to the procedure of Section 1.4.1.2 and returns to step 2, or otherwise proceeds to step 4.

(4) If BPC, BC, or DC are zero, the station sets CW and DC based on the value of BPC and the priority of the queued MPDU according to Table 8.2, increments BPC, and then sets BC = Random (CW). Otherwise, the station decrements BC and DC. (Note: The function Random (CW) returns an integer uniformly distributed between 0 and CW, inclusive). At each subsequent Slot Time, if the Medium State has become busy, the station goes to step 6. Otherwise, (a) if BC is not equal to 0 and PCS is not active, the station decrements BC, or (b) if BC = 0, the station begins transmission of the MPDU and then proceeds to step 5.

(5) After the station transmits a MPDU, if a collision occurs then the station proceeds to step 6. Otherwise, the backoff procedures are completed.

(6) The station uses the carrier sense mechanism described in Section 1.4.1 to determine the Medium State as determined by the VCS and PCS. While the Medium State is busy, the station does not transmit the MPDU and waits for the Medium State to change. When the Medium State goes to PRS0, the station resolves priority contention according to the procedures of Section 1.4.2.1 and then proceeds to step 3. When the Medium State goes to idle, the station proceeds to step 4.

One of the unique features of HomePlug AV backoff procedure is the use of the Defer Counter. The Defer Counter tracks the number of times a node has to defer to transmissions from other nodes. This enables the node to estimate the level of congestion (or contention) on the medium. A large number of defers indicates high

TABLE 8.2 CW and DC as a Function of BPC and Priority

BPC	DC	CW	
		CA3 and CA2	CA1 and CA0
0	0	7	7
1	1	15	15
2	3	15	31
>2	15	31	63

TABLE 8.3 Channel Access Priority versus Priority Resolution

Channel Access Priority	PRS0 State	PRS1 State
CA3	1	1
CA2	1	0
CA1	0	1
CA0	0	0

probability of collision, thus causing the node to increase its contention window. This enables HomePlug AV to reduce the collision overhead [19].

8.4.2.1 Priority Contention Priority contention is resolved in the Priority Resolution Slots (PRS0 and PRS1). Table 8.3 maps the channel access priority to the Priority Resolution Slots. A Priority Resolution Symbol followed by 5.12 μs of silence is transmitted in the Priority Resolution Slot for which a 0b1 is indicated in the table, with the following exception. If PCS is active (Priority Resolution Symbol detected) in a PRS in which the station did not assert the signal, then priority contention is considered lost and the station will not transmit in any remaining PRS and will not transmit in the immediately succeeding Contention State.

Table 8.3 shows the four possible levels of priorities: the highest priority is indicated by CA3 = 0b11 and the lowest priority is indicated by CA0 = 0b00. Packet Classification, described in Section 9.2, is used by HomePlug AV to determine the Channel Access Priority of CSMA/CA traffic.

Packet Classification described in Section 9.2 is used by HomePlug AV to determine the Channel Access Priority of CSMA/CA traffic.

8.5 TDMA CHANNEL ACCESS

HomePlug AV stations use Beacon-based TDMA (or Contention Free Access) in both Uncoordinated and Coordinated Modes for traffic requiring guaranteed QoS. TDMA allocations can be provided either on a per-Beacon Period basis (i.e., once per Beacon Period) or multiple times within each Beacon Period. Contention Free sessions go through the admission-control procedure before periodic access to the medium is granted. The admission-control procedure is controlled by the CCo.

A session with a regular CFP allocation can always start its transmission at its start time, as defined by the schedule, and must end its transmission by its end time, as defined by the schedule. If a session does not receive the Beacon, but the session has the current effective schedule information (due to schedule persistence), it may start its transmission during its allocated time.

8.5.1 Admission Control and Scheduling (Persistent and Nonpersistent)

The admission-control procedure in HomePlug AV deals with long-term allocation of network resources. All Contention-Free Sessions go through the admission-control procedure during session establishment. Scheduling procedure deals with the short-term (on a Beacon-period basis) allocation of network resources.

Stations may find that they require more time to transmit their data than they have been allocated in a Beacon Period. This may be due to a change in the source rate or changes in channel characteristics. To ensure that stations obtain the required allocation promptly, special fields in the Frame Control are used to convey allocation requirements to the CCo. The CCo may respond with changes to the persistent part of the schedule. The persistent part of the schedule is not very responsive and cannot change rapidly, since it is based on the maximum Schedule Countdown value used in the network. The CCo can also use the Nonpersistent Allocation fields in the Beacon to provide immediate allocation (refer to Section 5.2.2.8).

Note: Nonpersistent Allocations can be changed for every period and hence provide reasonably reliable and very efficient access with small delay to sessions with acute needs.

8.6 SUMMARY

This chapter provided a description of the hybrid CSMA/TDMA HomePlug AV channel access mechanisms, with a focus on the Beacon Period structure and operation for allocations and schedules, as well as for managing Coordinated and Uncoordinated or CSMA-only modes.

Chapter 9 defines the *connections* and *links* as well as the classification tools used in HomePlug AV to support the QoS requirements for multimedia traffic.

9

CONNECTIONS AND LINKS

9.1 INTRODUCTION

HomePlug AV is designed to support various multimedia applications within the home networks. It is therefore critical for HomePlug AV stations to be able to identify these applications and provide them with the required Quality of Service (QoS) (i.e., guaranteed bandwidth, latency, jitter, and packet error rates). For example, applications such as IP telephony require very low latency but are fairly tolerant to packet loss, while video streaming applications require bandwidth guarantees as well as low packet loss. On the other hand, basic Internet web browsing traffic is fairly tolerant to variations in bandwidth and latency. This section provides details on the Connections and Links that play important roles in provisioning QoS based on application requirements. This section is organized as follows:

- Section 9.2 provides details of the packet classification mechanisms that enable HomePlug AV stations to classify MSDUs based on their QoS requirements.
- Section 9.3 provide details of Connection Specification (CSPEC) that is used to define the QoS requirements of a Connection.
- Section 9.4 gives some fundamental facts about Connection and Links used in HomePlug AV.
- Section 9.5 explains Connection setup, modify, and teardown procedures.
- Section 9.6 discusses the Central Coordinator (CCo) functions for bandwidth management.

HomePlug AV and IEEE 1901: A Handbook for PLC Designers and Users, First Edition.
Haniph A. Latchman, Srinivas Katar, Larry Yonge, and Sherman Gavette.
© 2013 by The Institute of Electrical and Electronics Engineers, Inc. Published 2013 by John Wiley & Sons, Inc.

9.2 PACKET CLASSIFICATION

The packet classification function in HomePlug AV is a part of the convergence layer and is responsible for processing incoming MSDUs (i.e., Ethernet frames) from the host and mapping them into various transmit queues (or links) based on QoS requirements. This function is typically accomplished based on various packet classification fields in the MAC Service Data Unit (MSDU), such as

- Ethernet Source Address, Ethernet Destination Address, VLAN Tag,
- IPv4 Type of Service, IPv4 Protocol, IPv4 Source Address, IPv4 Destination Address,
- IPv6 Traffic Class, IPv6 Flow Label, IPv6 Source Address, IPv6 Destination Address,
- TCP Source Port, TCP Destination Port,
- UDP Source Port, UDP Destination Port.

All HomePlug AV stations are required to support packet classification based on Ethernet Source Address, Ethernet Destination Address, and VLAN Tag. HomePlug AV also supports classification rules that enable using multiple fields in the MSDU (e.g., Ethernet Source, Destination Address, and VLAN User priority) to map an MSDU into a link (or queue).

The VLAN User priority field in VLAN Tag is commonly used by HomePlug AV stations to prioritize connectionless MSDUs. The recommended mapping from the eight user priorities indicated in the VLAN Tag to the 4 HomePlug AV CSMA channel access priorities is shown in Table 9.1. This mapping follows the recommendation in IEEE 802.1D [20] (subclause 7.7.3). The rationale behind the choice of values in Table 9.1 is discussed in subclause H.2 of IEEE 802.1D. A consequence of the mapping shown is that frames carrying the default user priority (0) are given preferential treatment relative to user priorities 1 and 2. In addition, Table 9.2 (also from H.2 of IEEE 802.1D) defines the user priorities that should be assigned to application classes.

TABLE 9.1 Recommended VLAN User Priority-to-CSMA Priority Mapping

	HomePlug AV CSMA Priority	
User priority	0 (default)	1
	1	0
	2	0
	3	1
	4	2
	5	2
	6	3
	7	3

TABLE 9.2 Recommended Application Class-to-User Priority Mappings

User Priority	Application Class
7	Network Control—characterized by a "must-get-there" requirement to maintain and support the network infrastructure
6	"Voice"—characterized by >10-ms delay, and hence maximum jitter (one-way transmission through the LAN infrastructure of a single campus)
5	"Video" or "Audio"—characterized by >100-ms delay
4	Controlled Load—important business applications subject to some form of "admission control," such as preplanning of the network requirement at one extreme to bandwidth reservation per flow at the time the flow is started at the other
3	Excellent Effort—or "CEO's best effort," the best effort type services that an information-service organization would deliver to its most important customers
0	Best Effort—LAN traffic as we know it today (this user priority is actually serviced at a higher priority than user priorities 1 and 2 to accommodate legacy entities)
1, 2	Background—bulk transfers and other activities that are permitted on the network, but that should not impact the use of the network by other users and applications

The exact packet classification rules and capabilities of HomePlug AV stations are implementation dependent, with classification engines defined based on the applications for which the HomePlug AV stations are used. For example, implementations designed to support IPTV services have different classification rules and capabilities than devices intended for basic home networking.

9.3 CONNECTION SPECIFICATION (CSPEC)

The Connection Specification (CSPEC) contains the set of parameters that define the characteristics and QoS expectations of a Connection. Connections can be either unidirectional or bidirectional. For bidirectional Connections, the Connection specification for both the Forward Link and Reverse Link is contained in the CSPEC. The CSPEC of each Link is composed of two parts:

- The Connection Information (CINFO),
- The QoS and MAC parameters (QMP).

CINFO identifies the attributes of the Connection and the MAC and convergence layer operations required by the Connection at the source and destination STAs. CINFO contains the following information:

- *MAC Service Type:* indicates whether the link requires contention-based (CSMA) or contention-free services (TDMA).

- *User Priority:* indicates the CSMA channel access priority for transmission during Contention Period.
- *ATS to HLE:* indicates whether Arrival Time Stamp (ATS) should be passed to the host.
- *Smoothing:* indicates whether smoothing (or jitter control) needs to be performed for MSDUs belonging to this link.

The QoS and MAC parameters (QMP) of the link identify the QoS requirements (delay, jitter, data rates), as well as MAC parameters that are specific to the particular link. The QMPs exchanged between the Higher Level Entity (HLE) and Connection Manager (CM) are shown in Table 9.3.

The QoS and MAC parameters exchanged between two CMs include the parameters shown in Table 9.4. The QoS and MAC parameters exchanged between the CM and CCo are shown in Table 9.5. These parameters are used by the CCo to determine the size of Time Division Multiple Access (TDMA) allocation required to meet the QoS requirements of the application.

TABLE 9.3 QoS and MAC Parameter Exchanged between HLE and CM and Between CMs

CSPEC Parameter	Descriptions
Delay bound	Maximum amount of time allowed to transport an MSDU
Jitter bound	Maximum difference in the delay experienced by an MSDU
Average MSDU size	Average MSDU payload size in octets
Maximum MSDU size	Maximum MSDU payload size in octets
Average Data Rate	The average application data rate that is required for transport of MSDUs belonging to this Link
Minimum Data Rate	The minimum application data rate that is required for transport of MSDUs belonging to this Link
Maximum Data Rate	The maximum application data rate that is required for transport of MSDUs belonging to this Link
Maximum Inter-TXOP time	Maximum time allowed between two transmission opportunities (TXOPs) on the medium for this Link
Minimum Inter-TXOP time	Minimum time allowed between two transmission opportunities (TXOPs) on the medium for this Link
Maximum burst size	Maximum size of a single contiguous burst of MSDUs that is generated by the application at the maximum rate
Exception policy	Indicates whether the Connection needs to be terminated or reconfigured when the application characteristics are different than that indicated in the CSPEC
Inactivity interval	Maximum duration of time a Connection is allowed to remain inactive without transporting any application data before the CM can release the allocation
MSDU error rate	MSDU error rate requested
CLST	Convergence layer SAP type. This is set to indicate Ethernet SAP

(continued)

TABLE 9.3 (*Continued*)

CSPEC Parameter	Descriptions
CDESC	Connection Descriptor contains a set of fields that defines the Connection to the HLEs. It contains IP version, IP source and destination addresses, protocol type, source and destination ports. It is used by UPnP QoS and possibly by other HLEs
Vendor specific	Vendor-specific QoS and MAC information
ATS tolerance	Measured variance in value of Arrival Time Stamp (ATS) from the synchronized Network Time Base at the time the ATS is applied to the MSDU arriving at the convergence layer SAP of the transmit station
Smallest tolerable Average Data Rate	Smallest tolerable Average Data Rate indicates the smallest average data rate at which the application is able to operate. This is used for application that can support operation at different data rates (e.g., video stream that can operate at low quality when bandwidth is limited)
Original Average Data Rate	Original Average Data Rate indicates the average data rate at which the application intends to operate when sufficient station and network resources are available

TABLE 9.4 Additional QoS and MAC Parameter Fields Exchanged Between Two CMs

QoS and MAC Parameter Field (CM–CM)	Descriptions
Rx Window Size	Receive window size in number of 512-octet segments
Smoothing Buffer Size	The smoothing buffer size in octets that is required to support the Link at the transmitter and receiver
Bidirectional Burst	This parameter is only used for Bidirectional Connections when the traffic on the link is intended to be transmitted as a part of Reverse SOF of the Bidirectional Bursts initiated by the other Link in the Connection

9.4 CONNECTIONS AND LINKS

In HomePlug AV, a *Connection* is a data flow (a set of related MSDUs) between the HLE of the STA that establishes the Connection and HLE(s) of one or more destination STAs. A Connection can be either unidirectional or bidirectional. A *Link* is a unidirectional data flow (a packet or set of related packets) from the convergence layer (CL) of the source of the Link to the CL of one or more destinations of the Link. Links can be categorized as unicast or broadcast/multicast, depending on the number of destinations of the Link. A unicast Link may be either a Forward Link or a Reverse Link. The Forward Link is identified as originating at

TABLE 9.5 QoS and MAC Parameter Fields Between CM and CCo

QoS and MAC Parameter Field (CM–CCo)	Descriptions
TXOPs per Beacon Period	The number of uniformly spaced Transmit Opportunities (TXOPs) requested per Beacon Period
Average number of PBs per TXOP	The average number of 520-octet PHY Blocks per TXOP required for transporting MSDUs belonging to this Link
Minimum number of PBs per TXOP	The minimum number of 520-octet PHY Blocks per TXOP required for transporting the MSDUs belonging to this Link
Maximum number of PBs per TXOP	The maximum number of 520-octet PHY Blocks per TXOP required for transporting the MSDUs belonging to this Link
PPB_Threshold	The Pending PHY Block (PPB) threshold indicates the threshold of Pending PBs at which the Link requires extra bandwidth to clear the backlog. If there is sufficient bandwidth available, the CCo should provide extra allocation when the PPB threshold is exceeded
Surplus bandwidth	Surplus bandwidth indicates the excess amount of bandwidth required to support the Link relative to the average number of PBs per transmit operation
CDESC	Connection Descriptor contains a set of fields that defines the Connection to the HLEs. It contains IP version, IP source and destination addresses, protocol type, source, and destination ports. It is used by UPnP QoS and possibly by other HLEs
Vendor specific	Vendor-specific QoS and MAC information
Smallest tolerable average number of PBs per TXOP	Smallest tolerable Average number of PBs per TXOP indicates the smallest average number of PBs per TXOP at which the application is able to operate
Original average number of PBs per TXOP	Original average number of PBs per TXOP indicates the average number of PBs per TXOP at which the application intends to operate when sufficient station and network resources are available
Bidirectional burst	This parameter indicates that this Global link will be used for bidirectional bursting for a Local Link that is a part of this Connection

the STA that initiates the connection establishment procedure and terminating on the STA(s) responding to the connection establishment request. The Reverse Link is in the opposite direction to the Forward Link.

A Connection can be one of the following combinations of Links:

- a single unicast Link from the station that initiated the Connection to the terminating station of the Connection (i.e., a single Forward Link),
- a single unicast Link from the terminating station of a Connection to the initiating station of the Connection (i.e., a single Reverse Link).

- the combination of the above two Links (i.e., a bidirectional Connection composed of both a Forward Link and a Reverse Link),
- a single multicast/broadcast Link from the station that initiated the Connection to the terminating stations of the Connection.

The distinction between Connections and Links is made because, at the physical layer, each direction between two stations is likely to have different characteristics and must be allocated separately. By comparison, it is much easier for the HLE to request a bidirectional Connection and have the CMs and CCo set up in both directions of the Connection.

HomePlug AV supports the following three types of Links:

- *Global Links:* these are established and controlled by the CCo at the request of a CM. The CCo assigns the Global Link a TDMA allocation and a Global Link Identifier (GLID) (refer Section 9.5.1).
- *Local Links:* these are established and controlled by the CM and CCo is not involved in Local Link setup. Traffic belonging to local links is transmitted during the Contention Period (CP). The CM on the transmitting side of the Link assigns a Local Link Identifier (LLID) to identify the Link (refer Section 9.5.1).
- *Priority Links:* priority links are used for transmitting MSDUs that do not belong to any Connection (i.e., Local Link to Global Link). Each connectionless data packet is assigned a Priority Link ID (PLID = 0, 1, 2, 3). These PLIDs identify the priority of the packet being transported but do not uniquely identify a particular data flow. Traffic belonging to local links is transmitted during the Contention Period. Priority Links are also referred to as "Connectionless" links.

9.4.1 Link Identifiers

In HomePlug AV, an 8-bit Link Identifier is used to identify Global Links, Local Links, and Priority Links.

- *Priority Link ID:* The PLID uses values in the range from **0x00** to **0x03**. The PLID indicates the traffic class for priority resolution during CSMA. PLID of **0x00** uses channel access priority 0, PLID of **0x01** uses channel access priority 1 and so on.
- *Local Link ID:* The LLID is used for Connection-oriented traffic carried within the Contention Period and uses value in the range from **0x04** to **0x7F**.
- *Global Link ID:* The GLID is assigned by the CCo and is unique in a network and is used to identify different types of allocations. For contention-free allocation, it further identifies the unique Link that can use the medium.
  - **0xFF** = identifies a Local CSMA allocation.
  - **0xFE** = identifies a Shared CSMA allocation, which can be used only when the network is operating in Coordinated Mode.
  - **0xFD** = identifies an allocation used by a designated STA to transmit a Discover Beacon.

- **0xFC** = identifies an allocation for Contention-Free Period Initiation (CFPI).
- **0xF8–0xFB** = reserved.
- **0x80–0xF7** = identifies a unique contention-free Link in the network.

Note: The type of the LID can always be distinguished by its value.

Using Link Identifiers to uniquely identify a MAC Frame Stream is described in Section 6.3. A primary function of the LID is to identify the MAC Frame Stream associated with the PHY Block (PB) contained in a received MPDU. It is carried in the Frame Control (FC) of each MPDU that transports PBs.

9.4.1.1 Assignment of LIDs All LIDs are assigned at Connection setup. The originating STA assigns a unique LLID immediately. This LLID will be used as a part of the CID (refer Section 9.4.2) and, if the forward LID will be local, this is the same value used for LID-F. The terminating STA assigns the LLID-R field if there is either a Local or a Global Reverse Link. Each STA selects an LLID value that is unique within its transmitter. This allows the PBs in the MPDU to be routed to the correct MAC stream solely based on the LLID and the Source Terminal Equipment Identifier (STEI) in the Frame Control.

The CCo assigns the GLIDs used on Links in the Contention-Free Period (CFP). Upon receipt of the **CC_LINK_NEW.REQ**, the CCo assigns GLIDs to one or both Links associated with the Connection. The CCo assigns these GLIDs to be unique within its network (i.e., the GLIDs alone identify the Connection without reference to the STEI). This allows the PBs in the MPDU to be routed to the correct MAC stream solely based on the GLID in the Frame Control.

9.4.2 Connection Identifiers

When a Connection is first requested by an HLE, the CM assigns a Connection Identifier (CID) to the Connection. This CID is unique for each Connection within the AVLN. It is assigned in a manner to ensure that it is globally unique with the network. The CID is used in communications between the STA and CCo, as well as between the STAs. It can also be used for communication between the HLE and CM.

The CID is a 16-bit value constructed by placing the Terminal Equipment Identifier (TEI) of the originating STA in the upper eight bits and placing the LLID-F initially generated for the Connection by the originating STA in the lower eight bits. The STA always generates an LLID-F for a Connection to construct the CID, even if the Connection does not have a Forward Link or if the Forward Link is to be a GLID-F. The CID does not change when and if the CCo assigns a GLID-F to the Connection. Table 9.6 presents a summary of Link and Connection Identifiers.

9.5 CONNECTION SERVICES

HomePlug AV offers two transport services to the HLEs, namely a Connectionless Service (CLS) and a Connection-Oriented Service (COS).

The Connectionless Service (CLS) transports individual data packets between HLEs on different stations. CLS is used by an HLE prior to the time it sets up a

TABLE 9.6 Summary of Link and Connection Identifiers

Name	Contents/Value	Assigned by	Description / Notes
Connection ID (CID)	8-bit Originating STA TEI + 8-bit LLID-F	Originating STA	Used by the CCo and the STAs to identify a Connection. May also be used in primitives between HLE and CM
Link ID (LID)			Used by the MAC to identify the MAC Frame Stream.
Local Link ID (LLID)	**0b0XXXXXXX**		Values of **0x00** to **0x03** are not permitted for LLIDs
Forward (LLID-F)		Originating STA	Unique within the originating STA
Reverse (LLID-R)		Terminating STA	Unique within the terminating STA
Global Link ID (GLID)	**0b1XXXXXXX**		Unique within the network
Local CSMA allocation	**0xFF**	Fixed in Spec	Used by CCo to identify Local CSMA allocation.
Shared CSMA allocation	**0xFE**	Fixed in Spec	Used by CCo to identify Shared CSMA allocation.
Discover Beacon allocation	**0xFD**	Fixed in Spec	Used by CCo to identify Discover Beacon allocation
CFPI allocation	**0xFC**	Fixed in Spec	Used by CCo to identify Contention-Free Period Initiation allocation (refer Section 9.6.1)
Reserved for future use	**0xF8–0xFB**	Fixed in Spec	Reserved for future use
Forward (GLID-F)	**0x80–0xF7**	CCo	Contention-Free allocation used by a designated STA
Reverse (GLID-R)	**0x80–0xF7**	CCo	Contention-Free allocation used by a designated STA
Priority Link ID (PLID)	**0x00–0x03**	Fixed in Spec	Used for connectionless traffic

Connection and selectively thereafter. It is also used by legacy applications that are unaware of the availability of Connection services. Connectionless traffic always uses the Contention Period (CP). It does not offer guaranteed QoS, but does support prioritization of packets as they flow through the system.

The Connection-Oriented Service (COS) allows two or more HLEs to set up a logical Connection between themselves. A Connection can be either unidirectional (data flows in only one direction) or bidirectional (data flows in both directions). Connection-oriented traffic can use either the Contention-Free Period (CFP) or the CP, depending on the QoS requirements of the Connection. CFP Connections are used to provide guaranteed QoS.

A Connection is typically created when the HLE in a given station initiates a messaging sequence to set up the Connection. Based on the CSPEC provided by the HLE, the CM in this station determines how many Links are required and whether each Link should be a Global Link or Local Link. The CM then communicates with the CM in the destination station, and possibly with the CCo, to establish the one or more Links required to realize the Connection.

Connection setup can also be initiated by the convergence layer. This functionality is referred to as Auto-Connect. Auto-Connect eliminates the need for HLEs to play a proactive role in Connection setup. Auto-Connect also enables provisioning of QoS for traffic generated by HLEs that do not have the ability to initiate Connection setup. Some of the mechanisms that can be used by convergence layer for Auto-Connect are as follows:

- The convergence layer can be preconfigured with specific classifier rules and CSPECs for Auto-Connect purpose. When the convergence layer detects MSDUs matching these predefined classifier rules, it automatically initiates the indicated Cnnection and provides QoS based on the associated CSPEC.
- The convergence layer may also perform heuristic analysis of the MSDUs it receives and based on the characteristics of the flow, determine if a particular Connection needs to be established. For example, if there is persistent flow of MSDUs at high data rate, the convergence layer can choose to treat this as a video stream and provide a Connection-oriented service with appropriate QoS.

Once the Connection is established, the STA is responsible for monitoring the QoS performance of each of its Links. If a Link is not performing according to its CSPEC, the CM may initiate a Link reconfiguration with a new CSPEC or it may tear down the Connection.

It is possible to have several Connections between two STAs. Each of these Connections will have either Global or Local Links along with its own, possibly unique, CSPEC.

9.5.1 Connection Setup

The steps involved in setting up a Connection are as follows (Figure 9.1):

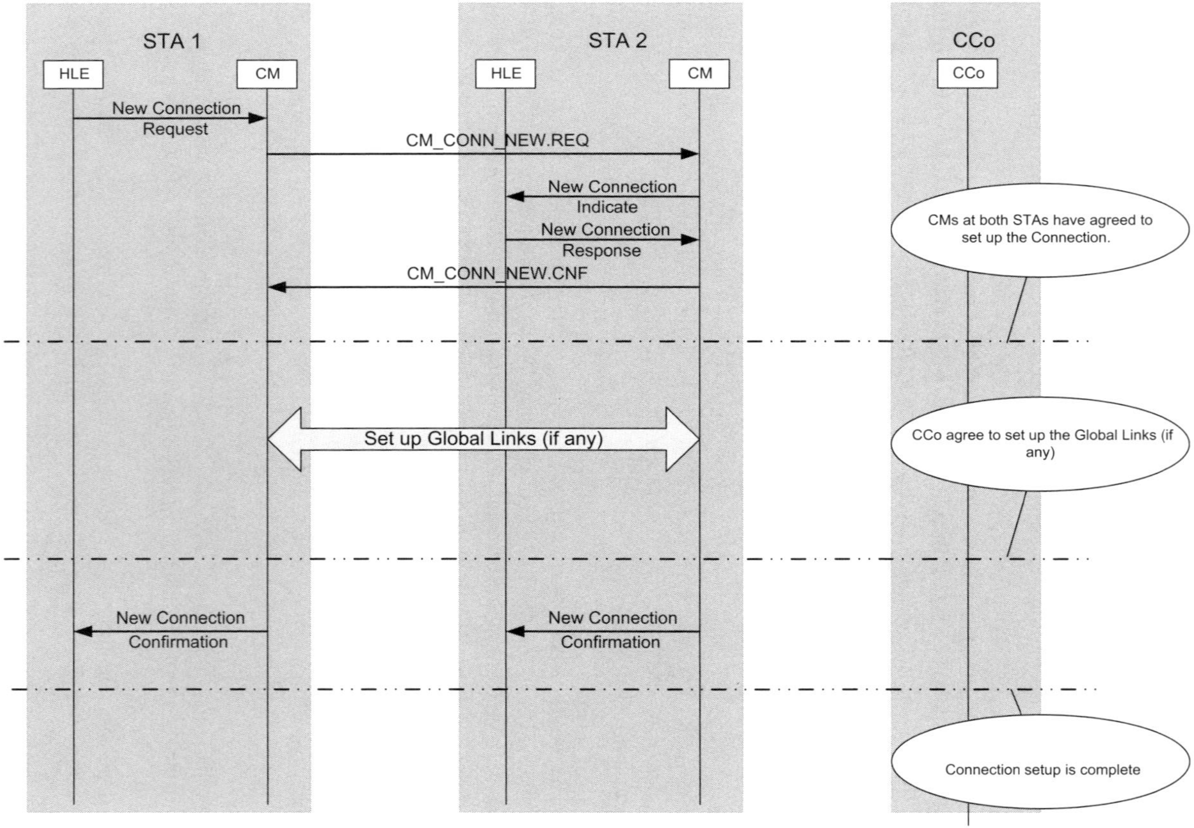

FIGURE 9.1 Connection setup.

- The HLE (or convergence layer, in case of Auto-Connect) tells the CM the CSPEC for the new Connection, the destination STA's MAC address, and the classifier rules that will match the messages sent by the HLE.
- The CM on the STA that initiated the Connection sends a **CM_CONN_NEW. REQ** message to the CM(s) on the terminating STA(s).
- The CM at the terminating STA informs its HLE of the new Connection and the HLE responds indicating whether the Connection can be added. This step is optional.
- The CM at the terminating STA sends a **CM_CONN_NEW.CNF** message to the CM of the initiating station indicating whether the Connection request is accepted or rejected.
- If negotiation between CMs is successful and no Global Links are required, the Connection setup is complete and the HLE (or convergence layer, in case of Auto-Connect) is notified.
- If the negotiation between CMs is successful and if Global Links are required, the initiating CM sends a **CC_LINK_NEW.REQ** to the CCo. CCo sends the **CC_LINK_NEW.CNF** message to all stations belonging to the Connection to indicate the success or failure of the Connection setup procedure (Figure 9.2).
- If negotiating between CMs is unsuccessful, the Connection setup is deemed to have failed.

When a Connection setup request is rejected by the CM due to insufficient station resources or rejected by the CCo due to insufficient bandwidth, a Proposed CSPEC containing the fields of the CSPEC that can currently be supported is communicated to the stations belonging to the Connection. This enables the application or the convergence layer to set up a new Connection for the streams, within the limits of the Proposed CSPEC.

9.5.2 Connection Monitoring

Once a Connection is setup, the CM will gather statistics for each Link. The CM will monitor the QoS being delivered to each Link and will reconfigure or tear down any Link that fails to meet its CSPEC, depending on the action specified by the CSPEC's Violation Policy parameter. If a Link is torn down, the corresponding Connection will also be torn down. The HLE may also monitor the Connection by requesting Link statistics from the MAC layer.

9.5.3 Connection Teardown

The CM of either STA will tear down a Connection at the request of the HLE or, if exception policy dictates, because the Connection's performance (i.e., the performance of one of its Links) is not meeting its CSPEC. When either station decides to tear down the Connection, the teardown process depends on whether the Connection includes a Global Link.

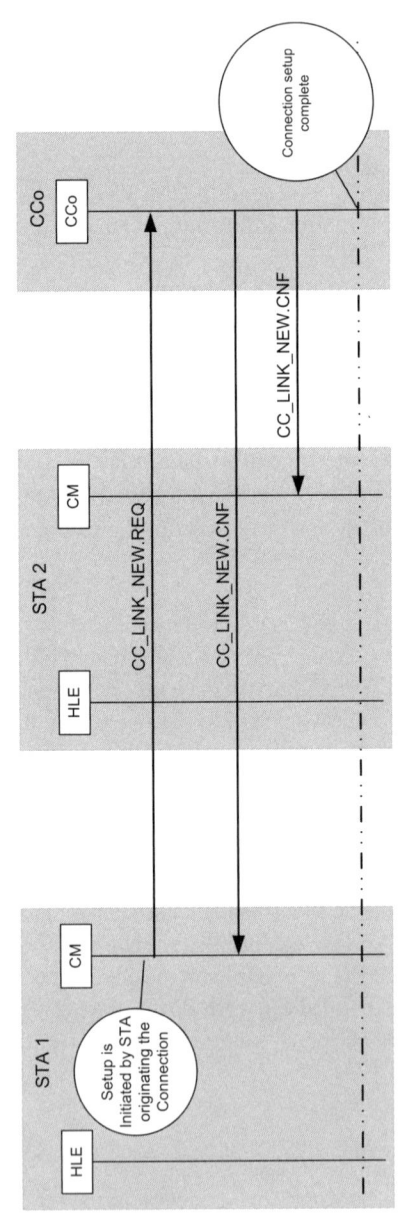

FIGURE 9.2 Global Link setup.

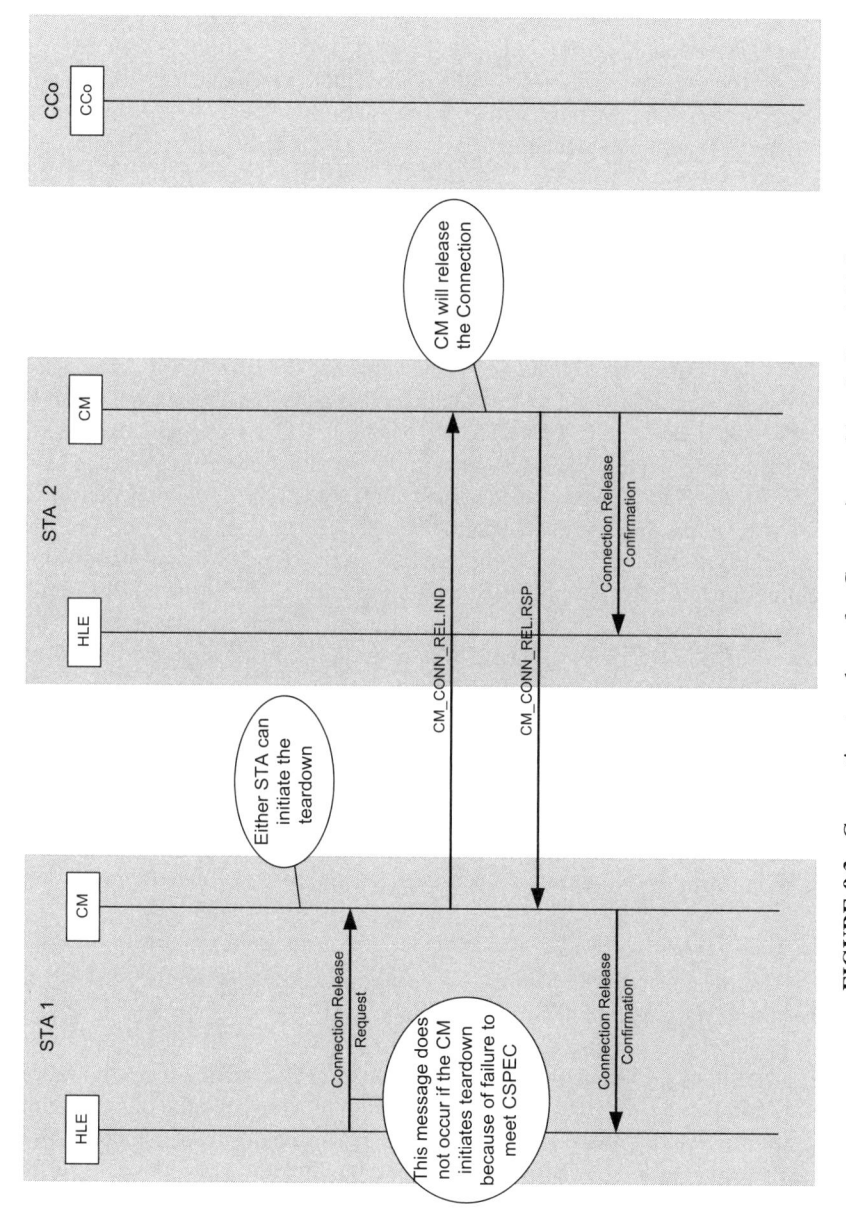

FIGURE 9.3 Connection teardown for Connections with only Local Links.

- For a Connection that contains only Local Links, the station initiating the teardown notifies the other station using the **CM_CONN_REL.IND** message. Reception of a **CM_CONN_REL.IND** causes the CM to release the Connection and respond with a **CM_CONN_REL.RSP** (Figure 9.3).
- For a Connection that contains Global Links, the station initiating the teardown notifies the CCo by send the **CC_LINK_REL.REQ** message. Upon reception of this message, the CCo indicates the release of this Connection to all stations that are part of this Connection by sending **CC_LINK_REL.IND** message (Figure 9.4).

The CCo may also initiate the teardown of a Connection due to bandwidth limitations. In such cases, the CCo sends the **CC_LINK_REL.IND** message to notify the teardown of the Connection to all stations that are part of the Connection. When the CM or CCo initiates a Connection teardown due to violation of the CSPEC, the CSPEC fields that are violated and their values at the time when teardown is initiated may be communicated to other stations as a part of the Violated CSPEC in **CM_CONN_REL.IND, CC_LINK_REL.IND,** and **CC_LINK_REL.REQ**. For example, if the CM tears down a Connection due to violation of the Average Data Rate, the Average Data Rate observed for the Link at the time at which Connection teardown was initiated is communicated.

When the CCo initiates a Connection teardown due to insufficient bandwidth, it may communicate a Proposed CSPEC containing the values of the CSPEC fields that can be supported to all stations belonging to this Connection as a part of the Proposed CSPEC in **CC_LINK_REL.IND**. Proposed CSPECs enable application-sable to support the stream at multiple data rates to setup a new Connection within the limits of the Proposed CSPEC.

9.5.4 Connection Reconfiguration

Connection reconfiguration occurs when the CSPEC parameters of the Connection change. Connection reconfiguration occurs for several reasons, including

- the HLE initiates a change in CSPEC for reasons specific to the application.
- The Auto-Connect or CM initiates a change in CSPEC when they determine that the CSPEC parameters have been changed.

Figure 9.5 shows a Connection reconfiguration. The Connection reconfiguration process is similar to the new Connection setup process. When a Connection reconfiguration request was rejected by the CM due to insufficient Station resources or by the CCo due to insufficient bandwidth, a Proposed CSPEC containing the fields of the CSPEC that can currently be supported can be communicated to the stations belonging to the Connection. This enables the application or the CM setup a new Connection for the streams within the limits of the Proposed CSPEC.

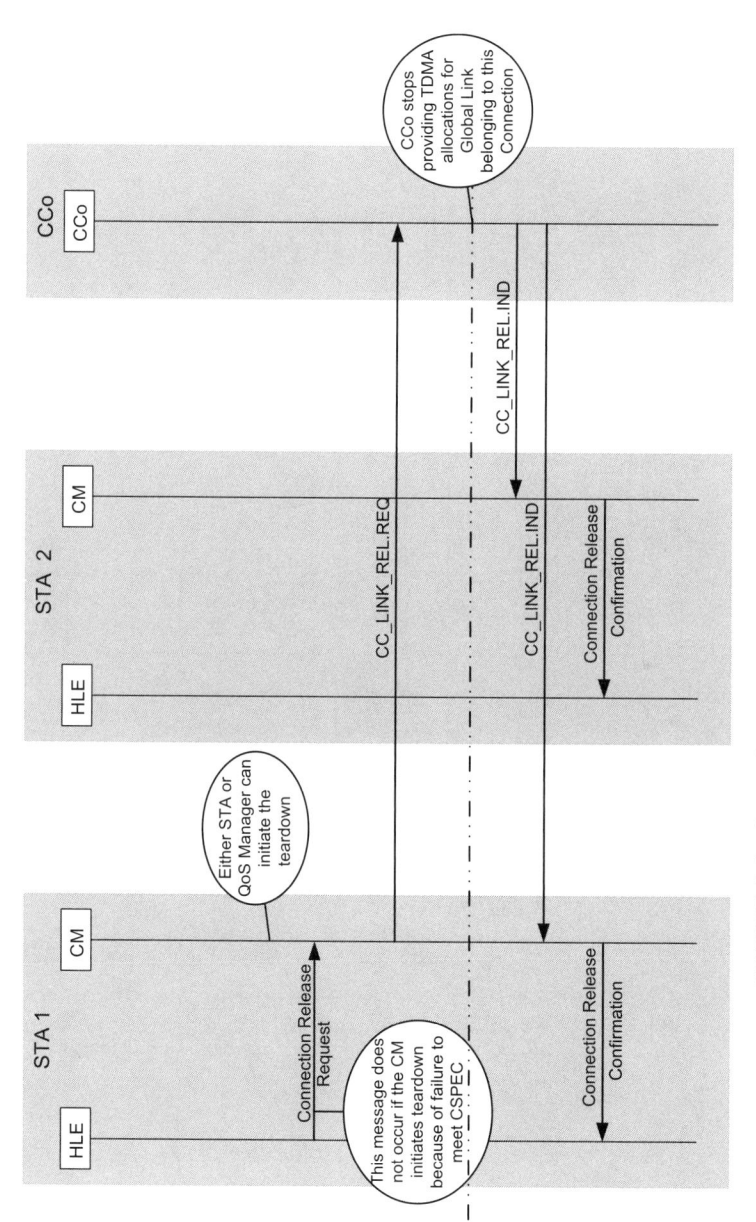

FIGURE 9.4 Connection teardown for Connections with Global Links.

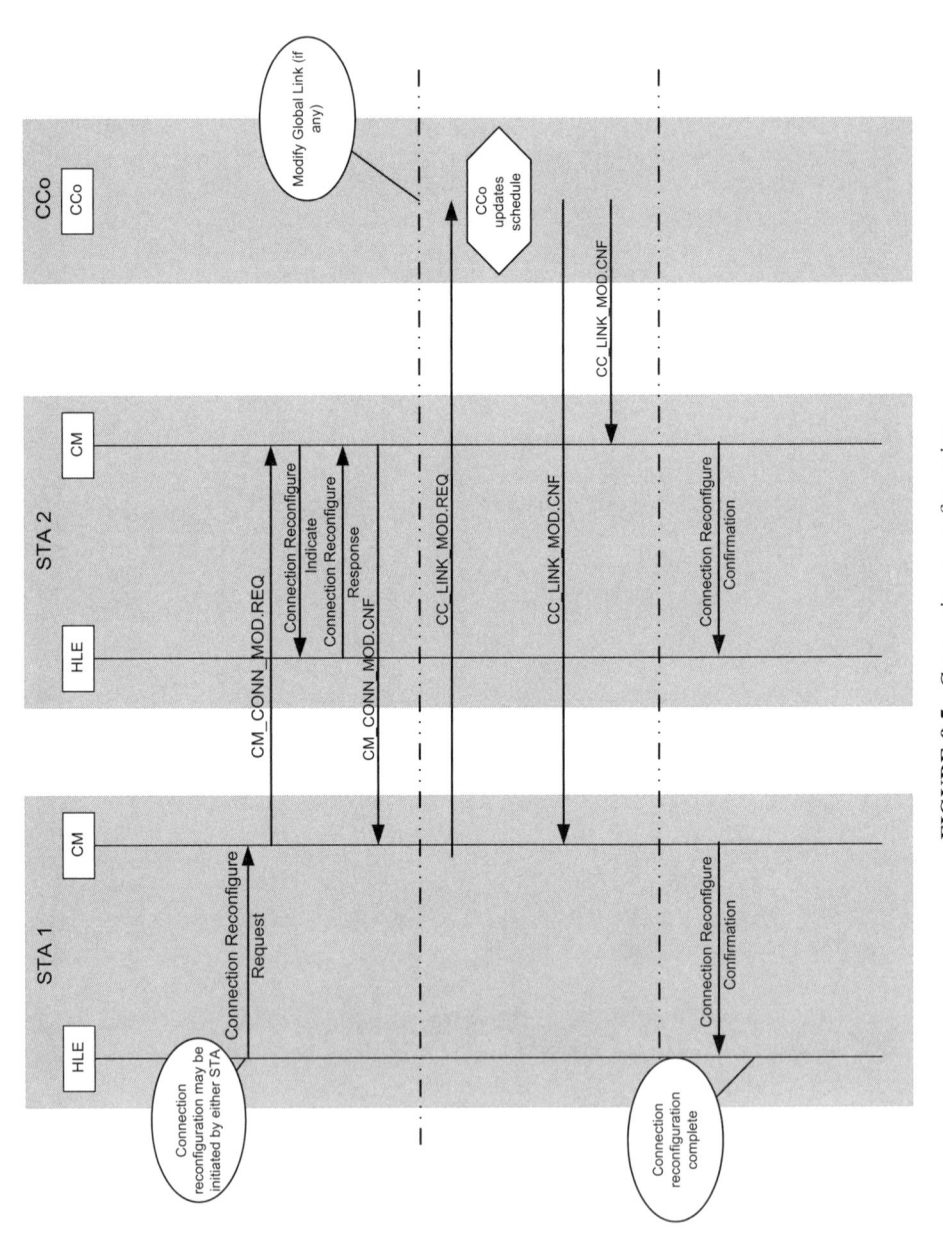

FIGURE 9.5 Connection reconfiguration.

9.5.5 Global Link Reconfiguration Triggered by CCo

The CCo can trigger reconfiguration of the Global Link because of

- consolidation of TDMA assignments,
- changes to channel characteristics,
- bandwidth-triggered Squeeze/De-Squeeze operations.

Global Link reconfiguration required to support the current CSPEC (which includes consolidation of TDMA assignments and adapting to changes in channel characteristics) and is performed without explicit messaging via Frame Control and the Beacon. The CM at the source of a Global Link provides continuous estimates of the channel characteristics to the CCo by providing Bit-load Estimates (BLEs) in the Frame Control. The CM may also send a **CC_BLE_UPDATE.IND** message to the CCo when it notices significant changes to the Bit-loading Estimates. These estimates, along with the Pending PHY Blocks (PPB) information (also provided in the Frame Control), are processed by the CCo. This results in CCo-initiated changes to Global Link allocations to support the negotiated CSPEC. The CCo notifies the two STAs about the new allocation by updating the schedule information in the Beacon.

9.5.5.1 Squeeze and De-squeeze The CCo of an AVLN may also reconfigure the existing Connections by requesting Connection(s) to reduce their bandwidth usage (i.e., request to Squeeze) when bandwidth is scarce or by allowing them to increase their bandwidth usage (i.e., request to De-Squeeze) when bandwidth becomes available.

The CCo may request a Connection to squeeze when there is insufficient bandwidth available to support all the ongoing Connections (due to degradation in channel characteristics) or to accommodate a new Connection. The CCo may request a squeezed Connection to de-squeeze when more bandwidth becomes available (e.g., when channel conditions improve). The CM may also continuously monitor the available bandwidth and request connection reconfiguration (to de-squeeze) when it determines that more bandwidth is available.

The following procedure is used by the CCo to Squeeze or De-Squeeze a Connection:

- When a CCo determines that a Connection needs to be squeezed or de-squeezed, it sends a **CC_LINK_SQZ.REQ** request to one of the stations belonging to this Connection.
- Reception of the **CC_LINK_SQZ.REQ** causes the CM to indicate the Connection reconfiguration request to the HLE.
- The HLE responds with indication on whether the Connection reconfiguration is successful.
- If the HLE accepts the Connection reconfiguration, the CM will negotiate with the peer CM for Connection reconfiguration. The procedure for performing this step is the same as described in Section 9.5.4.

- If the peer CM accepted the Connection reconfiguration (as indicated by **CM_CONN_MOD.CNF**), the CM will send a **CC_LINK_SQZ.CNF** indicating successful reconfiguration along with **CC_LINK_MOD.REQ** to the CCo.
- If the peer CM rejected the Connection reconfiguration, the CM will send a **CC_LINK_SQZ.CNF** indicating that the reconfiguration has failed.

The Connection Squeeze/De-Squeeze procedure is shown in Figure 9.6.

9.6 BANDWIDTH MANAGEMENT BY CCo

The main functions performed by the Bandwidth Manager in the CCo are

- *Scheduling and Bandwidth Allocation to Connections:* The CCo receives requests from STAs in the network, requesting bandwidth assignments for Connections. In response, the CCo assigns a Global LinkID (GLID) and schedules allocations to the Connection. The traffic characteristics, QoS guarantees, MAC services, and MAC parameters specific to a Connection are defined in the Connection Specification (CSPEC). Sounding and channel estimation results are used by the Bandwidth Manager in making allocations to Connection requests.
- *Admission Control:* When the CCo receives the connection establishment or Connection reconfiguration requests from a STA, the Bandwidth Manager must determine whether there is adequate bandwidth available to support the request, without compromising QoS of the existing Connections. The Bandwidth Manager is responsible for either accepting or rejecting the requests.
- *Beacon Period Configuration and Beacon Transmission:* The Bandwidth Manager must determine the allocations within a Beacon Period. It assembles and broadcasts the Beacon once for every Beacon Period.

9.6.1 Scheduler and Bandwidth Allocation

The CCo implements scheduling algorithms for the CFP. These algorithms make bandwidth assignments in the form of time grants. The assignments are carried in the persistent and nonpersistent Schedule BENTRIES that are broadcast in the Beacon.

The scheduling algorithms make updates to the schedules based on the following events:

- requests for new Links from STAs within the network,
- request for Link reconfigurations from the existing Links within the network,
- changes to the capacity of the existing Links as a result of changes to the physical channel.

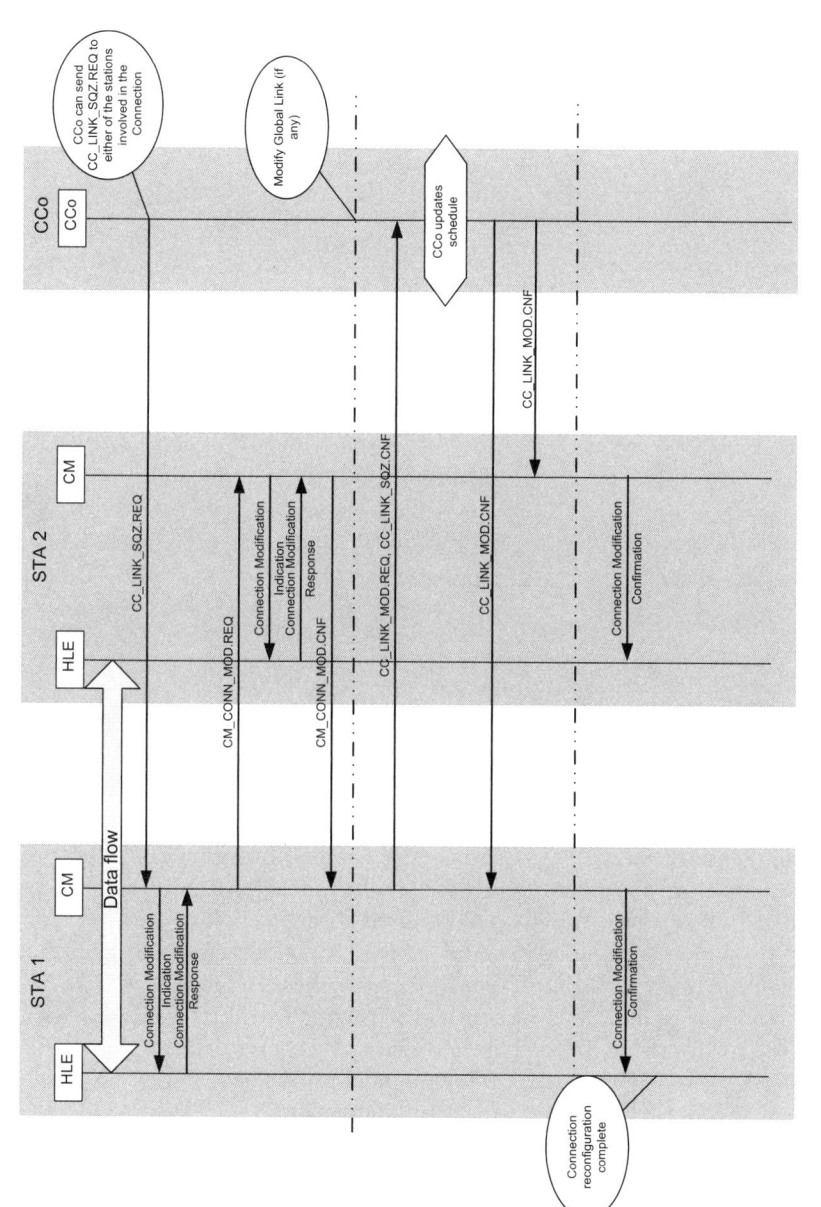

FIGURE 9.6 Connection Squeeze/De-Squeeze.

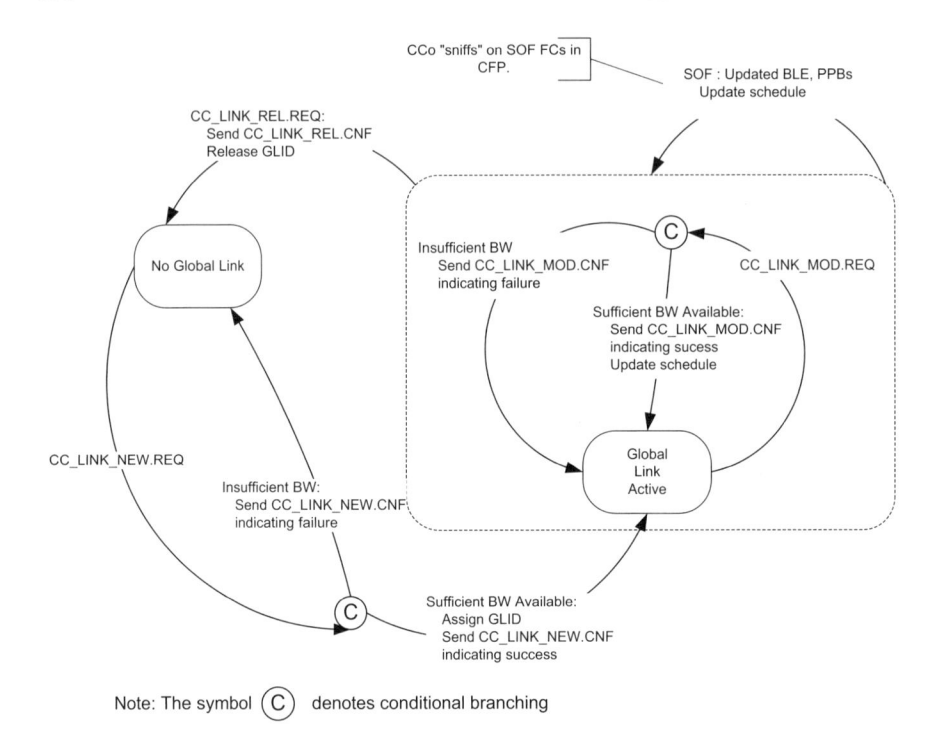

FIGURE 9.7 Global Link life cycle.

Figure 9.7 contains a finite state machine diagram describing the life cycle of a Global Link. The life cycle of the Link typically begins when the CCo receives a **CC_LINK_NEW.REQ** message from a station in the network. The **CC_LINK_NEW.REQ** message contains the CSPEC and CINFO parameters for a Connection that requires use of the CFP. It also includes Bit-load Estimates (BLEs) for the physical channel between the STAs requesting the Link. These BLEs are based on communications between the STA that occurred during the CP.

The CCo performs admission control on the Link based on the BLE. If there is sufficient bandwidth available to support the Link, the CCo sends the **CC_LINK_NEW.CNF** message indicating success and containing the GLID for the Link(s) to both STAs. At this point, the Link is active and carries user data.

If the admission control function returns failure to allocate, the CCo sends a **CC_LINK_NEW.CNF** message indicating failure along with a Proposed CSPEC to both STAs involved in the Connection setup.

While the Link is active, the CCo "sniffs" on the FC fields transmitted in the delimiter by the STA. The Start-of-Frame (SOF) FC contains a BLE field and a PPB field. The CCo uses the contents of the BLE field to update its estimates of the channel capacity versus line cycle. It uses the PPB field to determine whether to increase or decrease persistent allocation and/or to provide nonpersistent allocation for the Link.

The CCo may update its schedules (i.e., persistent and nonpersistent allocations) based on one of the following events:

- The STA may request a change in the QoS parameters of a Link. In this case, if sufficient bandwidth is available, the CCo updates the schedule; otherwise, the modification request is rejected.
- The CCo decides that a change to the schedule is needed based on the PPB and BLE fields.

Schedule updates are indicated to all stations in the network using persistent and nonpersistent BENTRIES in the Beacon.

9.6.2 Connection Admission Control

The Connection Admission Control procedure ensures that the station and network resources are not overallocated, thus ensuring the QoS guarantee on the admitted Connections.

CMs execute the Connection Admission Control procedure whenever a new Connection is requested or an existing Connection is modified. CMs may also continuously monitor the existing Connections for adherence to the negotiated traffic characteristics (traffic policing) and QoS guarantees. Violation of the CSPEC parameters may cause the CM to reconfigure or tear down an existing Connection.

The CCo executes the admission control procedure when a new Global Link is requested or an existing Global Link is modified. The CCo also continuously monitor all the existing Global Links and it may modify or terminate Connections if the available bandwidth is not sufficient to satisfy the CSPEC requirements.

9.6.3 Beacon Period Configuration

The Bandwidth Manager determines the allocations within a Beacon Period. The Bandwidth Manager assembles the portion of the Beacon payload dealing with bandwidth allocation.

9.7 SUMMARY

This chapter discussed the packet classification mechanism that enables HomePlug AV to distinguish the QoS needs of various MSDUs. The details are also provided on Connection Specification (CSPEC) and its associated Connection setup, modification and teardown procedures that enable the provisioning of parameterized QoS. The role of the CCo in bandwidth management was also explained.

Chapter 10 focuses on the question of security and network formation in HomePlug AV, from power-on to association, authentication, and authorization.

10

SECURITY AND NETWORK FORMATION

10.1 INTRODUCTION

This chapter examines the methodologies used in HomePlug AV to form a logical network and to ensure adequate security and privacy for devices within the logical network. The behavior of a station after power-on is described in Section 10.2. Forming a logical network, adding new stations and removing stations from an AVLN is discussed in Section 10.3. The various security keys used in HomePlug AV, including the various key entry modes (direct, remote, and push button) are discussed and explained in Section 10.4.

10.2 POWER-ON NETWORK DISCOVERY PROCEDURE

An AV STA performs the Power-on Network Discovery Procedure to determine whether another HomePlug network is active and if a new AVLN can be instantiated. Before initiating the Power-on Network Discovery Procedure, the AV STA will select a BeaconBackoffTime (BBT). If the STA was the CCo of an AVLN before it was powered down, BBT is chosen as a random value in the interval (MinCCoScanTime [1 s], MaxCCoScanTime [2 s]). Otherwise, BBT is chosen as a random value in the interval (MinScanTime [2 s], MaxScanTime [4 s]). BBT is the maximum duration of time for which an STA will execute the Power-on Network Procedure.

HomePlug AV and IEEE 1901: A Handbook for PLC Designers and Users, First Edition.
Haniph A. Latchman, Srinivas Katar, Larry Yonge, and Sherman Gavette.
© 2013 by The Institute of Electrical and Electronics Engineers, Inc. Published 2013 by John Wiley & Sons, Inc.

During the Power-on Network Discovery Procedure, if one or more AVLNs are detected, the STAs attempts to transmit the **CM_UNASSOCIATED_STA.IND** MME using multinetwork broadcast approximately once per unassociated STA Advertisement Interval (USAI, 1 s) to provide information to other stations that may be performing the same procedure. The transmission time of each **CM_UNASSOCIATED_STA.IND** MME is randomly chosen. Furthermore, the STA synchronizes its PhyClk to one of the detected AVLNs, and uses the SNID of that AVLN in the MultiNetwork Broadcast transmissions. This provides a way for other AV STAs that also hear the same AVLN to apply the appropriate PhyClk correction for reception.

During the Power-on Network Discovery Procedure, while no AVLNs are detected, the STA search for beacons and other unassociated STAs using Hybrid mode. After an AVLN is detected, the STA may adopt the HomePlug 1.0/1.1 coexistence mode of that AVLN. Optionally, the STA may continue to search in Hybrid mode (irrespective of the mode of the AVLN that it detected). This optional behavior will enable the STA to continue to detect beacons of noncoordinating AVLNs (if any). It is important to note that proper execution of the CSMA/CA channel access mechanism requires the STA to perform virtual carrier sense in the mode of the Shared CSMA region (i.e., AV-only or Hybrid mode) of the AVLN(s) being tracked. Thus, if the Shared CSMA region is in AV-only mode and the STA is searching in Hybrid mode, it has to transition to AV-only mode during the AVLNs Shared CSMA region as soon as it has a pending transmission. The STA should also transition back to Hybrid mode after the transmission is completed.

During the Power-on Network Discovery Procedure, detection of an AVLN with matching NID causes the unassociated STA to follow the procedure described in Section 10.3.5 for joining the AVLN. If the STA successfully joins the AVLNs, it abbreviates the power-on network procedure and become an STA in the AVLN (refer to Section 10.2.3). Failure to join successfully causes the STA to continue with the Power-on Network Procedure.

On the expiration of the BBT, the STA processes all the received unassociated STA information (if any) as follows:

- If the STA detects other unassociated STAs with a matching NID, it uses the procedure described in Section 7.2.1 to determine whether it has to become the CCo and instantiate a new AVLN.
- If the STA detects no other AVLNs and there are no unassociated STAs with matching NID, it becomes an unassociated CCo (refer to Section 10.2.1).

In all other cases, the STA operates as an unassociated STA (refer to Section 10.2.1).

An STA that failed to form or join an AVLN during the power-on procedure operates in unassociated STA mode or unassociated CCo mode. Both these modes are generically called unassociated STA mode. Section 10.2.1 and Section 10.2.2 provide details on the unassociated STA mode and unassociated CCo mode, respectively.

Figure 10.1 shows the basic functional flow for Power-on Network Discovery Procedure.

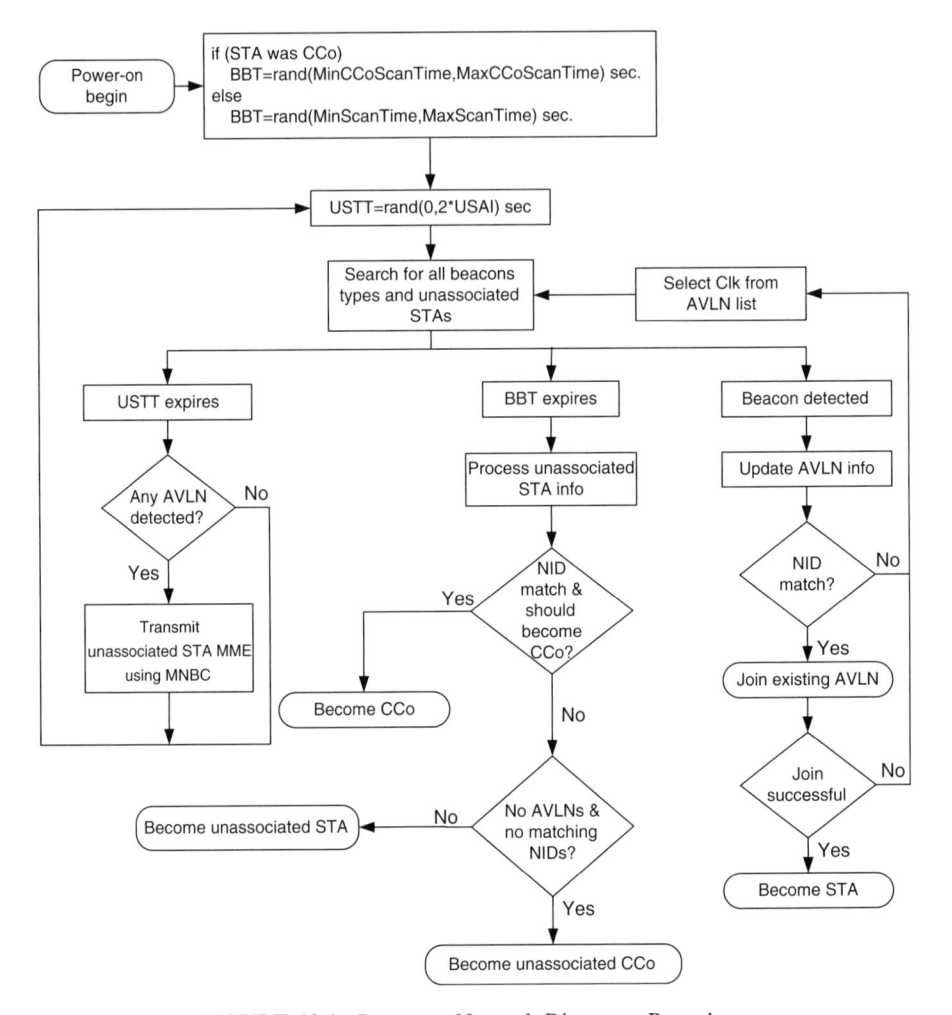

FIGURE 10.1 Power-on Network Discovery Procedure.

10.2.1 Unassociated STA Behavior

An unassociated STA that detects other AVLNs continue to send **CM_UNASSOCIATED_STA.IND** MME using multinetwork broadcast approximately once per MaxDiscoverPeriod (10 s) to provide information to other stations. The transmission time of each **CM_UNASSOCIATED_STA.IND** MME is randomly chosen. Unassociated STAs continues to operate in this mode until one of the following events occur:

- Detection of an AVLN with matching NID causes the unassociated STA to follow the procedure described in Section 10.3.5 for joining the AVLN. If the STA successfully joins the AVLN, it becomes an STA in the AVLN. Failure to join successfully causes the STA to continue operating as an unassociated STA.

POWER-ON NETWORK DISCOVERY PROCEDURE 175

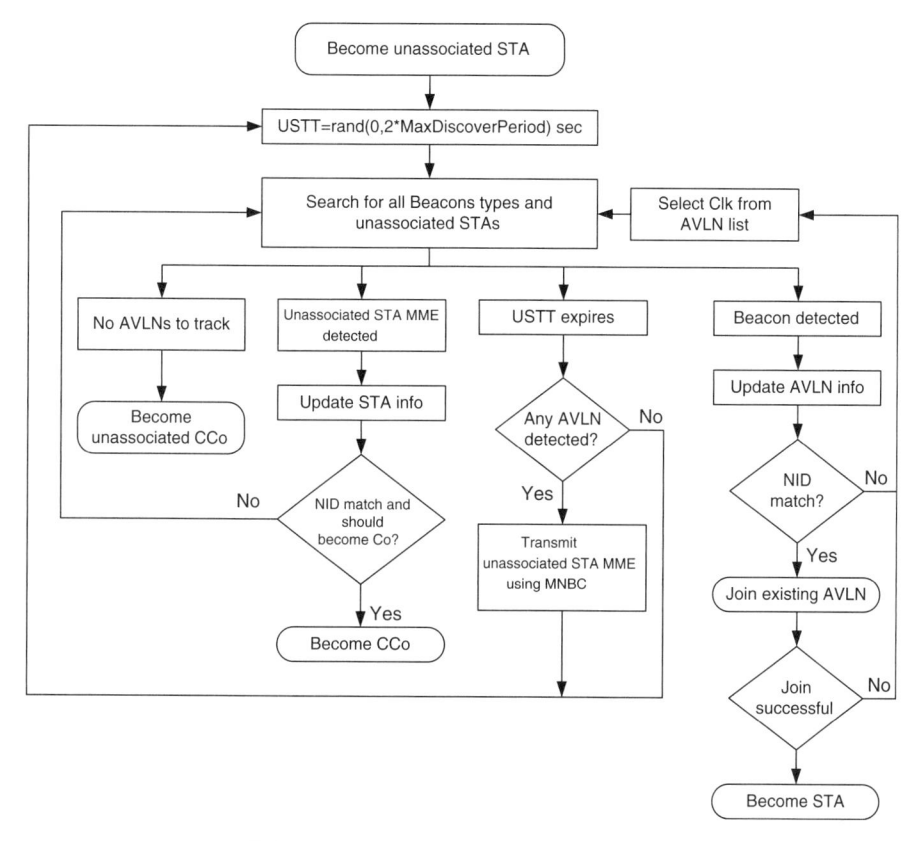

FIGURE 10.2 Unassociated STA behavior.

- If an unassociated STA determines that there are no AVLNs to track (i.e., AVLN(s) that it is tracking no longer exist), it becomes an unassociated CCo.
- Detection of a **CM_UNASSOCIATED_STA.IND** MME with matching NID causes the STA to follow the procedure described in Section 10.3.4.1.

Figure 10.2 shows the basic functional flow for unassociated STA behavior.

10.2.2 Unassociated CCo Behavior

Unassociated CCo continues to operate in this mode until one of the following events occur:

- Detection of another AVLN cause the unassociated CCo to start operating as an unassociated STA.
- Reception of valid **CC_ASSOC.REQ** causes the unassociated CCo to become a CCo.

Figure 10.3 shows the basic functional flow for unassociated CCo behavior.

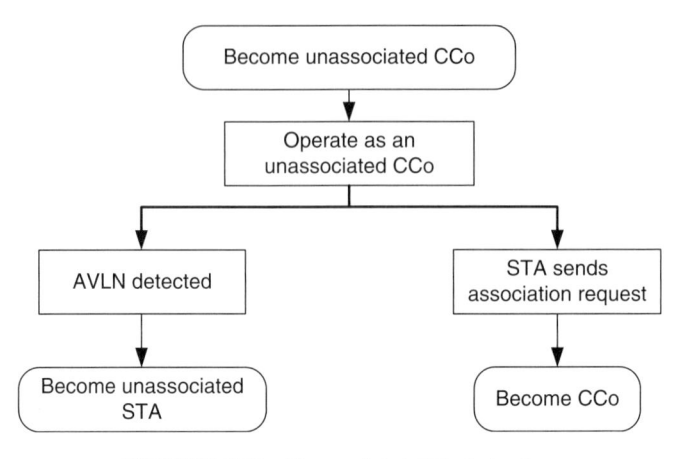

FIGURE 10.3 Unassociated CCo behavior.

10.2.3 Behavior as an STA in an AVLN

On joining the AVLN, if the STA is a user-appointed CCo, the STA sends a **CC_CCO_APPOINT.REQ** to the existing CCo to handover the CCo functionality. Successful handover of the CCo functionality will cause the STA to become a CCo.

If an STA in the AVLN fails to detect the Central or Proxy Beacons of the AVLN that it is part of for at least ten Beacon Periods and it is not the backup CCo, it restarts the power-on procedure.

If an STA in the AVLN is requested to leave the AVLN, it becomes an unassociated STA.

Figure 10.4 shows the basic functional flow for the behavior as an STA in an AVLN.

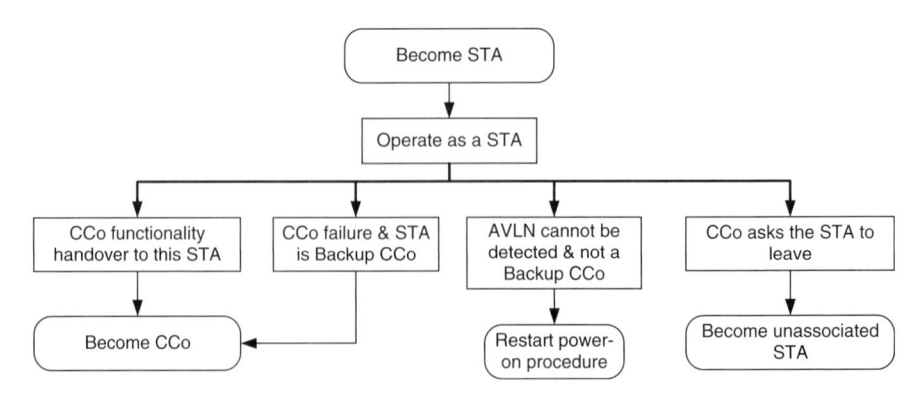

FIGURE 10.4 Behavior as an STA in an AVLN.

POWER-ON NETWORK DISCOVERY PROCEDURE   177

10.2.4 Behavior as a CCo in an AVLN

Once an STA becomes a CCo, if it determines that there are no STAs that have successfully joined the AVLN, it starts a Join Wait Timer. Join Wait Timer is the duration of time for which the STA will act like a CCo if no other STA has successfully joined the AVLN. It is recommended that Join Wait Timer be set to at least MaxDiscoverPeriod (10 s) to provide sufficient time for STAs to join the AVLN. If an STA joins the AVLN before the expiration of Join Wait Timer, the timer is cleared. Expiration of a Join Wait Timer causes the CCo to operate as an unassociated CCo if there are no other AVLNs present; otherwise, it starts operating as an unassociated STA.

Handing over of the CCo functionality to another STA in the AVLN causes the CCo to become an STA in the AVLN.

If all the STAs associated with the AVLN leave the AVLN and there are no other AVLNs detected, the CCo operates as an unassociated CCo. If all the STAs associated with the AVLN leave the AVLN and there is at least one other AVLN detected, the CCo becomes an unassociated STA.

Figure 10.5 shows the basic functional flow for the behavior as a CCo in an AVLN.

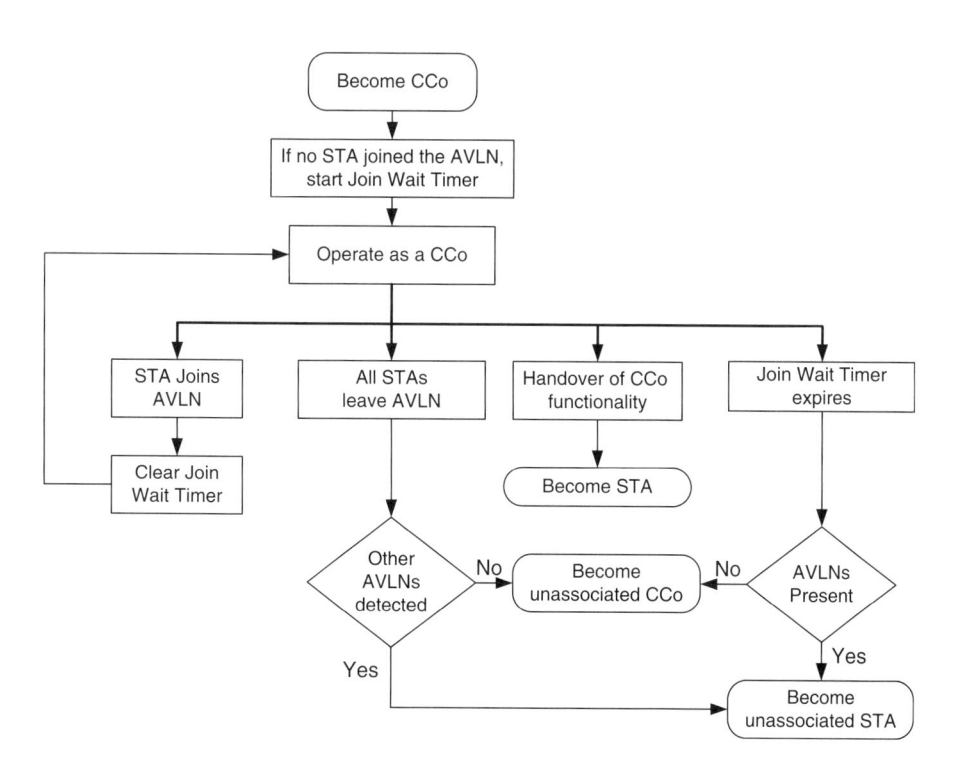

FIGURE 10.5 Behavior as a CCo in an AVLN.

10.3 FORMING OR JOINING AN AVLN

10.3.1 AVLN Overview

An AVLN is formed by STAs that possess a common NID and CCo. STAs in an AVLN will typically possess a common NMK and Security Level (SL), but a CCo may divide an AVLN into sub-AVLNs using multiple NEKs (possibly using multiple NMKs). All NMKs that are associated with the same NID have the same SL. When no AVLN exists and an STA discovers one or more other stations with the same NMK and SL, the STAs forms an AVLN. If an STA discovers an existing AVLN with the NMK and SL it possesses, it joins the existing AVLN. An AVLN is formed or joined using Association and Authentication.

To join (participate in) an AVLN, an STA must have the following:

- A valid NMK and Security Level (NMK-SL, obtained by Authorization—a.k.a. NMK Provisioning, as described in Section 10.4.2).
- A unique TEI (obtained by Association, as described in Section 10.3.2).
- The current NEK (obtained by Authentication, as described in Section 10.3.3).

Usually, the NMK-SL and NID are stored in nonvolatile memory.

10.3.1.1 Network Identification Each AVLN has a Network Identifier (NID) that is provided with or generated from the NMK-SL and is used to help an STA identify another STA or AVLN with the same NMK-SL.

It is also permitted for the CCo to use the same NID for multiple NMKs, forming a sub-AVLN with each NMK. As the NIDs are the same, an STA with a (NID, NMK) pair whose NID matches the NID of an AVLN attempts to join that AVLN. The CCo disambiguates the NMK depending on its knowledge of the STA's MAC address.

10.3.1.2 Human-Friendly Station and AVLN Names STA manufacturers provide default "Human Friendly" Identifiers (HFIDs) to the STAs. An HFID is a string of up to 64 ASCII characters chosen from the range of ASCII [32] to ASCII [127]. The HFIDs is stored in nonvolatile memory.

The HFID of an AVLN is obtained by sending a **CC_WHO_RU.REQ** MME and receiving the associated **CC_WHO_RU.CNF** reply.

10.3.1.3 Get Full AVLN Information The NID is contained in each Beacon transmitted by the CCo. Ordinarily, this will be sufficient, but there will be circumstances when the STA needs additional information about the AVLN (e.g., to identify the AVLN to the user, it will need to get the HFID of the AVLN). The Message Sequence Chart (MSC) shown in Figure 10.6 allows the new STA to get full AVLN information before obtaining a TEI. If the STA already has a TEI, that value is used within the messages.

10.3.1.4 Get Full STA Information The SNID and TEI are contained in each MPDU transmitted by an STA. Ordinarily, this will be sufficient, but there will be

FORMING OR JOINING AN AVLN   179

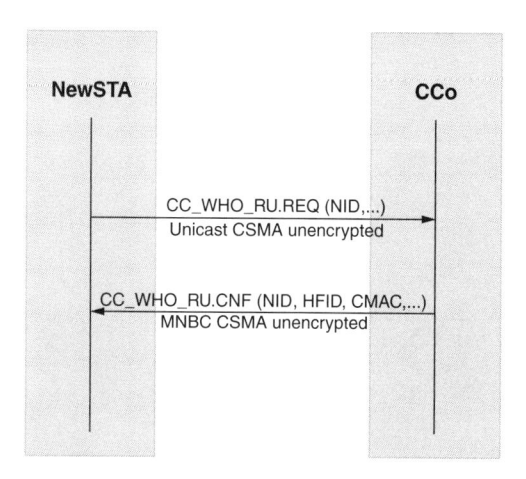

```
     NewSTA                              CCo
        |                                 |
        |------CC_WHO_RU.REQ (NID,...)--->|
        |      Unicast CSMA unencrypted   |
        |                                 |
        |<---CC_WHO_RU.CNF (NID, HFID, CMAC,...)---|
        |      MNBC CSMA unencrypted      |
        |                                 |
```

FIGURE 10.6 Getting Full AVLN information.

circumstances when a user needs additional information about the STA (e.g., the HFID of the STA is useful to identify the STA to the user).

The **CM_GET_HFID.REQ** MME may be used to get the HFIDs of the STA (manufacturer-set HFID or user-set HFID) or the HFID of the network. The **CM_STA_CAP.REQ** MME may be used to get detailed information about the STA including the optional features the STA supports, the HomePlug AV version, the OUI, and the product manufacturer's version number.

10.3.2 Association

Association is a process by which an STA obtains a valid TEI from the CCo of the AVLN with which it wants to associate. Disassociation is a process by which an STA should stop using a valid TEI for an AVLN with which it was once associated. On disassociation, the TEI becomes invalid until reassigned.

There is a single method of Association, but there are several methods of Disassociation.

When an STA initially wants to communicate with an AVLN, it performs Association as shown in Figure 10.7. When an STA joins an AVLN, the CCo provides the complete TEI Map to that STA and update all other STAs in the AVLN with the new TEI Map information using the **CC_SET_TEI_MAP.IND** MME.

10.3.2.1 TEI Assignment and Renewal The CCo assigns a Terminal Equipment Identifier (TEI) to each STA when it successfully associates with the AVLN. The TEI is 8-bit long and is unique within the AVLN. Table 10.1 shows the possible values of the 8-bit TEI.

The CCo maintains the list of assigned TEIs and the corresponding mapping of MAC addresses. The CCo sends the **CC_SET_TEI_MAP.IND** MME to STAs in the AVLN. After an STA is disassociated (or has been expelled), its TEI will be

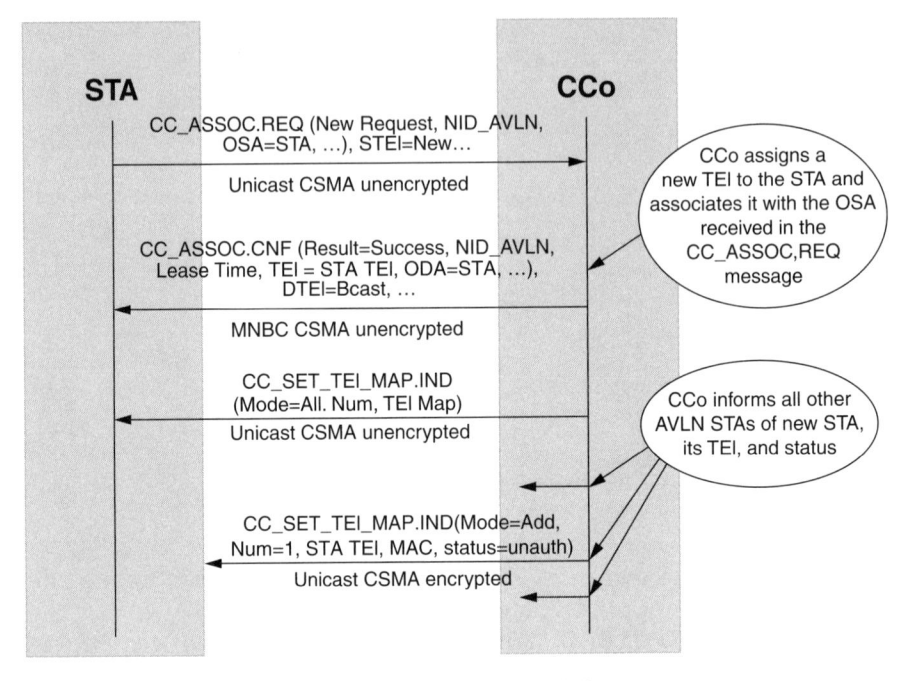

FIGURE 10.7 STA Association.

TABLE 10.1 TEI Values

TEI Value	Interpretation
0x00	TEI not yet assigned. This value may be used by a CCo to communicate with another CCo or by a new STA that has not yet associated with the AVLN.
0x01–0xFE	Identifies an STA within the AVLN.
0xFF	Broadcast TEI—All STAs treat message as addressed to them.

reclaimed by the CCo. The reclaimed TEI is not reused for a period of at least 5 min and the CCo sends the **CC_SET_TEI_MAP.IND** MME to update all the stations.

An STA may use the **CC_SET_TEI_MAP.REQ** MME to query the CCo for a full copy of the TEI Map. The CCo responds to the STA with the **CC_SET_TEI_MAP.IND** to provide the TEI Map.

10.3.2.1.1 Disambiguated TEIs Although the CCo ensures that the TEIs it assigns are unique within the AVLN, the same TEI value may be assigned to a different STA in a neighbor AVLN, so it is important to disambiguate the TEI by associating it with the network where it is generated. The NID or the SNID can be used for this purpose.

10.3.2.1.2 TEI Leases and Renewals When the CCo assigns or renews a TEI, it specifies a lease time. This is the length of time for which the STA may use the TEI. If the lease time expires before the TEI is renewed, the STA stops using the TEI and apply for another. Lease time can range from 15 min to 48 h.

Lease time is measured from the time the CCo generated the corresponding **CC_ASSOC.CNF**. Variable amount of delay can be incurred from the time the CCo generated the **CC_ASSOC.CNF** and the STA receives and processes the message. Implementations should consider this when determining the expiry time of the TEI lease and the time at which the STA starts renewing its TEI lease. Before the lease time has expired, an STA should go through the association process again to renew its TEI. It is recommended that the STA starts its TEI renewal process at least 5 s before the expiration of its lease time.

10.3.3 Method for Authentication

Once an STA has associated and has a valid NMK, it uses this NMK to join the AVLN. If the CCo verifies the STA's NMK, it will give the STA an NEK. Once an STA is authenticated successfully, the CCo maintains the STA's authentication status as long as the STA's association status is maintained. Thus, an STA is not required to reauthenticate subsequent to TEI renewal.

There is one method for Authentication; it is used by all STAs. The joining STA sends a **CM_GET_KEY.REQ** containing KeyType = NEK, the STA's TEI and MAC address. It also contains a freshly generated nonce. The message is placed in an Encrypted Payload of a **CM_ENCRYPTED_PAYLOAD.IND** MME encrypted by the NMK, with PID = **0x00** and PMN = **0x01**.

If the CCo confirms that the correct NMK was used to encrypt the message, it sends a **CM_GET_KEY.CNF** message (in a **CM_ENCRYPTED_PAYLOAD.IND** with the payload encrypted with the NMK) to the STA. This message contains the NEK and EKS along with the nonce sent in the request and is placed in an Encrypted Payload encrypted using the NMK. The procedure is shown in Figure 10.8.

If the CCo cannot decrypt the request encrypted by that NMK (indicated by **CM_ENCRYPTED_PAYLOAD.RSP** with Result = Failure), the new STA flags the NMK as invalid on this AVLN (NID and SNID) and either restart the process of obtaining a (valid) NMK on this AVLN or try to join a different AVLN (same NID but different SNID).

The new STA can begin using the NEK as soon as it successfully receives it from the CCo.

10.3.4 Forming a New AVLN

Two unassociated STAs can form a new AVLN when one of four conditions is met:

- They have the same (NID, NMK) pair, and one or both receives the other's **CM_UNASSOCIATED_STA.IND**MME (refer to Section 10.3.4.1).

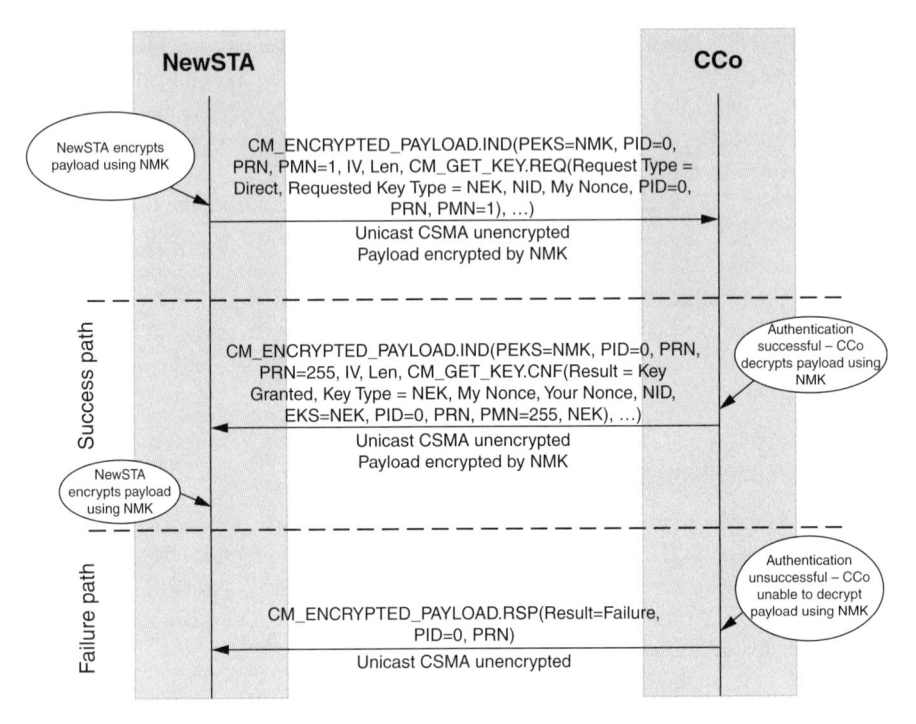

FIGURE 10.8 Provision NEK for a new STA (Authentication).

- One STA sends the other STA its NMK encrypted with the other STA's DAK (refer to Section 10.3.4.2).
- They both have NMK-SCs, and one's HLE indicates that it should enter the SC-Join state and the other's HLE indicates that it should enter the SC-Add state (refer to Section 10.3.4.3).
- They both have NMK-SCs and the HLE of each indicates they should enter the SC-Join state (refer to Section 10.3.4.4).

In each of these cases, two unassociated STAs exchange MMEs that includes the CCo capability of each STA, and from these MMEs, each STA recognizes that a new AVLN needs to be formed. One of the STAs will become the CCo (as described in Section 7.2.1) and the other will associate with the new CCo, and possibly perform the NMK key exchange and ultimately authenticate.

CM_SET_KEY.REQ/CNF, CM_UNASSOCIATED_STA.IND, and **CM_SC_JOIN.REQ/CNF** MMEs contain CCo capability information; this information allows the recipient to determine the STA that can become the CCo of the new AVLN. The STA with greater CCo capability becomes the new CCo. Once the AVLN is formed, the CCo may be changed due to other factors.

An STA that determines that another STA should become the waits to try to associate with it until it starts receiving the Central Beacon from that STA or possibly another STA. Such STAs may only repeat the MMEs that the other STA must receive

FORMING OR JOINING AN AVLN

to prompt it to form an AVLN. The other STA should reach the same conclusion and become a CCo and start issuing Central Beacons, perhaps in Coordinated Mode if a neighboring network is present. An STA that becomes a CCo begins transmitting the Central Beacon and continues to do so as long as at least one STA has successfully associated with it or while it hears MMEs from other STAs that should associate with it.

When the STA has or receives an NMK associated with an NID matching that of the new CCo (possibly as a result of one of the processes described earlier), it tries to authenticate using the protocol described in Section 10.4.3. If authentication fails, the NMKs are not the same and the protocol aborts. In this case, if the CCo has no associated STAs, it ceases AVLN operation and returns to being an unassociated STA (unless it cannot detect any other AVLN). If authentication failed for reason other than different NMKs, the STAs should try again to form a new AVLN. If the authentication succeeds, the initial AVLN formation is complete and the STAs may go on to add more STAs to the AVLN, select a new CCo, and so on.

10.3.4.1 Two Unassociated STAs with Matching NIDs

Two STAs with identical NMKs and identical security levels will also have identical default NIDs. Identical NIDs do not guarantee that the NMKs are identical, but the probability against this is very small (one in 2^{54}). The NID is advertised in the **CM_UNASSOCIATED_STA.IND**MME, so when one STA receives the other STA's **CM_UNASSOCIATED_STA.IND**MME, it will observe the matching NID.

At this point, one of the STAs (or a third STA) becomes a CCo. The **CM_UNASSOCIATED_STA.IND**MME contains the CCo capability and the MAC address of the sender (as part of the generic MME format). The receiver can then decide whether it or another STA should become the CCo.

If the STA determines that it should become the CCo, it forms a new AVLN and start sending Central Beacons, and then wait for other STA(s) to associate.

If the STA determines that another STA should become the CCo, it waits until it detects a Central Beacon with matching NID. In the meanwhile, the STA continues to send **CM_UNASSOCIATED_STA.IND** MMEs periodically, in case the other STA did not correctly receive its earlier advertisements and hence does not know to become the CCo of a new AVLN. After the nonCCo STA has detected the new CCo's Beacon with matching NID, it sends the CCo a **CC_ASSOC.REQ**MME asking it for a TEI.

If more than one AVLN with the same NMK is formed, the STA may attempt to join any AVLN with matching NID that it detects. Once the STA has associated with the CCo, it tries to authenticate as described in Section 10.4.3. This entire process is shown in Figure 10.9.

10.3.4.2 Two Unassociated STAs Form an AVLN Using a DAK-encrypted NMK

When the HLE provides an STA with the DAK of another station and tells it to send the other STA an NMK and NID, the STA broadcasts a **CM_SET_KEY.REQ** MME containing a TEK and encrypted with the DAK as the payload of a **CM_ENCRYPTED_PAYLOAD.IND** MME, sent unencrypted using multinetwork broadcast. All STAs that receive the MME tries to decrypt it; if one succeeds, the

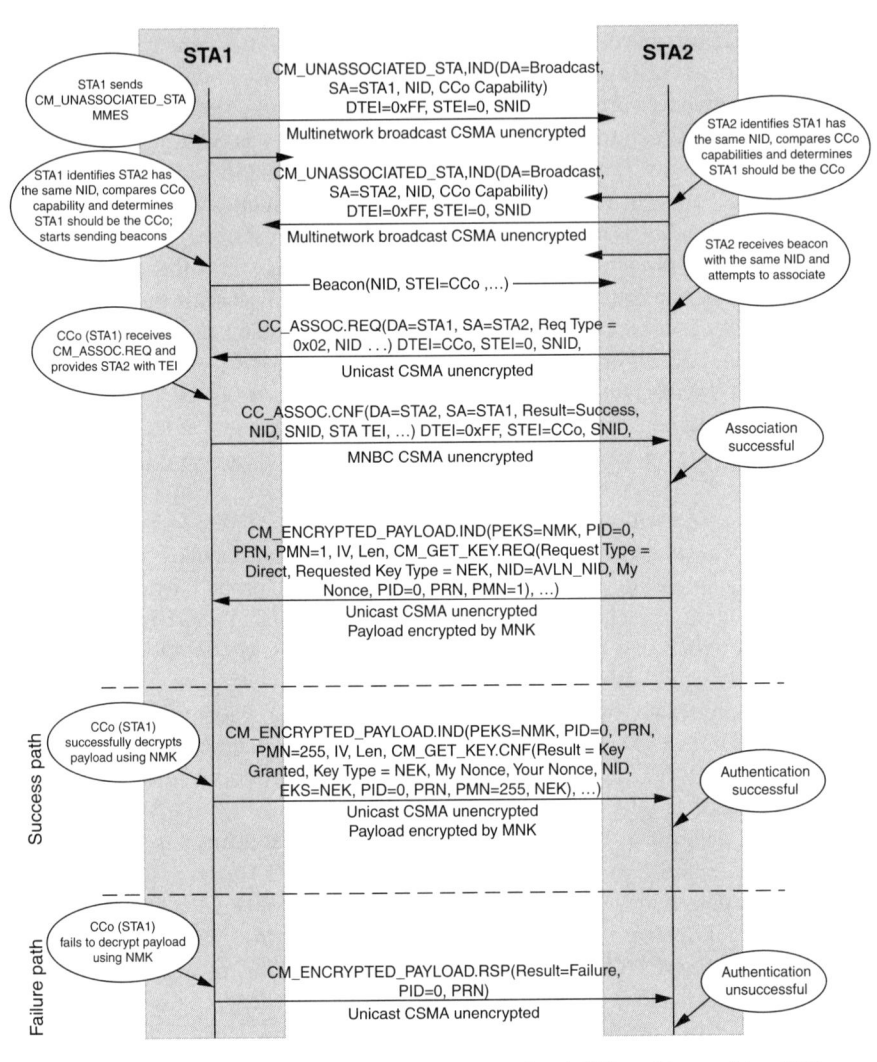

FIGURE 10.9 AVLN formation by two unassociated STAs with matching NIDs.

successful STA responds with a **CM_SET_KEY.CNF** MME encrypted with the TEK as the payload of a **CM_ENCRYPTED_PAYLOAD.IND** MME sent unencrypted using unicast.

Each STA compares the CCo capability fields (present in the first two messages) and determines the STA that should become the CCo, as defined in Section 7.2.1. Once one STA becomes the CCo and the other STA has associated with it, the STAs complete the protocol as described in Section 10.4.2.4. When the protocol is completed successfully, the STA that is not the CCo authenticates with the CCo to obtain the NEK. The new CCo uses in its Central Beacon the NID that was sent with the NMK; the other STA must also use the NID associated with the

FORMING OR JOINING AN AVLN

NMK and wait for a Central Beacon with the matching NID before it authenticates with the CCo.

The STA sending the DAK-encrypted payload continues to transmit periodically until it either receives a response or until it times out.

If the STA that was given the DAK by its HLE later joins with some other STA to form an AVLN, it may restart this protocol by sending the DAK-encrypted payload as an AVLN STA.

An associated (and even authenticated) STA that receives a new NMK via DAK-encrypted payload leaves its current AVLN and form an AVLN with the STA that initiated the DAK-based protocol, even if that STA is not initially a part of any AVLN.

The entire process (omitting failure paths) is shown in Figure 10.10. See also Section 10.4.2.4.

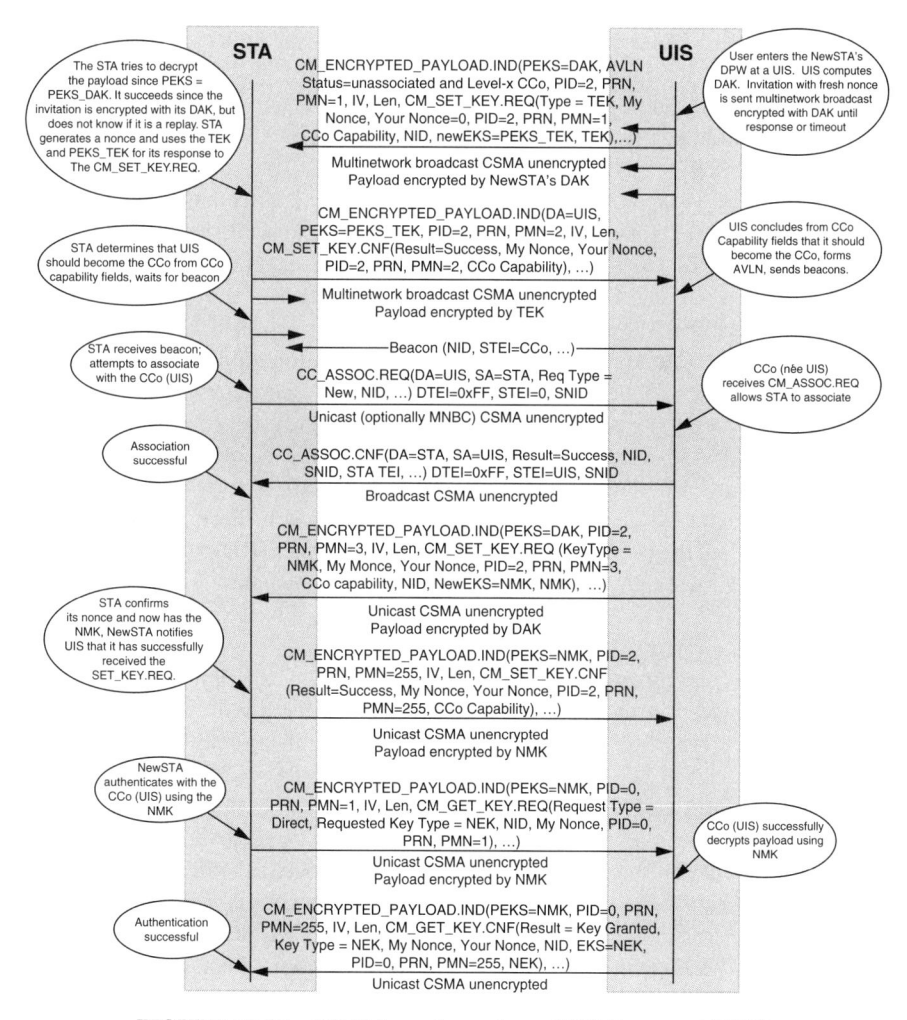

FIGURE 10.10 AVLN formation using a DAK-Encrypted NMK.

10.3.4.3 Two Unassociated STAs: One in SC-Add and One in SC-Join When the HLE places an STA into the SC-Join state, it transmits **CM_SC_JOIN.REQ** MMEs using multinetwork broadcast periodically until it either joins an AVLN or times out. If the HLE places the STA into the SC-Add state, however, it does not advertise this, but waits to hear another STA transmitting **CM_SC_JOIN.REQ** MMEs until it either adds a new STA or times out. Optionally, an STA in the Simple Connect SL may cache recently received **CM_SC_JOIN.REQ** MMEs in anticipation of its HLE placing it into the SC-Add state.

When an STA in the SC-Add state detects a **CM_SC_JOIN.REQ** MME, it responds to it with a **CM_SC_JOIN.CNF** MME. The STA that was in the SC-Add state then becomes the CCo of a new AVLN and starts issuing Beacons. When the STA in the SC-Join state detects the Beacons, it associates with the CCo, regardless of relative CCo capabilities. This case is distinguished from the case below in which two unassociated STAs are in SC-Join by the STA in SC-Add setting the CCo Status field to **0b1**. If the joining STA has greater CCo capability, it will later become the CCo through autoselection of the CCo and the CCo Handover process (refer to Section 7.2.2 and Section 7.4).

After the new STA has associated with the new CCo, the two STAs optionally perform Channel Adaptation (refer to Section 3.6) to have channel adapted tone maps. Finally, the CCo starts the UKE protocol. This first establishes a shared TEK, which is used by the CCo to provide the new STA with its NMK (refer to Section 10.4.2.5). When the new STA has the NMK, it authenticates and joins the AVLN. This entire process is shown in Figure 10.11.

Channel-adapted tone maps are generated based on the unique attenuation and noise characteristics between a transmitter and receiver. Since power line channels are highly frequency selective and can vary significantly between stations, it is very unlikely that messages transmitted using channel adapted tone maps are properly decoded by stations other than the intended receiver. Using channel adapted tone maps while executing the UKE significantly limits the ability for stations other than the (new) CCo and STA involved in the UKE to properly decode the exchanged message. This option significantly enhances security of UKE protocol.

10.3.4.4 Two Unassociated STAs: Both in SC-Join It is possible that both STA's HLEs can be placed in the SC-Join state. In this case, both will begin to transmit **CM_SC_JOIN.REQ** MMEs with their CCo capability using multinetwork broadcast. When an STA in SC-Join mode receives another STA's **CM_SC_JOIN.REQ** MME, it determines the STA that should become the CCo as defined in Section 7.2.1. If it is the one to become the CCo, it changes its state to SC-Add, generate a new random NMK-SC, send a **CM_SC_JOIN.CNF** MME to the other STA with its NID, establish itself as a CCo, and start issuing Central Beacons, forming a Neighbor Network if necessary. When the other STA receives the **CM_SC_JOIN.CNF** MME and detects the Beacon with the same NID, it associates with the new CCo. The two STAs optionally perform channel adaptation before commencing the UKE protocol as above (refer to Section 10.3.4.3).

FORMING OR JOINING AN AVLN

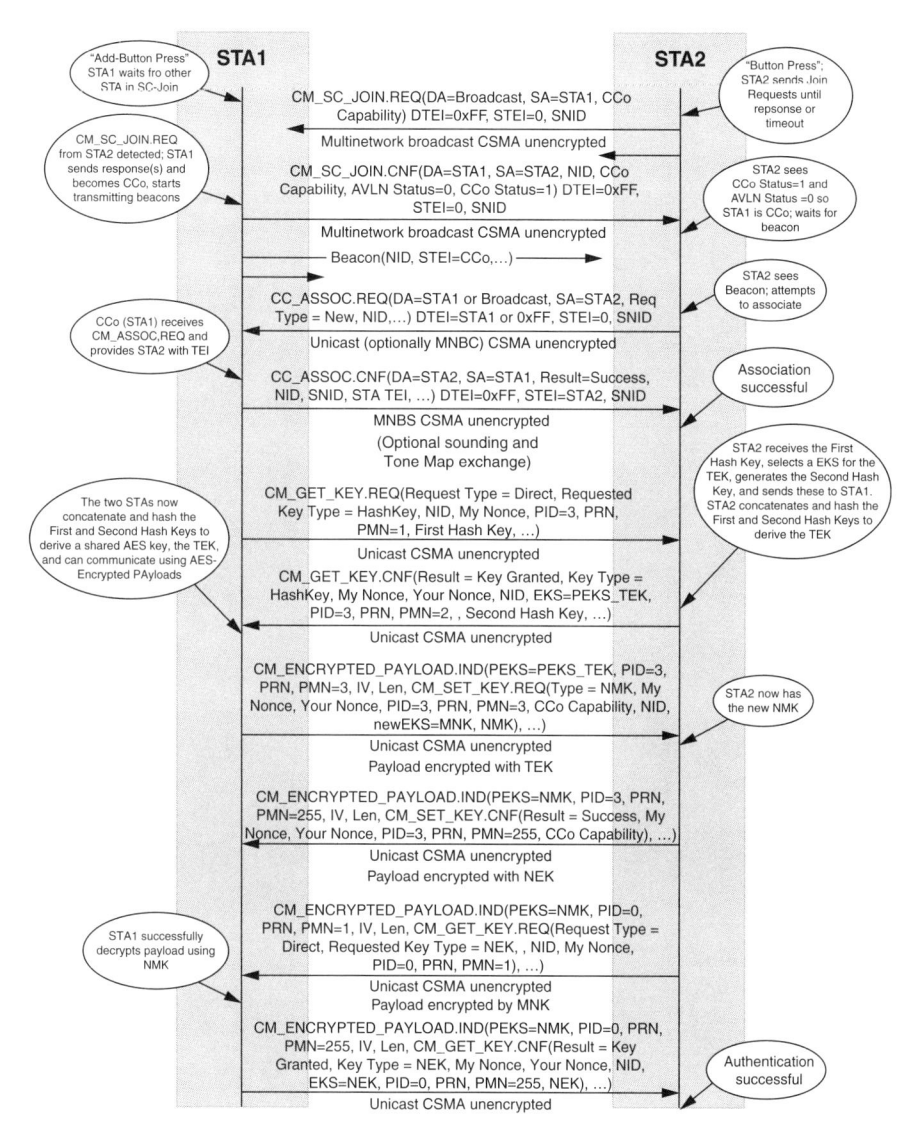

FIGURE 10.11 AVLN formation using UKE by one STA in SC-Add and one STA in SC-Join.

The STA that determines that it should not become the CCo must wait for the other STA to send the **CM_SC_JOIN.CNF** MME and for that STA to begin issuing Central Beacon. In the meanwhile, the STA continues to send **CM_SC_JOIN.REQ** MMEs periodically, in case the other STA did not correctly receive its earlier transmissions and hence does not know to send the **CM_SC_JOIN.CNF** MME and become the CCo of a new AVLN.

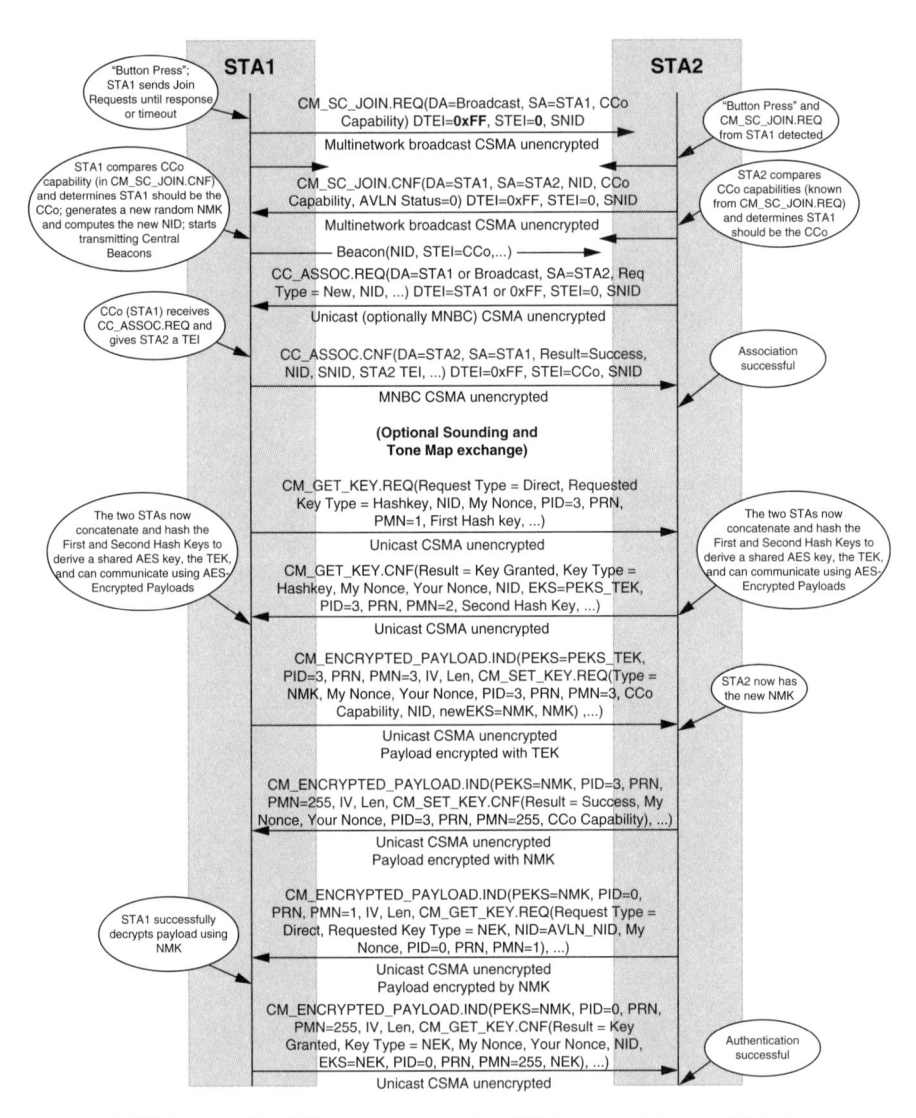

FIGURE 10.12 AVLN formation using UKE by two STAs in SC-Join.

This entire process is shown in Figure 10.12.

10.3.5 Joining an Existing AVLN

An unassociated STA may join an existing AVLN when one of the following three conditions is met:

- It has the same NMK and Security Level and detects the AVLN's Central Beacon or the Discover Beacon of one of the STAs in the AVLN (refer to Section 10.3.5.1).

- One of the AVLN STAs sends the unassociated STA its NMK encrypted with the unassociated STA's DAK (refer to Section 10.3.5.2).
- The AVLN has an NMK-SC, the unassociated STA's HLE indicates that it should enter the SC-Join state, and the HLE of an STA in the AVLN indicates that it should enter the SC-Add state (refer to Section 10.3.5.3).

In each of these cases, based on the initial information that is received or exchanged, the unassociated STA will recognize that it needs to associate with an existing AVLN. After association, the STA proceeds with the protocol to receive the NMK for the AVLN if it does not already possess it. On successful reception of the NMK for the AVLN, the STA then authenticates using the NEK distribution protocol described in Section 10.3.3.

10.3.5.1 Matching NIDs Two STAs with identical NMKs and identical Security Levels will also have identical default NIDs. Identical NIDs do not guarantee that the NMKs are identical, but the probability against this is very small. The NMK held by an STA may be associated with a nondefault NID; in this case, the nondefault NID associated with the NMK is used for matching purposes. The NID is advertised in the Central Beacon, Proxy Beacon, and Discover Beacons, so when an unassociated STA receives one of these, it observes the matching NID.

The unassociated STA must take the first step; the AVLN STAs (including the CCo) must wait for it to initiate the process. The unassociated STA sends the CCo a **CC_ASSOC.REQ** MME asking for an initial TEI within the AVLN. The CCo replies with a **CC_ASSOC.CNF** MME with STATUS = Success and a TEI assignment, unless it is out of TEIs or is in the process of transferring CCo status to another STA. In the latter cases, the CCo replies with STATUS = Defer, and the new STA will have to retry later.

Once the STA has associated with the CCo, it tries to authenticate as described in Section 10.4.3.

This entire process is shown in Figure 10.13.

10.3.5.2 DAK-Encrypted NMK When an STA with a suitable User Interface (the UIS) already on an AVLN is provided with the DAK of another STA and told by the HLE to send the other STA its current NMK, the UIS transmits a **CM_SET_KEY.REQ** MME containing a TEK and encrypted with the DAK as the payload of a **CM_ENCRYPTED_PAYLOAD.IND** MME, sent unencrypted using multinetwork broadcast. The STA need not be the CCo to do this; it is sufficient that it is in an AVLN.

All STAs that receive the MME tries to decrypt it; if one succeeds, that successful STA responds with a **CM_SET_KEY.CNF** MME encrypted with the TEK as the payload of a **CM_ENCRYPTED_PAYLOAD.IND** MME, sent unencrypted, then it associates with the AVLN's CCo and complete the protocol as described in Section 10.4.2.4. An STA that fails to successfully decrypt a payload encrypted with a DAK ignores the message.

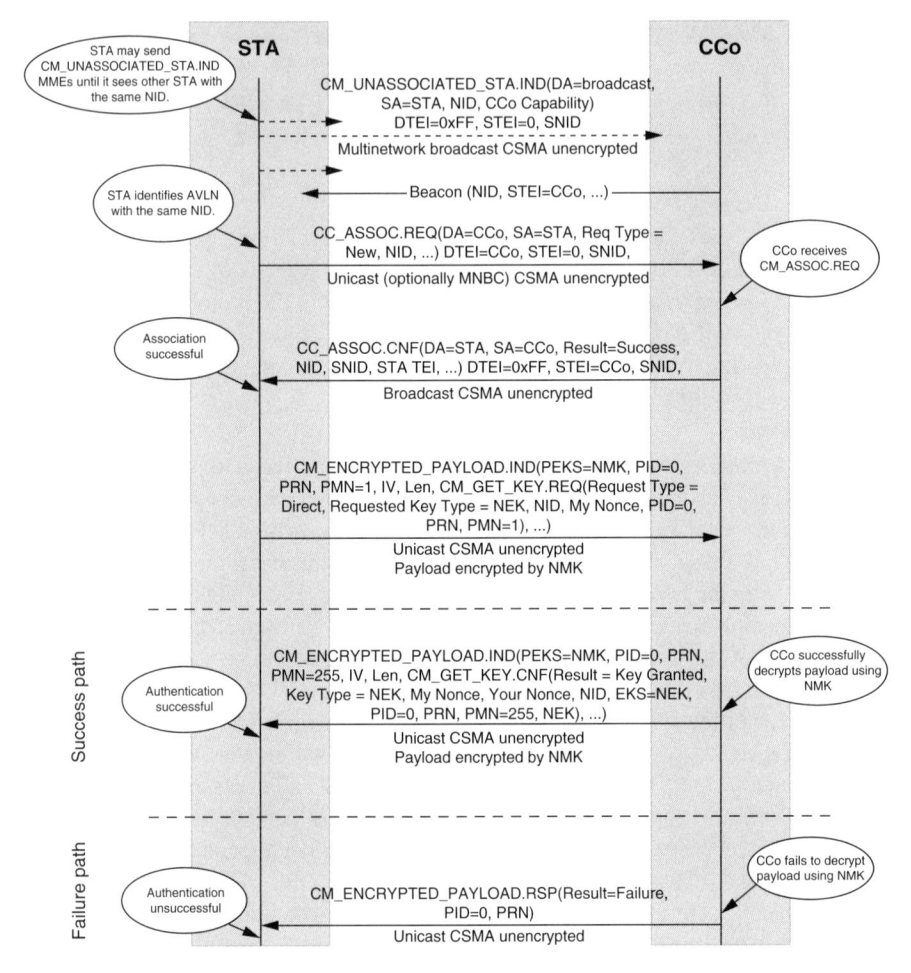

FIGURE 10.13 New STA joins existing AVLN with matching NID.

When the DAK-encrypted NMK provisioning protocol is completed successfully, the new STA accepts the NMK and SL, then tries to authenticate with the CCo as described in Section 10.4.3 and joins the AVLN.

This entire process is shown in Figure 10.14.

10.3.5.3 SC-Join and SC-Add When the HLE places an STA into the SC-Join state, the STA transmits **CM_SC_JOIN.REQ** MMEs using multinetwork broadcast periodically until it either joins an AVLN or times out. If the HLE places the STA into the SC-Add state, however, the STA should not advertise this, but waits to hear another STA transmitting **CM_SC_JOIN.REQ** MMEs until it either adds a new STA or times out. Optionally, an STA in the Simple Connect SL may cache recently received **CM_SC_JOIN.REQ** MMEs in anticipation of its HLE placing it into the SC-Add state.

FORMING OR JOINING AN AVLN

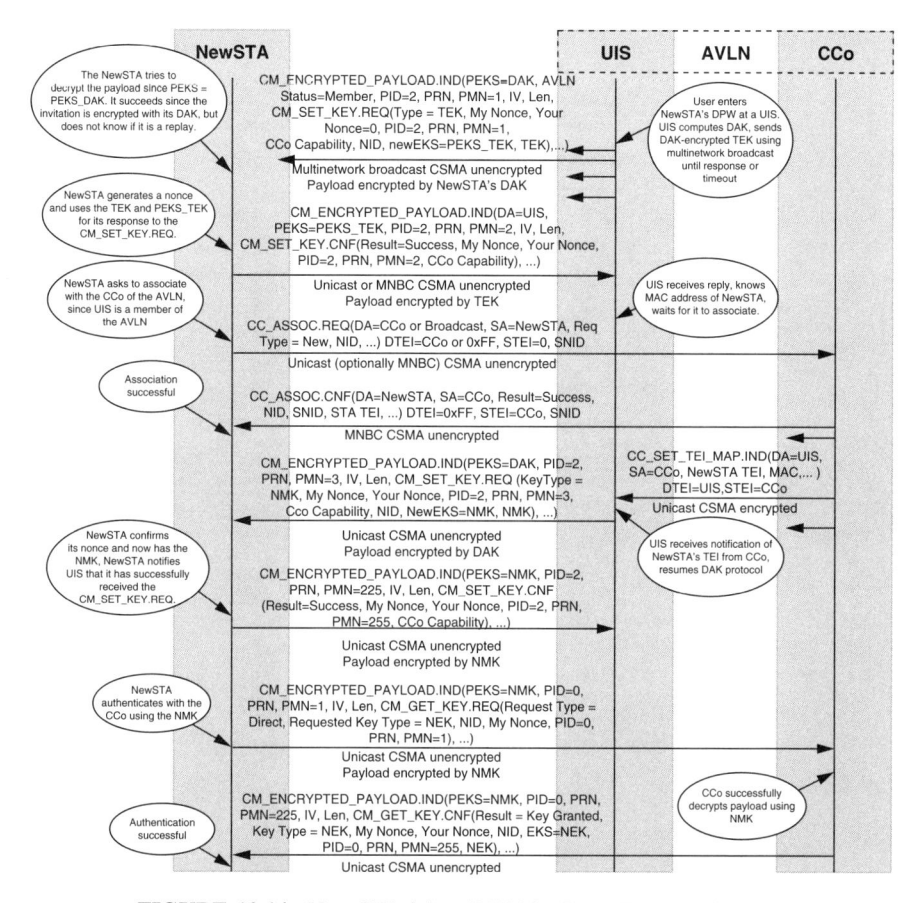

FIGURE 10.14 New STA joins AVLN by DAK-Encrypted NMK.

When the AVLN STA in the SC-Add state detects a **CM_SC_JOIN.REQ** MME, it responds with a **CM_SC_JOIN.CNF** MME. The STA need not be the CCo to do this; it is sufficient that it is in an AVLN. The new STA in the SC-Join state associates with the AVLN. At this point, the two STAs optionally perform Channel Adaptation to have channel-adapted tone maps.

Once the new STA has associated with the AVLN (and optionally established channel-adapted tone maps), the AVLN STA starts the UKE protocol. The AVLN STA knows the new STA has associated due to the updated TEI Map received from the CCo. The UKE protocol first establishes a shared TEK, which is used by the AVLN STA to provide the new STA with its NMK-SC (refer to Section 10.4.2.5). When the new STA has the NMK-SC, it tries to authenticate with the CCo as described in Section 10.4.1.5 and join the AVLN.

This entire process is shown in Figure 10.15.

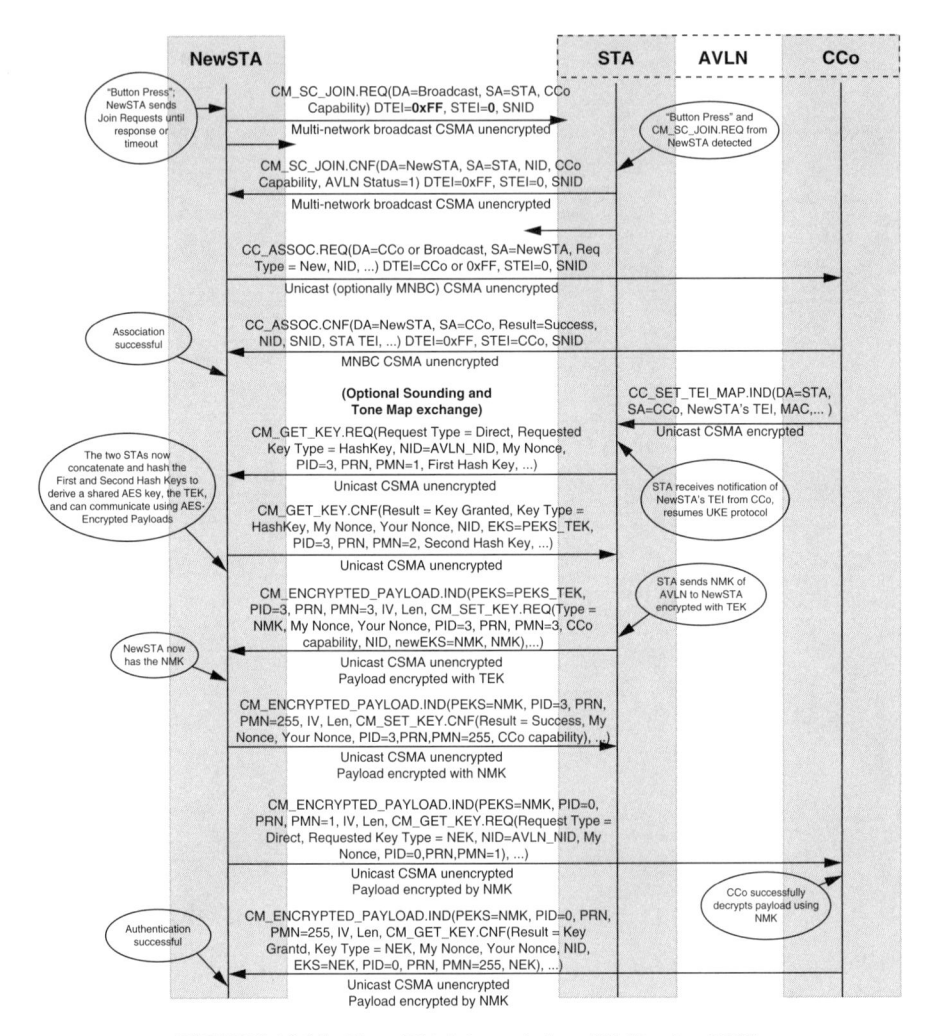

FIGURE 10.15 New STA joins existing AVLN using UKE.

10.3.6 Leaving an AVLN

If the STA is powered down or instructed to leave the AVLN by the user, it notifies the CCo of its departure as shown in Figure 10.16.

The STA waits until it receives an acknowledge response from the CCo before actually leaving. If it does not receive an ACK within three Beacon Periods, it will try to send the message a second time, after which it should not use the TEI it had been assigned for any further communications with the AVLN (the CCo or any member STA).

If the user has overtly requested disassociation (note that power down is an implicit request, not an overt request), the user may also want to tell the STA not

SECURITY OVERVIEW

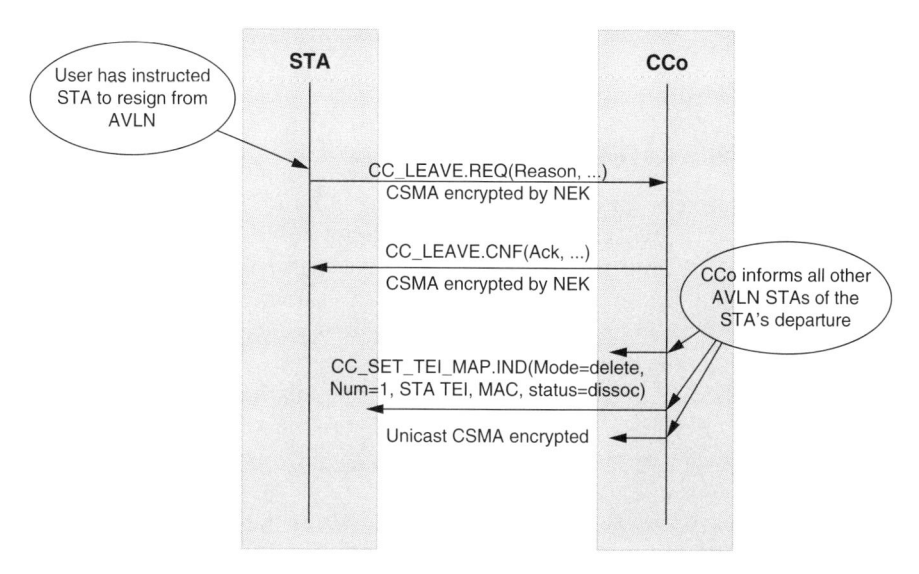

FIGURE 10.16 Disassociation—STA leaves AVLN.

to try to reassociate with the AVLN in future. In this case, the STA must also change the NMK.

10.3.7 Removing a Station from an AVLN

The only secure way to remove an STA from an AVLN is to change the NMK (refer to Section 10.4.2.7). The DAK of the removed STA should be discarded.

A CCo may send a **CC_LEAVE.IND** MME to an STA to remove the STA from the AVLN. The CCo may also change the NEK and not provide the new NEK to the STA it wants to remove.

10.4 SECURITY OVERVIEW

HomePlug AV security performs two functions, namely, controlling access to the AVLN and securing the privacy of data transferred on the AVLN.

Security is provided by means of a single encryption algorithm and a single hash function. They are 128-bit AES encryption (refer to Section 10.4.5) and SHA-256 secure hash function [21], respectively. Security is also enhanced through the use of nonces (refer to Section 10.4.6.3), which help to prevent unauthorized replays of MMEs. Access to the AVLN and ability to participate in the AVLN are provided by encryption keys and passwords.

With the exception of the encryption of PBs in the MAC/PHY using the NEK, all security-related activities (e.g., payload encryption and key management) take place in an entity that may be thought of as the "Security Layer." This entity is implementation-dependent and is assumed to lie in the Control Plane. The Security

Layer is responsible for generating the **CM_ENCRYPTED_PAYLOAD.IND** messages.

10.4.1 Encryption Keys, Pass Phrases, Nonces, and Their Uses

All the encryption keys used in HomePlug AV are 128-bit AES keys. These may be machine-generated or they may be based on pass phrases. Both the terms "Password" and "pass phrase" describe the same object. The term "pass phrase" is preferred, but the term "password" is also preserved because of its use in acronyms such as "DPW" and "NPW." Keys based on pass phrases come in two varieties: Device Password (DPW) and Network Password (NPW).

In addition, HomePlug AV uses Hash Keys (longer, machine-generated strings) for the Unicast Key Exchange (UKE) protocol, and nonces for freshness and association between messages in a protocol run.

10.4.1.1 Device Access Key (DAK) The Device Access Key (DAK) is unique to an STA. Each STA is provided with an unique DAK during manufacture. Another STA may, if it knows a particular STA's DAK, use that DAK to encrypt a message intended only for the particular STA. On receipt, such a message is treated as equivalent to the direct entry of the NMK from HLE. The DAK cannot be reset, and it should never be sent over the medium, even in encrypted MMEs. Support for the DAK is mandatory in all implementations of HomePlug AV.

10.4.1.2 Device Password (DPW) As part of the packaging of the product (perhaps as a label on the back or bottom of the product), the user is provided with a Device Password (DPW)—a password that will uniquely generate the new STA's unique DAK via the standard hashing algorithm defined in Section 10.4.6.1. The DPW is the value that is actually entered by the user during Authorization using the DAK (refer to Section 10.4.2.4).

10.4.1.3 Network Membership Key (NMK) The Network Membership Key (NMK) is used by an STA to prove its membership in an AVLN (or a sub-AVLN); that is, its right to join (participate in) an AVLN or a sub-AVLN (refer to Section 2.3). Thus the NMK defines the sub-AVLN. The user may designate the NMK(s) by entering an NPW or may elect machine generation (which is more secure). Hashing the NMK produces the default NID offset, but the NMK may be associated with a nondefault NID in some cases.

10.4.1.4 Network Password (NPW) The Network Password (NPW) is the value that generates the NMK when it is run through the hashing function described in Section 10.4.6.1.

10.4.1.5 Network Encryption Key (NEK) During normal operation, most messages are encrypted using the Network Encryption Key (NEK), which is generated only by the CCo and is never exposed to the user. Support for two NEKs (one present and one future) is mandatory.

The NEK can only be set by a CCo that internally generates it randomly (refer to Section 10.4.6.2), or by the NEK-provisioning processes, whereby the CCo provides an STA with the NEK in an Management Message Entry (MME) encrypted with the NMK. The NEK is not known by an STA until the STA has completed the authentication process (refer to Section 10.4.3) and joined the AVLN.

10.4.1.6 Temporary Encryption Key (TEK) The Temporary Encryption Key (TEK) is an AES key that is used to encrypt messages on a temporary private channel between two STAs. It may be distributed using the receiver's DAK, or generated by the Unicast Key Exchange (UKE) protocol (protected from standard equipment by unicast, and possibly at the signal level by tone map modulation). It can be used over unauthenticated channels (i.e., it may be distributed without proof of freshness using the DAK, or it may be generated using UKE). It should not be distributed to more than one STA, and both sender and receiver must discard it after no more than Max_TEK_Lifetime (120 s). If the TEK was exchanged using UKE, it may be used only until a protocol message with PMN = **0xFF** is sent or until the protocol aborts (by either STA). Once a message with PMN = **0xFF** has been sent, the TEK established at the beginning of that protocol run should no longer be used. Support for at least one TEK is mandatory. Refer to Section 10.4.6.2 for generation of random AES keys.

10.4.1.7 Nonces Nonces are pseudo-random numbers that are used only once (i.e., the sequence of nonce values for a given station are unpredictable). Practically, as the number of bits in a nonce is finite, they may repeat, but the same nonce should never be used with the same encryption key; that is, changing the encryption key in essence clears the set of unusable nonces.

Nonces are used to prevent replay attacks (when an STA receives a nonce that it recently generated, it can be assured that the message it received was composed recently also).

10.4.2 Methods for Authorization (NMK Provisioning)

An AVLN is defined as a set of stations that share a common NID and CCo with which they share a common NMK and Security Level. Typically, all STAs in an AVLN can communicate with each other and share a common NMK and Security Level, but the CCo may form separate sub-AVLNs within an AVLN, each with its own NMK.

Before a new STA can participate in an AVLN, it must determine whether to join an existing AVLN or to form a new one. The determination of whether to join a given AVLN consists of three decisions:

1. At what level of security does the user wish the AVLN to operate?
2. Does the STA want to authorize the AVLN? (More precisely: does the STA's owner want to have the STA join the AVLN?)
3. Does the AVLN want to authorize the STA's membership request? (More precisely: does the AVLN's owner want to allow the STA to join the AVLN?)

These decisions must occur before a new STA can fully participate in the selected AVLN. The first decision determines the operating mode of the AVLN and the ways in which its NMK(s) may be distributed. All three decisions implicitly require user participation and approval. It is possible for all three decisions to be made simultaneously by a single action of the user. Several alternative methods of NMK Provisioning are provided to enable these User Experience (UE) scenarios. These methods are specified in this section.

Regardless of whether the three processes occur separately or concurrently, the end result, if successful, is that the STA possesses a Network Membership Key (NMK) that it can use now and in the future to join the AVLN.

In general, NMK Provisioning will occur once and the authorization will be permanent. It is possible for the STA to be expelled from an AVLN by providing a different NMK to all stations except the STA being expelled, which would invalidate the STA's NMK. Subsequent distribution and use of a new NEK(s) then excludes the expelled STA from secure communication within the AVLN.

It is the implementer's responsibility to inform the user if an STA cannot connect to an AVLN because of a Security Level mismatch. If the user needs to change the Security Level for an STA to join an AVLN, the user should be so advised. This situation can arise when the user enters the NPW on multiple devices or uses a mix of mechanisms to distribute NMKs.

There are three methods for NMK Provisioning; each has its own merits and shortcomings, which are described with the UEs that these methods support. These methods are covered in three sections:

- 10.4.2.3 Direct Entry of the NMK,
- 10.4.2.4 Distribution of NMK Using DAK, and
- 10.4.2.5 Distribution of NMK Using Unicast Key Exchange (UKE).

Support for all methods is mandatory. Note, however, that the user experience(s) supported by a method will not be available to the user in the absence of a suitable user interface on the STA. Depending on the method used, the STA might or might not need to engage in AVLN communication before obtaining the NMK. The Security Level of the AVLN may also restrict the methods that may be used for NMK distribution.

The choice of the preferred method for obtaining an NMK from among those supported by an implementer is left with the user. NMK Provisioning is closely linked to Security and Privacy. It is expected that the user will determine the method to be used based on the method's ease of use and the user's perceived need for security and privacy.

10.4.2.1 Security Level The Security Level (SL) of the NMK defines both the Security Level of the AVLN and the methods that can be used to distribute the NMK. The SL is encoded as part of the NID so that AVLNs at different SLs cannot be considered to be the same AVLN even if the rest of their NID is the same.

Table 10.2 shows the interpretation of the 2-bit SL value.

SECURITY OVERVIEW

TABLE 10.2 Security-Level Interpretation

Security-Level Value	Interpretation
0b00	Simple connect. The NMK may have been exchanged using UKE.
0b01	Secure Security Level. The NMK must not have been exchanged using UKE.
0b10–0b11	Reserved.

The Security Level applies primarily to Encryption Key Management protocols. It is assumed that all data plane traffic and, with a few exceptions specified herein, all MMEs will be encrypted with the NEK. STAs are not required to accept and should not process any unencrypted data plane traffic and should not be assumed to do so. STAs are not required to accept unencrypted control plane traffic outside of the exceptions specified herein, and should not be assumed to do so. STAs that do accept other unencrypted traffic do so at their own risk.

10.4.2.1.1 Secure Security Level Table 10.3 shows the Security Levels used in NMK provisioning. An NMK-HS is associated with Secure Security Level. An STA will only accept a new NMK-HS that is either set by direct entry by the host or sent using the DAK-based distribution method (refer to Section 10.4.2.4). These are termed Secure Key distribution methods. The STA should not distribute an NMK-HS in its possession except by using the DAK of another STA.

An STA should neither accept nor distribute an NMK-HS using Unicast Key Exchange (UKE). An STA with an NMK-HS should ignore local Add/Join button presses.

10.4.2.1.2 Simple Connect Security Level An NMK-SC is associated with Simple Connect Security Level. An STA should only accept a new NMK-SC that it generates, or is sent using UKE or its DAK, or that it receives from the HLE using

TABLE 10.3 Security Level and NMK Provisioning

Key Status	Description	NMK Provisioning Allowed
NMK-SC	Simple Connect, NMK randomly generated by MAC or set by HLE	Direct (from HLE), UKE, or DAK
NMK-HS	Secure Security Level, NMK hashed from password—Password can be application-generated for the first station, but the user would have to record the password or use DAK provisioning for all additional stations. The STA ignores local Add/Join button presses (i.e., Add, Join button presses and UKE distribution require an explicit Security Level change, or reset).	Direct (from HLE) or DAK

the **CM_SET_KEY.REQ** message. The NMK-SC should be generated randomly by the first STA to become the CCo of a Simple Connect Security Level AVLN. An STA in Simple Connect may use UKE or the DAK of another STA to distribute NMK-SCs.

Just because an STA is in Simple Connect SL does not mean that it will provide its NMK-SC or accept a new NMK-SC from another STA using UKE. The HLE must place the STA in the SC-Join state to accept an NMK-SC from another STA using UKE, and must place the STA in the SC-Add state to provide its NMK-SC to another STA using UKE.

10.4.2.1.3 Changing the Security Level An STA that changes Security Level should discard the previous NMK.

A Station in Secure Security Level may only be changed to Simple Connect SL by receipt of an NMK-SC in the DAK-based distribution method (refer to Section 10.4.2.4), or by the HLE. If it is changed to Simple Connect SL by the HLE (e.g., by resetting the STA), it should generate a new NMK. In either case, it should discard the previous NMK-HS.

An STA in Simple Connect SL may only be changed to Secure SL by receipt of an NMK-HS in the DAK-based distribution method (refer to Section 10.4.2.4) or by the HLE. If it is changed to Secure SL by the HLE, the HLE should supply the new NMK-HS. In either case, the STA should discard the previous NMK-SC.

10.4.2.2 Preloaded NMK All devices are required to have an NMK at all times, so at a minimum, each device must have a random NMK. However, a vendor may replace the random NMK with an NMK derived from an NPW of the vendor's choice. This NMK must be set at the Secure Security Level. Alternatively, a vendor may cause several devices shipped as a set to have the same randomly generated NMK set at the Simple Connect Security Level. These options allow vendors to ship products that connect to each other out of the box for user convenience.

10.4.2.3 Direct Entry of the NMK Direct entry only implies that the HLE somehow obtains an NMK that it intends for the STA to use, and passes it to the Convergence Layer (CL) along with the NID (including the Security Level) using the **CM_SET_KEY.REQ** message over the H1 interface. The CL should store it for current and future use in joining the specified AVLN. The HLE may obtain the NMK from a set of (NID, NMK) pairs that it has stored, from user entry of an NPW (and generation of the default NID), from key exchange using some higher layer authentication mechanism or by other means (refer to Section 10.4.2.6) If a HomePlug AV Device is provided with a suitable user-interface mechanism, it may permit the user to enter the Network Password directly into the STA, from which the NMK-HS and default NID are generated. The user-interface mechanism may also allow the user to enter a nondefault NID. The HLE will obtain the Network Password (NPW) and Secure Security Level from the user, generate the NMK-HS (as described in Section 10.4.6.1) and default NID offset, and pass the NMK-HS with Security Level of the NID set to Secure to the CL via the **CM_SET_KEY.REQ** message. Whether the HLE retains the NPW and NID for future use is an implementation issue.

An STA that changes the NMK in this manner will leave its current AVLN (if any) and become an unassociated STA. It may join an existing AVLN or form a new one if it detects another STA with the same NMK-SL (refer to Section 10.3.5).

10.4.2.4 Distribution of NMK Using DAK The user may enter the DPW of an STA into any User Interface Station (UIS) on the AVLN (or into an unassociated STA). The HLE will derive the DAK from the DPW and pass it to the STA, possibly with the NMK and NID to distribute. The STA will use the DAK to perform Payload Encryption of a **CM_SET_KEY.REQ** MME containing a Temporary Encryption Key (TEK) and transmit the encrypted payload to the STA in a **CM_ENCRYPTED_PAYLOAD.IND** MME using multinetwork broadcast (refer to Section 6.9.1). The protocol ID used should be PID = **0x02** and the PMN should be **0x01**. The User Interface Station (UIS) should periodically broadcast this message until it either receives a response or times out.

When the new STA receives and correctly decrypts the payload of this message using its DAK, as indicated in the PEKS, it cannot tell if the message is a replay or not. The new STA should use the TEK supplied to respond with a **CM_SET_KEY.CNF** MME containing a new nonce in a **CM_ENCRYPTED_PAYLOAD.IND** MME unicast to the sender indicating that it has a matching DAK.

By receiving the TEK-encrypted payload in the response, the UIS knows that the new STA has a matching DAK and has the TEK. The UIS should wait until both it and the New STA are associated with the same AVLN (known from the TEI Map information) before resuming the DAK-encrypted NMK key distribution protocol.

Note: An STA receiving a DAK-encrypted TEK does not need to disassociate from its current AVLN until it has correctly received the NMK later in the protocol. This is to facilitate its return to the current AVLN should the protocol fail.

Consider the case in which two STAs have obtained CCo capability and AVLN status information from the invitation messages exchanged. If the UIS is part of an AVLN, the new STA (i.e., the one whose DAK was used to encrypt the first MME) should associate with the AVLN of the UIS. If the UIS determines that it will become the CCo, it should begin transmitting Central Beacons and the new STA should associate with the CCo (UIS). If the UIS is an unassociated STA and the new STA is determined to become the CCo, the new STA will become the CCo and the UIS should associate with it.

Once the two STAs are associated with the same AVLN, the key distribution protocol resumes. The UIS should send the new STA the NMK in a **CM_SET_KEY.REQ** MME as an encrypted payload encrypted with the DAK in a unicast **CM_ENCRYPTED_PAY LOAD.IND** MME, including the nonce received in the previous **CM_SET_KEY.CNF** MME. The new STA will receive this and obtain the NID (including the Security Level) and NMK, which is now knows to be fresh because of the nonce. The new STA will then confirm its correct receipt of the NMK by using it to encrypt the **CM_SET_KEY.CNF** MME sent as an encrypted payload in a unicast **CM_ENCRYPTED_PAYLOAD.IND** MME in response.

By virtue of receiving the third message in the protocol correctly (encrypted with its DAK and containing a new NMK and NID), the STA will know that the user wants it to join the AVLN from which the NMK was received and should join it immediately, even if it is already participating in another AVLN. An STA should set its security mode to the Security Level provided in the NID sent in the **CM_SET_KEY.REQ** MME when it receives the NMK via DAK-based encryption and should discard its current NMK.

Note: The STA should not discard its current NMK(s) on receipt of a DAK-encrypted **CM_SET_KEY.REQ** MME, as this may be a replay. Should the protocol abort before reception of the UIS' NMK, the new STA will disassociate from the UIS AVLN, reassociate (if necessary) with its previous AVLN, and resume use of its current NMK.

If the protocol completes successfully, the STA (i.e., the one that does not have the NEK) will disassociate and associate (if not already associated with the correct AVLN) and authenticate using the NMK, as described in Section 10.3.

The entire process for an unassociated STA forming a new AVLN is described in Section 10.3.4, and for an AVLN STA adding a new STA, the process is described, in Section 10.3.5. Absent receipt of a response to its initial invitation, the UIS need not send an abort message. An STA that receives an abort message with matching Protocol Run Number (PRN) from the other STA will terminate the protocol on its end unless it has already completed it. In particular, a new STA that correctly receives the NMK may attempt to authenticate with that NMK regardless of whether it receives an abort message later.

10.4.2.5 Distribution of NMK Using Unicast Key Exchange (UKE) If both the new STA and an existing STA have suitable user interfaces—alphanumeric entry is not a requirement—the new STA may, on invitation, establish a private channel with the UIS using a Unicast Key Exchange (UKE) method, establish a shared TEK and then obtain the NMK via the TEK-secured communications channel. This also requires that the AVLN be in Simple Connect (SC) Security Level, one STA be in the SC-Join state and the other be in the SC-Add state.

Note: The user may explicitly place an STA already in an AVLN in SC-Join state, causing it to leave its current AVLN and seek another to join (even if it later becomes the CCo as provided below). If the UKE protocol aborts, the STA may resume using the existing NMK. Likewise, an unassociated STA may be placed in the SC-Add state to continue to use its current NMK (e.g., when this is one previously used in an AVLN, but the other STAs are not currently present).

When an unassociated STA in the SC-Join state observes another STA in the SC-Join state and determines that it should become the CCo, it will enter the SC-Add state, generate a new random NMK-SC, and become the CCo of a new AVLN, as specified in Section 10.3.4.4. Once the user has directed the STA to join an AVLN (possibly with a single button press at the STA) and has directed the AVLN to adopt a new STA (possibly with a single button press at a designated STA already on the AVLN), the AVLN can give the STA the NMK. Two stages are involved in this process.

- First, a private channel is established using a Unicast Key Exchange (UKE) method.
- Second, the network STA (the one with the NMK to distribute) sends the new STA the NMK over the private channel.

There is some chance that a new STA joins the wrong AVLN, or that an AVLN adopts the wrong new STA. In the former case, the new STA may be directed to join a different AVLN. In the latter case, the AVLN may expel the incorrectly adopted new STA. In extreme cases, the user may have to abandon UKE and use one of the other, more secure methods to distribute the NMK to the desired stations. Once a shared AES key is established, it will be used until the protocol is aborted, or the TEK is superseded by additional AES encryption keys, subject to limitations (e.g., TEK timeout, refer to Section 10.4.1.6).

Note: Two STAs can use this method to establish a private AES key likely to be known only to themselves and distribute a new NMK-SC, even if they already share a common AES key known to other STAs (e.g., an NMK or an NEK).

To use UKE, the two STAs should receive some positive indication from the user that they are to join the same AVLN at the Simple Connect SL. This requires both STAs to be at the Simple Connect SL, and implies that they both have NMK-SCs. The two STAs first detect each other (using **CM_SC_JOIN.REQ/CNF** MMEs) and either associates with an existing AVLN or form a new AVLN (if both are unassociated STAs). Once the two STAs are associated with the same AVLN (known from the TEI Map updates from the CCo), they should establish channel-adapted tone maps (refer to Section 3.6) and then start the UKE protocol.

UKE requires that the STAs use unicast communications to derive the common TEK. Each STA contributes a secret Hash Key during an exchange of unencrypted, unicast **CM_GET_KEY.REQ/CNF** MMEs with Requested Key Type = HashKey with the PID = **0x03**, PMN = **0x01** and **0x02** respectively. The STA in SC-Add sends the **CM_GET_KEY.REQ** MME, and the STA in SC-Join responds with the **CM_GET_KEY.CNF** MME containing the PEKS chosen by the STA in SC-Join. The two Hash Keys are concatenated and hashed to produce the shared, unauthenticated TEK as defined in Section 10.4.6.2. The secrets (Hash Keys) exchanged by the STAs will be pseudorandom strings of length 384 octets.

The first two messages in UKE are sent unencrypted. These messages are unicast and should be sent using channel-adapted tone maps to minimize the likelihood of their communication being intercepted. These messages should not be retransmitted (that is, if the PB containing a Hash Key is not properly received, a new Hash Key should be generated before retransmitting; alternatively, if either of the two initial messages fails to deliver on the first attempt, then the protocol may be aborted and resumed with a different PRN).

Once both the STAs have the TEK derived from the Hash Keys, the STA in SC-Add should send the STA in SC-Join the NMK and the NID (which will indicate SC Security Level) in a **CM_SET_KEY.REQ** MME, containing the nonce received in the previous message, as an encrypted payload encrypted with the TEK with PID = **0x03** and PMN = **0x03** in a unicast **CM_ENCRYPTED_PAYLOAD. IND** MME. If the NID received with the NMK in the **CM_SET_KEY.REQ** MME does not indicate that the SL is SC, the receiver will abort the protocol by sending an unencrypted **CM_ENCRYPTED_PAYLOAD.RSP** with Result set to Abort. Otherwise, the STA in SC_Join should respond with a **CM_SET_KEY.CNF** MME, containing the nonce just received, encrypted with the NMK with PID = **0x03** and

PMN = **0xFF** in a unicast **CM_ENCRYPTED_PAYLOAD.IND** MME. This verifies that the NMK was received and concludes the UKE portion of the process.

The new STA will then go on to authenticate using the NMK to obtain the NEK from the CCo. The entire process for unassociated STAs is described in Section 10.3.4.4. For one unassociated STA and one STA in an AVLN, it is described in Section 10.3.4.3.

10.4.2.6 Distribution of NMK Using Other Key Management Protocols As an alternative to the methods described earlier for Authorization, HomePlug AV fully supports the use of HLE standards such as **802.1x**, EAP, and SNMP.

All messages sent by the HLE for key distribution should be transmitted as **CM_ENCRYPTED_PAYLOAD.IND** MMEs with PID = **0x04**. The 16-octet field used for the Initialization Vector (IV) in encrypted messages should be used for the UUID (see references [22,23]) of the protocol sending the messages. This field will remain constant for the duration of the protocol run. The UUID field may be used by the HLE to distinguish between application protocols using this lower-level transport mechanism.

The entire "encrypted" payload portion of the message should not be interpreted by the sending or receiving STA, and is thus available to the HLE to carry its protocol messages. In particular, the STA should not check the Protocol ID (PID), Protocol Run Number (PRN), or Protocol Message Number (PMN) fields across the "encrypted" and "unencrypted" parts of the MME. References to PID, PRN, and PMN in this section only concern the "unencrypted" fields of the MME.

The STA will use the ODA and its knowledge of TEIs to determine whether or not to broadcast a message it receives from the HLE with PID = **0x04**. The OSA and/or ODA in these messages might not belong to a HomePlug AV STA. All messages in the same protocol run should use the same PRN (randomly selected by the HLE application at the start of the protocol run), and the first message should have PMN = **0x01**. Subsequent messages in the protocol run will increment the PMN until the last message, which should have PMN = **0xFF**. An STA that receives a valid **CM_ENCRYPTED_PAYLOAD.IND** MME with PID = **0x04** from the power line medium will pass the entire **CM_ENCRYPTED_PAYLOAD.IND** MME to its HLE for processing.

If an STA receives (from its HLE or from the power line medium in response to a message it sent) a valid **CM_ENCRYPTED_PAYLOAD.IND** MME with PID = **0x04** and PMN = **0x02**, it may choose to associate with the same AVLN of the other STA or form a new AVLN as described in Section 10.3. In particular, this is desirable when one or both of the STAs are unassociated. Sufficient information is present from the messages (the SNID and TEI) that the STAs can communicate using unicast without association, but communication may be more reliable if they are associated with the same AVLN.

The HLEs that generate and process these messages are responsible for properly setting the PRN and incrementing the PMN. An STA that transmits one of these messages using MultiNetwork BroadCast (MNBC) may repeat that message unchanged multiple times for reliability purposes. An STA that receives duplicate messages (i.e., with the same PID, PRN, and PMN) may discard the duplicate copies silently. The HLE must be able to manage the reception of duplicate messages. It is up to the HLE to generate the response to any message received.

When the HLE sends the last message in the protocol, it will use PMN = **0xFF**, regardless of the success or failure of the protocol. If the protocol succeeds in providing an NMK to the new STA, the HLE for the STA receiving the NMK should use the **CM_SET_KEY.REQ** message to set the NMK and NID (including the SL) of its STA, as described in Section 10.4.2.3. When both STAs have the NMK, both should also have the same NID, and the STAs should form an AVLN as described in Section 10.3.4.1 or expand an existing AVLN as described in Section 10.3.5.1.

10.4.2.7 Changing the NMK If for some reason (e.g., a rogue STA has managed to get into the AVLN) the user suspects a hostile environment, the user may force a change to the NMK. Essentially, this redefines the AVLN and the NID may change along with the NMK. The user will change the NMK (and typically, the NID) for each STA that held the compromised NMK individually, using direct entry, the DAK, if it is available (Section 10.4.2.4), or UKE (Section 10.4.2.5). In the latter case, only an NMK-SC may be distributed.

Changing the NMK may change the NID, which is expected to be a disruptive event. That is, while the STAs in the AVLN are receiving the new NMK, leaving the old AVLN with the compromised NMK and NEK, and joining the new AVLN, they might not be able to sustain QoS levels for ongoing connections. In fact, they may have to terminate existing connections and establish new connections in the new AVLN. The degree to which the STA and the HLE shield applications and the user from these interruptions of services is both implementation- and vendor-dependent.

10.4.3 NEK Provisioning

The NEK is always provided by the CCo and always encrypted by the NMK. The NEK should never be set using any other encryption key or by an unencrypted MME. The NEK is never passed over the H1 interface (neither to the host from the STA nor from the host to the STA). NEK provisioning can be initiated in one of two ways, namely a new STA joins an AVLN and is given the NEK (with PID = **0x00**, Section 10.4.3.1) or the CCo determines to change the NEK for part or all of the AVLN ((with PID = **0x01**, Section 10.4.3.2).

10.4.3.1 Provisioning NEK for New STA NEK provisioning for a new STA is called Authentication and is described in Section 10.3.3.

10.4.3.2 Provisioning NEK for Part or All of the AVLN The CCo may update the NEK at any time, and is required to do so as specified in Section 10.4.1.5. It does so using the procedure shown in Figure 10.17. The PID should be set to **0x01** for all messages in this protocol. The CCo will provision each STA individually to get a positive ACK from each STA. After all active STAs have been provisioned with the new NEK and EKS, the CCo will cause the change to be effective by putting an

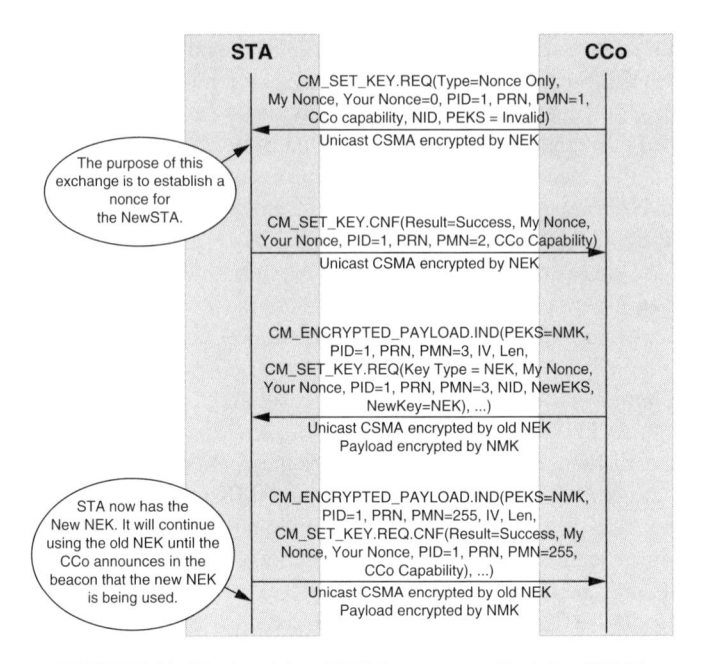

FIGURE 10.17 Provision NEK for part or all of the AVLN.

Encryption Key Change BENTRY in the Beacon for the appropriate number of Beacon Periods. If an STA that belongs to an AVLN detects an Encryption Change BENTRY in that AVLN's Beacon and the STA does not have the new NEK, it should request the new NEK from the CCo using the authentication procedure specified in Section 10.3.3. The STA should not use the new NEK until indicated by the CCo in the Encryption Key Change BENTRY in the Central or Proxy Beacon.

10.4.4 Encryption Key Uses and Protocol Failures

Encryption keys are used for specific purposes, and improper use of these keys is one of several forms of protocol failure.

The NEK is used exclusively for encrypting PBBs (i.e., at the PHY level); that is, only PBBs are encrypted with the NEK, and nothing else may be encrypted with the NEK. All data plane traffic and most control traffic will be between authenticated STAs in an AVLN, and those PBBs will be encrypted using the current NEK for that AVLN. A few messages may be sent unencrypted at the PHY level. For more information, see Table 11.5 in the HomePlug AV 1.1 Specification [24].

A failure to decrypt a PBB using the NEK cannot be detected by the PHY; it can only use the PBCS to check that the encrypted PBB was demodulated properly, then decrypt whatever it receives. Proper decryption of the PBB can only be checked by the ICV at the MAC Frame level. Incorrect MAC Frames should be discarded silently by the receiver, and it is up to the HLE to recover from this loss.

SECURITY OVERVIEW

All other keys (DAK, NMK, TEK) are used to encrypt the payload of **CM_ENCRYPTED_PAYLOAD.IND** MMEs, which are usually sent unencrypted at the PHY level. (Figure 10.17 is an anticipated exception.) How an STA responds to a **CM_ENCRYPTED_PAYLOAD.IND** MME depends on how it was sent as well as the ability of the STA to decrypt it and to validate it.

When the PID is **0x04**, only the unencrypted fields are checked, and the entire **CM_ENCRYPTED_PAYLOAD.IND** MME will be passed from the HLE to the PHY or from the PHY to the HLE without interpreting the "encrypted" (or "uninterpreted") portion (see Figures 10.18 and 10.19).

An STA that receives a **CM_ENCRYPTED_PAYLOAD.IND** MME should attempt to decrypt the payload using the key indicated in the PEKS, if it is valid and the PID is not **0x04**. Proper decryption is determined by using the last octet to ascertain the length of the Random Filler, then decoding the encapsulated MME for the MME length, and then checking the CRC as well as the Protocol Identifier (PID), PRN, and PMN (refer to Section 10.4.7). These last three should match their unencrypted values exactly. Lastly, if a TEK is used, it must not have expired (refer to Section 10.4.1.6). Failure to decrypt a **CM_ENCRYPTED_PAYLOAD.IND** MME properly will cause the recipient to respond with a **CM_ENCRYPTED_PAYLOAD.RSP**

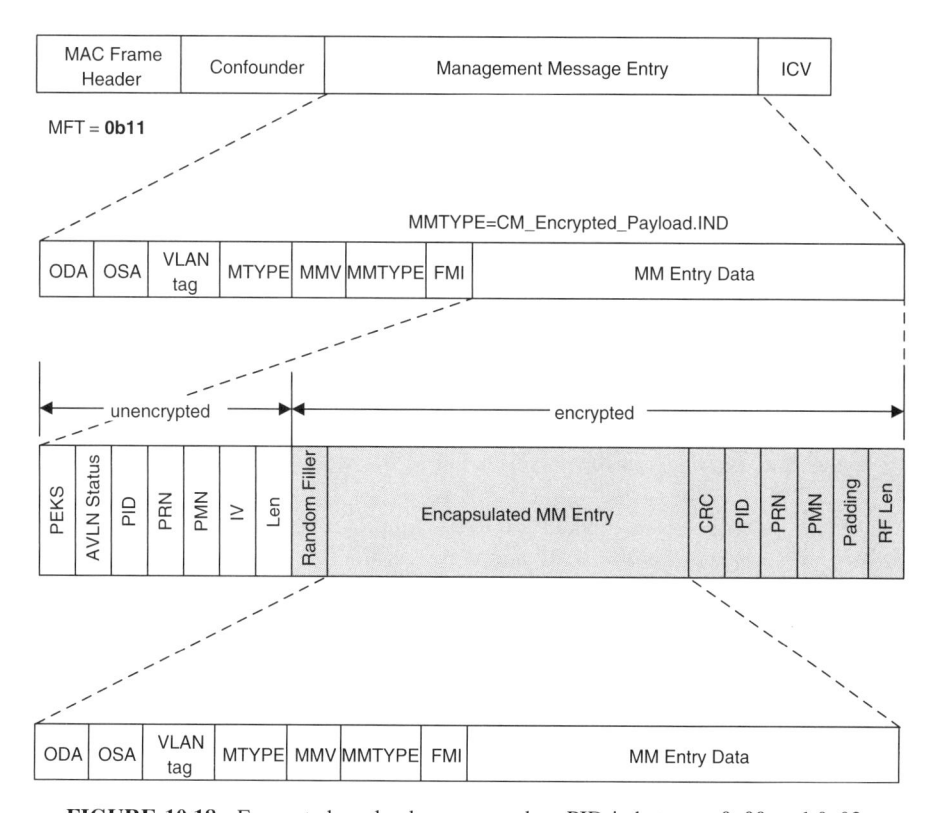

FIGURE 10.18 Encrypted payload message when PID is between **0x00** and **0x03**.

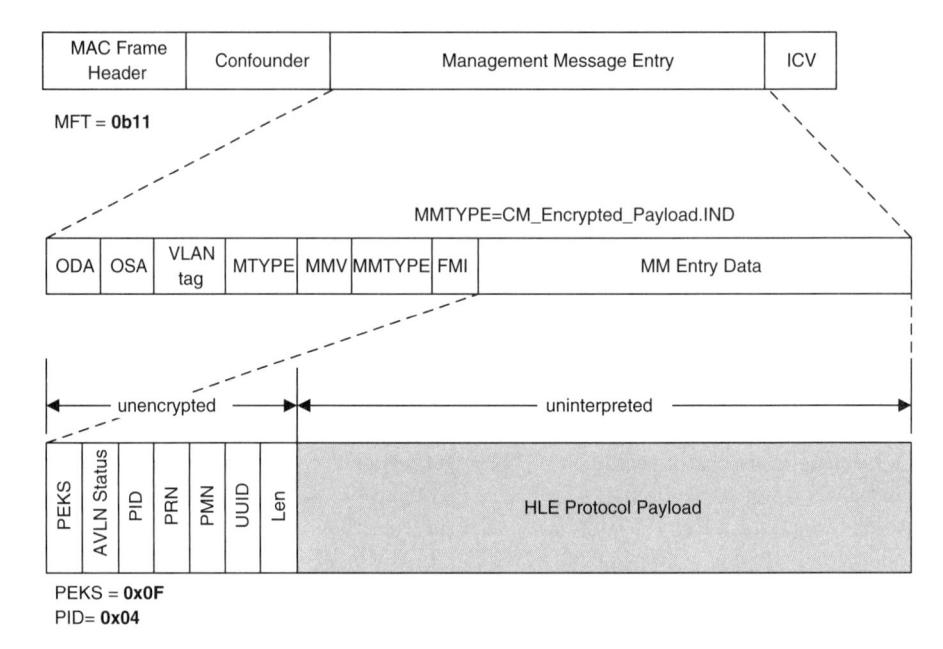

FIGURE 10.19 Encrypted payload message when PID = **0x04**.

MME indicating failure, unless the **CM_ENCRYPTED_PAYLOAD.IND** MME was sent with a broadcast address ODA. Note that a broadcast DTEI may be necessary even when the transmission is unicast in the ODA, so the DTEI cannot be used to determine whether a response is needed.

If the **CM_ENCRYPTED_PAYLOAD.IND** MME is decrypted correctly, other checks must be performed at the level of the protocol. At the minimum,

- the PID, PRN, and PMN must be appropriate,
- the key indicated in the PEKS and the encapsulated MME type must be appropriate for the PID and PMN, and
- the nonces, if any, in the encapsulated MME must be valid.

The PID must be between **0x00** and **0x04**. All protocol runs should start with the first message having PMN = **0x01** and end with the last message having PMN = **0xFF**. If the PMN is greater than **0x01**, the PRN must be already known and the Your Nonce field in the current message must match the My Nonce field in the previous message in that protocol run; otherwise, the protocol run is aborted. The PMN must be in the appropriate range for the PID, and must be the next PMN expected in that run of the protocol (the special PMN value of **0xFF** is the sole exception and is used to indicate the last message). Once an STA sends or receives a message with PMN = **0xFF**, it should ignore all further messages with that PID and PRN. It is recommended that at least 16 of the most recently used sets of {PID, PRN} pairs are stored. Except for PID = **0x04**, the key indicated by the PEKS must be of the correct type for that PID and PMN, and the MME must be of the correct type for that PID and PMN.

A **CM_ENCRYPTED_PAYLOAD.IND** MME with an incorrect PID, an unknown PRN with PMN greater than **0x01**, an out-of-sequence PMN, or an encapsulated MME of the wrong type for the PID and PMN should be ignored. An exception for duplicate MMEs is noted below. A correct MME type that has other defects may be ignored or, if it is a request, may cause a corresponding confirm to be sent indicating the failure. When a confirmed MME indicating the failure is sent, it should be sent as the body of an encrypted payload MME that does not encrypt its payload (indicated by PEKS = **0xF**).

The higher protocol must be able to withstand the loss of a message in the protocol sequence. This is done by setting a timer for a time by which a response is expected, and retransmitting the last message if no response has been received by that time. Since these protocol messages are exchanged using CSMA, and may also have to traverse several layers of processing by the recipient, the timer should be set to account for these latencies. When a duplicate of the previous received message in a protocol run is received, then a duplicate of the message sent in response to it (if any) must be sent.

10.4.5 AES Encryption Algorithm and Mode

AES Encryption may be performed either at the PHY Block level, as described in Section 5.4.7.1, and/or within a **CM_ENCRYPTED_PAYLOAD.IND** MME. Both modes are mandatory.

10.4.5.1 PHY Block-Level Encryption   HomePlug AV uses the 128-bit Advanced Encryption Standard (AES) algorithm in Cypher Block Chaining (CBC) Mode. The encryption method is described in Section 5.4.7.1.

The 4-bit Encryption Key Select (EKS) in the Frame Control is an index of the encryption key used for encrypting the PBBs. EKS values are defined in Section 4.4.1.5.2.8. The values instruct the STA about how to encrypt/decrypt the message, including use of no encryption at all.

10.4.5.2 Payload-Level Encryption   Payload-level encryption should be identical to PHY-level encryption with the following exceptions:

- Only a portion of the payload is encrypted.
- The PEKS (called the PEKS to distinguish it from the EKS) identifying the Encryption Key and, if AES encryption is used, the Initialization Vector, are carried in the unencrypted portion of the payload.
- The bit-ordering is different.

The payload should be padded to a 128-bit boundary as needed. The padding should be random bits.

10.4.6 Generation of AES Encryption Keys

10.4.6.1 Generation from Passwords The HomePlug AV privacy function may at times require the generation of an encryption key from a password. In all such cases, the mechanism for creating a key from a password should be the PBKDF1 function, as shown in the PKCS #5 v2.0 Password-based Cryptography Standard [25], using SHA-256 as the underlying hash algorithm. The iteration count used to calculate the key should be 1000. The salt value should be 0x0885 6DAF 7CF5 8185 for DAKs and 0x0885 6DAF 7CF5 8186 for NMK-HSs. After the 1000th iteration, the leftmost 16 octets of the SHA-256 output (as described in FIPS-180-2 change notice [21]) will be used as the AES encryption key. The first octet of the output corresponds to octet 0 of the AES encryption key. The bit ordering of the AES encryption key within an octet is dependent on where it is used.

HomePlug AV passwords (DPW and NPW) should be limited to strings of ASCII characters chosen from the range ASCII [32] to ASCII [127]. The length of a DPW should be between 16 and 64 characters inclusive. The length of a NPW should be between 8 and 64 characters inclusive. Users will be provided a warning by the user interface that the entered NPW may not be secure when the user enters a NPW less than 24 characters in length. Passwords are not sent to the Convergence Layer and it is recommended that they need not be retained. DPWs should be pseudo-randomly generated and recommended to be a minimum of 20 characters long. It is recommended that pseudo-random NPWs be a minimum of 16 characters long and user-selected NPWs a minimum of 24 characters long.

10.4.6.2 Automatic Generation of AES Keys All STAs should be able to generate Hash Keys (used in UKE) and AES encryption keys (e.g., NEK, NMK, and TEK) directly. The generation of these is implementation-dependent. Implementers should ensure that key and secret generation are based on a good random number generation algorithm. The method used should be at least as good as the following baseline method.

In the baseline method, a station uses a keyed cryptographic hash algorithm (such as the one used to generate keys from passwords) with a key distinct to the station and not the same as its DAK. It should maintain an internal counter that is never exposed. This counter value will be hashed to obtain a pseudorandom number, and the counter is then incremented each time a pseudorandom number is generated. RFC 4086 [26] and the book on Applied Cryptography by Schneier [27] may be used for guidance.

10.4.6.3 Generation of Nonces Generation of nonces should be done in a way so that the repetition of a nonce is very unlikely and that the nonces are not predictable. A method at least as good as the one described in Section 10.4.6.2 for automatic generation of AES keys should be used. A possible way to generate a nonce is to use a counter that is passed through a keyed secure hash function. The counter is incremented each time a nonce is generated and the key for the hash is kept private to the station. The same key and counter may be used for generating nonces and random numbers for different purposes, by using additional information specific to the purpose that is also passed through the hash function.

10.4.7 Encrypted Payload Message

The **CM_ENCRYPTED_PAYLOAD.IND** management message may be exchanged over the medium unencrypted. As such, they are most useful for the processes that establish or distribute keys. They may also be used by higher layer protocols to perform discovery or key distribution (with PID = **0x04,** refer to Section 10.4.2.6).

Within HomePlug AV, these are used for five distinct purposes:

- Distribution of the NEK using the NMK based on a request from an STA seeking to authenticate (PID = **0x00,** refer to Section 10.4.3).
- Distribution of the NEK using the NMK initiated by the CCo when rotating the NEK (PID = **0x01,** refer to Section 10.4.3).
- Distribution of the NMK using the DAK (PID = **0x02,** refer to Section 10.4.2.4).
- Distribution of the NMK using UKE (PID = **0x03,** refer to Section 10.4.2.5).
- Execution of higher-layer protocols (PID = **0x04,** refer to Section 10.4.2.6).

Each of these is given its own PID for use in the PID fields (one unencrypted, one encrypted, except when PID = **0x04**) of the **CM_ENCRYPTED_PAYLOAD.IND** message. The PID value of **0x04** is set aside for higher-layer protocols, so that the MAC knows to pass these messages across the H1 interface uninterpreted.

Figure 10.18 shows an entire Encrypted Payload Message when PID is between **0x00** and **0x03**, including all of its components contributed by MAC Framing and other processes.

The Payload Encryption Key Select (PEKS) indicates the AES encryption key used to encrypt the payload. When used with a higher-layer protocol or when a confirmed MME indicating the failure is sent, a PEKS of No Key (**0x0F**), may be used to indicate that the payload is not encrypted.

When PID = **0x04**, the 16-octet IV field will be used for a UUID (see references [22,23]), and the portion of the **CM_ENCRYPTED_PAYLOAD.IND** MME from Random Filler to RF Length will not be interpreted by the STA (see Figure 10.19). This entire portion will instead be HLE protocol payload (i.e., the eight fields defined for the encrypted portion are instead defined by the higher-layer protocol).

The limitation of MNBC transmissions to a single PB constrains the length of the **CM_ENCRYPTED_PAYLOAD.IND** MME to be no more than 502 octets, and the HLE Payload to be no more than 460 octets, unless the **CM_ENCRYPTED_ PAYLOAD.IND** MME is fragmented. Fragmentation at the MAC level is discouraged, however, due to the unreliability of MNBC, and if necessary, should be done by the HLE.

10.4.8 User Interface Station (UIS)

An UIS is any station that is capable of providing a suitable mechanism (preferably keyboard and display) to enable user interaction with the AVLN and capable of supporting network control and security functions for the AVLN. It might not be physically attached to the AVLN (i.e., it might connect through a bridging STA).

There is no limit to the number of UISs that may exist in the AVLN.

10.5 SUMMARY

This chapter examined in detail the security protocols used in HomePlug AV, given context-specific definitions and explanations of Encryption Keys, Passphrases, and Nonces. Details of NMK and NEK provisioning and recommended standards for generating keys and nonces were also given.

The next chapter describes additional useful and unique features of the HomePlug AV MAC, including facilities for channel estimation, bridging, features that guarantee coexistence with HomePlug 1.0.1 and the operations of Proxy Networks in handling hidden stations.

11

ADDITIONAL MAC FEATURES

11.1 INTRODUCTION

The HomePlug AV MAC supports the key features of (i) channel estimation, (ii) bridging to facilitate communication between the Powerline Communication (PLC) and other networks, (iii) coexistence with HomePlug 1.0.1 stations to facilitate harmony in the HomePlug Ecosystem, and (iv) Proxy Networking to handle hidden station problems. The mechanisms implemented in HomePlug AV to enable these features are described in the following sections of this chapter.

11.2 CHANNEL ESTIMATION

Channel estimation is the process of measuring the characteristics of the powerline channel to adapt the operation of the physical layer (PHY) to provide optimal performance. Channel estimation comprises:

- Selection of the modulation method(s) used on each carrier. Any given carrier may use different modulations at different times within the AC line cycle period.
- Selection of the forward error correction (FEC) rate.
- Selection of the guard interval length.
- Selection of the intervals within the AC line cycle where a particular Tone Map setting applies.

HomePlug AV and IEEE 1901: A Handbook for PLC Designers and Users, First Edition.
Haniph A. Latchman, Srinivas Katar, Larry Yonge, and Sherman Gavette.
© 2013 by The Institute of Electrical and Electronics Engineers, Inc. Published 2013 by John Wiley & Sons, Inc.

The FEC rate and guard interval length can vary over the AC line cycle period, but they are the same for all carriers at any given time.

A Tone Map consists of a unique set of modulation methods for each carrier, the guard interval for the Orthogonal Frequency Division Multiplexing (OFDM) Symbol, and the FEC Rate. In short, channel estimation is the process by which Tone Maps are selected for different intervals within the Beacon Period.

11.2.1 Channel Estimation Procedure

The channel estimation procedure enables a transmitter to obtain Tone Maps that can be used at various intervals of the AC line cycle while communicating with a particular receiver. Powerline channels are unique between each transmitter and receiver. Hence, the channel estimation procedure is executed independently between each transmitter and receiver. Tone Maps are exchanged between stations by means of **CM_CHAN_EST.IND** and/or **CM_TM_UPDATE.IND** Management messages. These messages are also referred to as Channel Estimate Indication (CEI) messages within the specification.

The channel estimation procedure can be divided into two phases depending on whether the transmitter has any valid Tone Maps:

- initial channel estimation; and
- dynamic channel adaptation.

The transmitter invokes initial channel estimation when it needs to transmit data to a particular destination and does not have any valid Tone Maps. During initial channel estimation, the transmitter sends one or more Sound MAC Protocol Data Units (MPDUs) to the receiver. The Sound Reason Code in these Sound MPDUs is used to indicate that the Sound MPDUs are transmitted as part of an initial channel estimation procedure. The receiver uses these Sound MPDUs to estimate the channel characteristics and designate a Default Tone Map that may be used by the transmitting station (STA) anywhere in the AC line cycle. In addition, the receiver may also provide one or more AC line cycle adapted Tone Maps during Initial channel estimation. This approach allows the STA to start communicating using Tone Map modulated data quickly, and avoids the complicated interactions between the channel access procedures and the channel estimation procedures.

Once the initial channel estimation is complete, the receiver continuously monitors the channel characteristics based on received MPDUs (either data or Sound) and provides dynamic updates to the Default Tone Map and/or to the Tone Maps that are valid at specific intervals of the AC line cycle. This process is referred to as dynamic channel adaptation. In contrast to initial channel estimation, the receiver is responsible for invoking dynamic channel adaptation. The transmitter provides passive support by behaving as indicated in the Channel Estimate Indication (CEI) messages and indicating error events using Sound MPDUs.

The channel estimation procedures also include mechanisms for negotiating the number of Tone Maps that can be used, maintaining lists of valid Tone Maps, and

CHANNEL ESTIMATION

maintaining the lists of the intervals within the AC line cycle where each Tone Map may be used. The procedures are described in the following subsections.

11.2.2 Initial Channel Estimation

The transmitter initiates the channel estimation procedure if it does not have any valid Tone Maps. Initial channel estimation may take place in either the CP or the Contention Free Period (CFP). If the initial channel estimation is performed in the CP, the transmitter contends for the channel prior to sending Sound MPDUs to the receiving STA. Conducting the initial channel estimation in the CP may preclude the transmitter from transmitting Sound MPDUs during certain parts of the AC line cycle. Similarly, if the initial channel estimation is performed in the CFP, the transmitter may lack sufficient CF allocation to span a complete AC line cycle. In either case, the receiving STA is required to provide a Tone Map referred to as the Default Tone Map that is valid for all portions of the Beacon Period (or AC line cycle). The receiver may also provide one or more AC line cycle adapted Tone Map during initial channel estimation.

During the initial channel estimation procedure, the transmitter uses the Max Tone Maps Requested (REQ_TM) field in the Sound Frame Control to indicate the maximum number of Tone Maps that it can support. REQ_TM is a value in the range 0 to **MAX_TONE_MAPS** (a value of 7). Once a REQ_TM value is advertised to the receiver, the transmitter will not change the value until it has reinitiated a new initial channel estimation procedure. The receiver stores the number of Tone Maps supported by the transmitter until all the Tone Maps for the transmitter have become invalid (i.e., have been explicitly invalidated by the receiver or have become stale). The receiver ensures that the number of Tone Maps it provides to the transmitter never exceeds this limit. ROBO Tone Maps are not included in this limit. The Default Tone Map is included in this limit if it is a non-ROBO Tone Map.

Initial channel estimation comprises the following steps (see Figure 11.1):

(1) The transmitting station has data to send and determines that it has no valid Tone Maps for communication with the destination STA.
(2) The transmitting station initiates the channel estimation procedure by sending a Sound MPDU with Sound Reason Code set to indicate Sounding for Initial Channel Estimation. This MPDU specifies the maximum number of Tone Maps that the transmitting STA can allocate to this Link (REQ_TM, refer to Section 5.6.1.10).
(3) The receiving STA responds with a Sound MPDU, with the Sound ACK Flag (SAF) bit in the FC set to **0b1** and no payload.
(4) The transmitting STA continues sending Sound MPDU (SAF = 0) until the receiving STA responds with SAF = **0b1** and SCF (Sound Complete Flag) = **0 b1**, indicating that it has received sufficient data to generate the Tone Maps.
(5) The receiving STA generates a new Tone Map and assigns a new TMI_AV to it, and sends them to the transmitting STA in the **CM_CHAN_EST.IND**

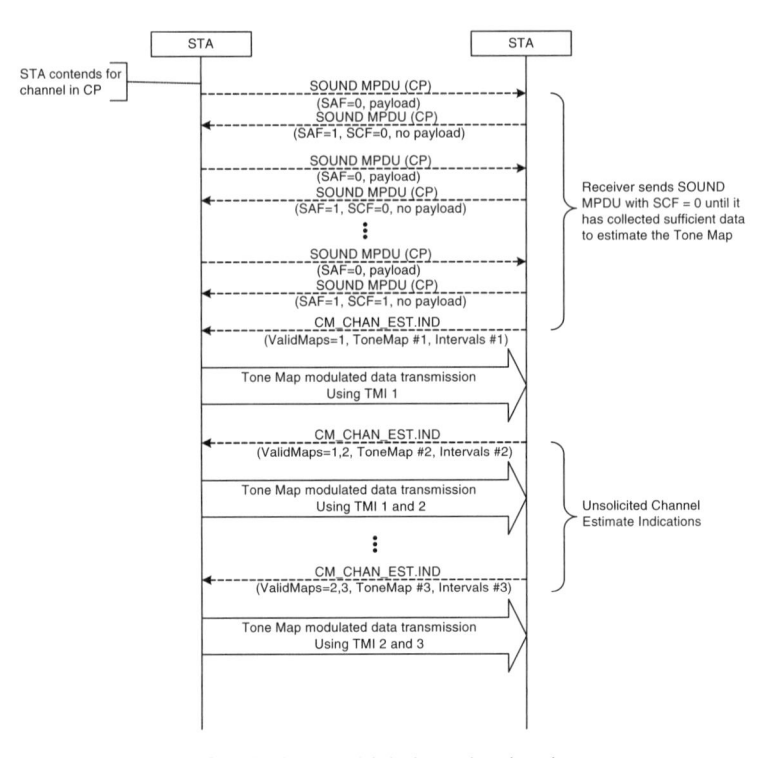

FIGURE 11.1 Initial channel estimation.

message. The newly assigned TMI_AV is the only entry in the valid TMI_AV list, and it indicates that the new Tone Map is the Default Tone Map. The Response Type field for this CEI message indicates that the message contains a Default Tone Map that is generated as a result of the initial channel estimation procedure.

(6) The receiving STA may also generate one or more AC line cycle adapted Tone Maps after generating the Default Tone Map.

(7) The transmitting STA should begin to use the newly assigned Tone Maps immediately after reception.

An implementation may use ROBO-AV modes for transmitting data or management messages before completing the initial channel estimation. It is recommended that the use of ROBO-AV modes be minimized as they are very inefficient.

11.2.3 Dynamic Channel Adaptation

Dynamic channel adaptation is performed by the receiver subsequent to the initial channel estimation. This process may result in dynamic updates to the Default Tone Map (i.e., replacing an existing Default Tone Map with a new Default Tone Map).

CHANNEL ESTIMATION

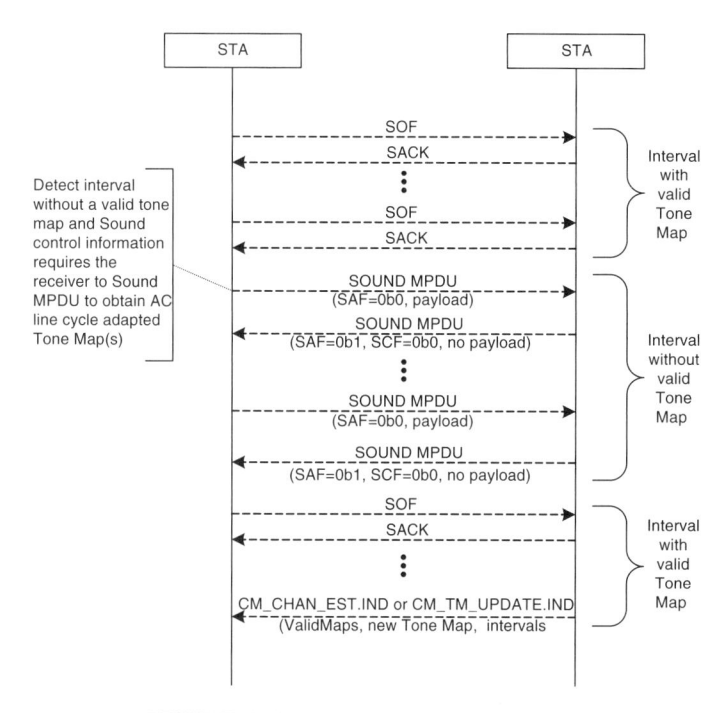

FIGURE 11.2 Dynamic channel adaptation.

This process may also result in the generation of AC line cycle adapted Tone Maps that are valid at various intervals of the AC line cycle, some of which may replace existing Tone Maps (Figure 11.2).

Channel adaptation during CP gets complicated due to the possibility of collisions. For this reason, the receiver may require the transmitter to not use a subset of the AC line cycle adapted Tone Maps in the CP. This information is conveyed independently for each Tone Map using the CPF field in the CEI messages. In contrast to the Default Tone Map, AC line cycle adapted Tone Maps are fine-tuned for channel characteristics within that specific interval of the AC line cycle. Hence, the transmitter should use the AC line cycle adapted Tone Map whenever one is available and can be used (based on CPF field in CEI).

The receiver uses the Sound Control field in the CEI messages to control the behavior of the transmitter during CP (SCL_CP) and CFP (SCL_CFP) when it encounters a region in the AC line cycle where there is no AC line cycle adapted Tone Map. When the Sound Control Field requires transmission of Sound MPDUs during intervals of the AC line cycle and there is no corresponding AC line cycle adapted Tone Maps, the transmitter should send Sound MPDUs.

The transmitter may choose to use the Default Tone Map or ROBO Tone Maps even when an AC line cycle adapted Tone Map is available and can be used during a specific interval. The transmitter may also choose to use the Default Tone Map or the ROBO Tone Map during intervals of the AC line cycle where the receiver requested

Sound MPDUs. In such cases, the receiver might not be able to provide a new AC line cycle adapted Tone Map for that interval. The transmitter should minimize such deviations from recommended behavior.

The Sound Reason Code (SRC) in the Sound MPDU is used by the transmitter to indicate the reason for transmitting Sound MPDUs. The transmitter sends Sound MPDUs for five different reasons, corresponding to different Sound Reason Code (SRC) values:

(1) Sound MPDU transmitted to obtain the Tone Map corresponding to a TMI_AV, which has been specified by the receiver in the intervals information, but is not recognized by the transmitter. Such conditions can occur when the CEI message carrying the Tone Map did not get delivered.
(2) Sound MPDU transmitted to indicate a Tone Map error condition detected at the transmitter. In this case, receiver resend all valid Tone Maps. The transmitter may use this Sound Reason Code when it determines that multiple Tone Map Indices identified in CEI are not available. Such conditions can occur when the CEI message(s) carrying the Tone Map(s) did not get delivered.
(3) Sound MPDU transmitted to obtain initial channel estimation.
(4) Sound MPDU transmitted in an interval where receiver has indicated that no AC line cycle adapted Tone Maps are available.
(5) Sound MPDU transmitted in an interval specified as Unusable by the receiver.

The Sound Complete Flag (SCF) in Sound ACK is used by the receiver to indicate that the receiver has received sufficient Sound MPDUs for completing the channel estimation. The interpretation of SCF should be based on the Sound Reason Code (SRC) contained in the corresponding Long Sound MPDU.

- For SRC Case (1), SCF indicates the completion of channel estimation in the interval where the corresponding TMI_AV is used.
- For SRC Case (2), SCF indicates the completion of channel estimation in all intervals of the AC line cycle.
- For SRC Case (3), SCF indicates the completion of initial channel estimation.
- For SRC Cases (4) and (5), SCF indicates the completion of channel estimation in the interval where the corresponding Long Sound MPDUs are transmitted.

The Dynamic Channel Adaptation procedure is described below:

(1) The transmitting STA receives a Default Tone Map and zero or more AC line cycle adapted Tone Maps following the initial channel estimation. The CEI message includes the Sounding Control information (SCL_CP and SCL_CFP) that indicate when Sound MPDUs need to be transmitted to obtain an AC line cycle adapted Tone Map(s).

(2) The receiver continuously monitors the channel characteristics based on Tone Map modulated MPDUs and provides updates to existing Tone Maps and/or invalidates an existing Tone Map.

(3) The transmitting STA detects a transmit opportunity (in either CP or CFP) in an interval that does not have a valid AC line cycle adapted Tone Map. It uses the Sound Control information received in prior CEI messages to determine whether it needs to send a Sound MPDU or data modulated MPDU using the Default Tone Map.

(4) If the transmitter sends a Sound MPDU, the receiving STA responds to the Sound MPDU with a Sound MPDU containing no payload and SAF set to **0b1**.

(5) The transmitting STA operates normally in intervals where valid Tone Maps are defined; that is, it sends one or more SOF MPDUs carrying user data and the receiving STA responds with a SACK MPDU.

(6) When the receiving STA has detected a sufficient number of Sound MPDUs, it responds with SCF set to **0b1** (i.e., sounding for the corresponding interval is complete) and generates a new Tone Map for the interval.

(7) The receiving STA sends the **CM_CHAN_EST.IND** (or **CM_TM_UPDATE.IND**) message to the transmitting STA. It provides a new Tone Map and indicates that this new Tone Map should be used in the interval in question.

(8) The transmitting STA should begin to use the newly assigned Tone Maps immediately after reception.

The receiving STA is not required to generate a new Tone Map. It may instead simply extend an existing Tone Map to cover the interval in question. In this case, it sends the **CM_CHAN_EST.IND** (or **CM_TM_UPDATE.IND**) message with an updated INTERVALS field, but without a new Tone Map.

The receiver ensures that a valid Default Tone Map is always available at the transmitter during the dynamic channel adaptation process. The receiver may choose the Default Tone Map to be the same as one of the AC line cycle adapted Tone Maps. The receiver may also choose either STD-ROBO_AV or HS-ROBO_AV as a Default Tone Map.

11.2.4 Maintenance of Tone Maps

The receiver is responsible for ensuring that one or more valid Tone Maps are available in intervals of the AC line cycle where a transmitter may potentially transmit. Subsequent to generating a Tone Map, the receiver continues to monitor the MPDUs modulated with that Tone Map to determine whether the Tone Map is still valid. If the Tone Map is deemed to be invalid (for example, due to high PB errors), the receiver can provide a new Tone Map to replace the existing Tone Map or it may force the transmitter to invalidate the existing Tone Map.

There are two mechanisms that a receiver can use to invalidate a Tone Map:

(1) The receiver can transmit a CEI message with a valid Tone Map list that does not include the invalidated Tone Map.
(2) The receiver can indicate that a Tone Map is invalid in the SACKI field (refer to Section 5.4.1.7). This causes the transmitter to invalidate the Tone Map without the need to send any CEI message.

It is recommended that the receiver use a combination of both mechanisms for invalidating Tone Maps (i.e., the receiver sends a CEI message as soon as the Tone Map is invalidated and subsequently use SACKI to indicate that the TMI_AV is invalidated if the transmitter continues to use the invalidated Tone Map).

The transmitter associates all valid Tone Maps with a stale timer set to 30 s. If 30 s elapse since the last CEI message was received from the receiving station, all valid Tone Maps are considered stale and has to be discarded. If a CEI message is received, the transmitter discard all Tone Maps that are not in the valid Tone Map list of the CEI message and restart the 30-s stale timer for valid Tone Maps.

The transmitter should not expire a Tone Map unless the Tone Map becomes stale or unless it is explicitly requested to do so by the receiver using a CEI message.

11.2.5 Tone Map Intervals

Tone Map Intervals are defined as time periods within the Beacon Period where a particular Tone Map may be used [28]. Since the Central Coordinator (CCo) locks the Beacon Period to the AC line cycle, intervals are synchronized to the AC line cycle.

Channel and noise characteristics over the powerline tend to be periodic with the underlying AC line cycle. In some cases, these impairments occur at twice the frequency of the AC line cycle (i.e., 100 or 120 Hz), while in other cases they may occur at the same frequency as the AC line cycle. The Tone Map intervals are defined over the entire Beacon Period and are therefore periodic with a frequency of half the line cycle. The receiver explicitly specifies the selected Tone Map on all intervals within the Beacon Period. It is anticipated that the receiver will typically define the Tone Maps to be used over the AC line cycle and then repeat this definition to fill out the entire Beacon Period.

The receiving STA specifies the intervals within which various Tone Maps may be used, subject to the following rules:

- The Default Tone Map may be used anywhere in the Beacon Period (refer to Section 11.2.7).
- With the exception of the Default Tone Map, intervals are disjoint (nonoverlapping).
- The transmitter is not allowed to transmit PHY Protocol Data Units (PPDUs) with the PPDU payload crossing the boundary between intervals using different Tone Maps, except when the PPDU payload starts or ends within 150 μs of the boundary. The following exceptions to this rule are allowed:

- PPDUs using Standard ROBO, Mini ROBO, High Speed ROBO, and Default Tone Maps are not restricted to Tone Map intervals.
- Reverse SOF MPDU are not required to strictly obey the Tone Map interval boundaries.
- PPDU transmissions that use Tone Maps having a PHY data rate of less than 15 Mbps at the input to the FEC encoder are not required to strictly obey the Tone Map interval boundaries.
- The receiver has to specify intervals that are large enough to carry a complete PPDU, including at least one PB, based on the indicated Tone Map.
- The current intervals definition is always carried in the CEI message.

11.2.6 Priority of Channel Estimation Response

The nominal channel access priority for a CEI message is set to CA2. Optionally, if the request for channel estimation was sent at CA3 (i.e., Sound MPDU that is transmitted as part of a Global Link or a Link with channel access priority equal to CA3), the CEI message should be sent at CA3. To minimize delays experienced by the CEI message, it is recommended that the message not be multiplexed with segments from other MAC frame streams having a lower priority.

11.2.7 Channel Estimation with Respect to the AC Line Cycle

Channel and noise characteristics over powerlines tend to be periodic with respect to the underlying AC line cycle. The HomePlug AV channel estimation mechanism should take this into consideration while generating the Tone Maps.

HomePlug AV channel estimation is performed with respect to the receiver. The receiver determines the number of Tone Maps (within the limits specified by the transmitter) and the intervals within the Beacon Period where they are valid. The interval information is sent to the transmitter along with the individual Tone Maps.

The transmitter then selects the Tone Map for a given PPDU based on the interval within the Beacon Period where the PPDU is to be transmitted. Since the Beacon Period is synchronous to the AC line cycle, the selection of Tone Maps by the transmitter is also synchronous to the AC line cycle.

A HomePlug AV station is required to support a minimum of two transmit and two receive Tone Maps for each station it is actively communicating with, up to two stations. Thus, a HomePlug AV station should be able to support a minimum of four transmit and four receive Tone Maps. In practice, HomePlug AV implementations support significantly larger number of tone maps).

11.3 BRIDGING

A device is acting as a Bridge if it is bridging traffic to a network that cannot natively receive said traffic. This is often the case when working with two different physical

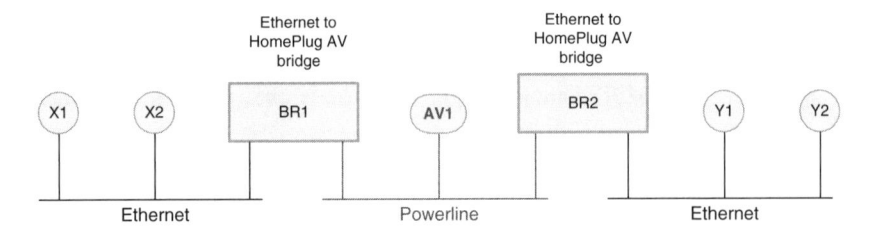

FIGURE 11.3 HomePlug AV bridging to Ethernet Networks.

networks, such as a powerline network and an Ethernet network. Figure 11.3 shows an example of a HomePlug AV network bridging two separate Ethernet networks. In this example, Bridges BR1 and BR2 enable bridging of traffic from {X1, X2} and {Y1, Y2} to the HomePlug AV network, respectively. It should be note that most HomePlug AV products available in the market are individual adapters with an Ethernet interface (i.e., Ethernet bridges).

Bridging in the HomePlug AV is accomplished through source-aware bridging, that is, each AV station maintains knowledge of the powerline destination to which an Ethernet packet needs to be sent to reach the bridged destination. For example, AV1 maintains information that enables it to transmit Ethernet packets address to Y1 as unicast transmissions to BR2. Source-aware bridging is necessitated in HomePlug AV as broadcast transmissions are very inefficient and unreliable over powerline channels. For example, if AV1 does not maintain knowledge that Y1 can be reached through BR2, any Ethernet packet that has to be set to Y1 needs be transmitted as broadcast using ROBO modulations that provide less than 10 Mbps physical layer rate. This results in significantly lower efficiency and will also incur lower reliability due to partial acknowledgments.

Source aware bridging enables AV stations to communicate directly to the AV bridge, thus enabling the use of channel adapted unicast tone maps and selective acknowledgment optimize the MAC efficiency and reliability.

There are two components to AV bridging:

- acting as a bridge;
- communicating through a bridge.

11.3.1 Acting as an AV Bridge

AV Bridges learn the source MAC address of all devices connected on the bridged network. The exact mechanism used by an AV bridge to identify devices in the bridged network is implementation dependent. One approach that can be used is by sniffing the source address of each Ethernet packet received from the bridged network. For example, BR1 can learn that X1 and X2 are bridged by sniffing at the source address in Ethernet packets received from the bridged network.

An AV Bridge maintains a Local Bridge Destination Address Table (LBDAT) that lists all source addresses of the devices that it is aware of on the network for which it

BRIDGING

```
LBDAT = {X1, X2}           RBAT For BR1 = {X1, X2}    LBDAT = {Y1, Y2}
RBAT For BR2 = {Y1, Y2}    RBAT For BR2 = {Y1, Y2}    RBAT For BR1 = {X1, X2}
```

(X1) (X2) — BR1 — AV1 — BR2 — (Y1) (Y2)

Ethernet — Powerline — Ethernet

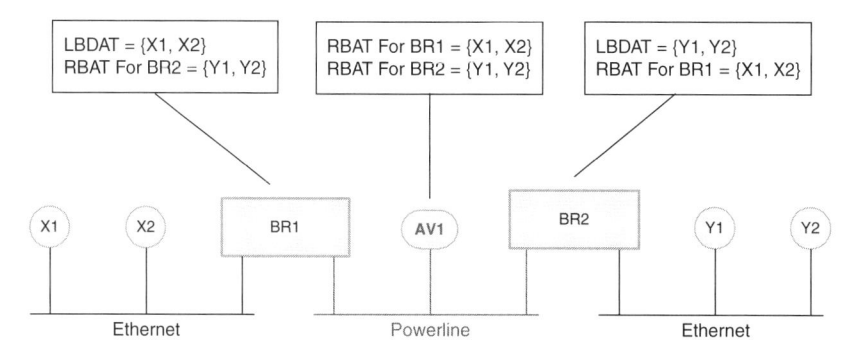

FIGURE 11.4 LBDAT and RBAT of HomePlug AV Stations.

is bridging. The LBDAT is communicated to all other AV stations via the **CM_BRG_INFO.CNF** message.

The **CM_BRG_INFO.CNF** message is regularly communicated to all of the AV devices for which it has an active Tone Map with a period of no less than 100 s and no greater than every 100 ms.

STAs in the HomePlug AV Local Network (AVLN) may send a **CM_BRG_INFO.REQ** message to obtain the LBDAT of any other STA in the AVLN. All STAs in the AVLN properly respond to a **CM_BRG_INFO.REQ** with a **CM_BRG_INFO.CNF**. STAs that are not capable of acting as a bridge respond to a **CM_BRG_INFO.REQ** with a **CM_BRG_INFO.CNF** that has the Bridging Station Flag (BSF) set to indicate that the STA does not perform bridging functions.

The AV Bridges age the Local Bridge Destination Address Table (LBDAT) such that devices on the bridged network that have not transmitted a packet for at least 100 s are removed from the LBDAT.

Figure 11.4 shows the LBDAT for BR1 and BR2 in the network illustrated in Figure 11.3.

11.3.2 Communicating through an AV Bridge

Communication through a bridge refers to the mechanisms used to communicate with a device whose DA is not known to exist on the AV network or any other known network. When this occurs, the traffic must be transmitted in such a way as to ensure that the intended devices will receive it, as long as the device's network is reachable.

AV Bridges will communicate their LBDATs to at least all stations for which the bridge has a current Tone Map. The **CM_BRG_INFO.CNF** message from an AV Bridge will contain the MAC addresses listed in the LBDAT for that bridge. The AV Station creates a Remote Bridged Address Table (RBAT) containing the MAC addresses in the **CM_BRG_INFO.CNF** and associates it with the Terminal Equipment Identifier (TEI) for the AV Bridge that sent it. The AV Station maintains a separate RBAT for each AV Bridge that the Station is communicating with. Figure 11.4 shows the RBAT for various stations in the network illustrated in Figure 11.3.

AV Stations independently age each RBAT so that if the station has not received a **CM_BRG_INFO.CNF** from the associated AV Bridge within at most 100 s, it deletes the RBAT for that bridge.

When an AV Station intends to transmit a packet containing a unicast ODA, it must determine whether that ODA is Known or Unknown. An ODA is known under the following three circumstances: (i) if it corresponds to the MAC address of an AV station which is associated with a TEI, (ii) if it is listed in the RBAT for an AV Bridge and, hence, is a bridged destination on a remote network, or (iii) if it is the multicast address of a known multicast group. Similarly an ODA is unknown if either it is (i) unicast and not a known AV Station or a known Bridged Destination, (ii) the broadcast address, or (iii) an unknown multicast address.

Communication with a destination differs, depending on whether its ODA is known or unknown.

11.3.2.1 Communication with a Known DA To determine whether a DA is known, the AV station scans through its list of AV stations and its RBAT. If the AV station maintains a separate RBAT for each bridge, it should ensure that the scan order for the RBATs is performed in chronological order, starting with the most recently received RBAT and continuing to the oldest. This recommendation is important to properly track mobile STAs.

11.3.2.1.1 Known AV Station When the ODA corresponds to another AV station, the sending station will know the destination station's TEI. When sending to a known AV Station, a sending station sets the DTEI to the TEI of the destination station. The Source Terminal Equipment Identifier (STEI) is set to the TEI of the sending station.

11.3.2.1.2 Known Bridged Destination When the Original Destination Address (ODA) is listed in the RBAT for a known AV Bridge the sending station will know that the destination is located on a bridged network and is a Bridged Destination. When sending to a known Bridged Destination, a sending station sets the DTEI to the TEI of the AV Bridge. The STEI is set to the TEI of the sending station.

11.3.2.1.3 Known Multicast Address When the ODA is multicast and the sending station is aware of the unicast destinations (Known destinations) participating in the multicast group, the sending station may send the traffic unicast to the known destinations.

When sending multicast traffic to the known multicast participants, the sending station follows the rules above for sending traffic to Known AV Stations and/or Known Bridged Destinations as is appropriate.

11.3.2.2 Communicating with an Unknown DA

11.3.2.2.1 Unknown Unicast Destination When a destination address is unicast and not a known AV Station or a known Bridged Destination, the sending station does not know whether the destination is on an AV network or a bridged network. Because of this, the traffic has to be sent in broadcast mode to ensure that it reaches its intended destination and for it to become known.

When sending to an unknown destination, the sending station sets the DTEI either to the TEI of the multicast proxy (refer to Section 6.13.2) or to broadcast TEI. The STEI is set to the TEI of the sending station.

11.3.2.2.2 Broadcast Address When a destination address is the broadcast address the sending AV station broadcasts the traffic. When sending to the broadcast MAC address, the sending station sets the Destination Terminal Equipment Identifier (DTEI) either to the TEI of the multicast proxy or to the broadcast TEI. The STEI is set to the TEI of the sending station.

11.3.2.2.3 Unknown Multicast Address An unknown multicast address is one in which the sending station is unaware of the unicast destinations participating in the multicast group. In this case, the sending station broadcasts the traffic.

When sending to an unknown multicast address, the sending station sets the DTEI either to the TEI of the multicast proxy or to the broadcast TEI. The STEI is set to the TEI of the sending station.

11.4   HOMEPLUG 1.0.1 COEXISTENCE

HomePlug AV STAs are likely to operate in environments containing legacy HomePlug 1.0.1 STAs. HomePlug 1.0.1 STAs transmit in the same frequency band as HomePlug AV STAs, resulting in interference [29]. Hence, a means for coexistence is essential. It should be noted that HomePlug 1.0.1 STAs always operate in CSMA/CA mode and do not understand the Beacon based medium access (with CPs and CFPs) used by HomePlug AV. Hence all HomePlug AV stations as well as HomePlug AV based stations (HomePlug AV2, HomePlug Green PHY, and IEEE 1901), implement the HomePlug 1.0.1 coexistence modes described below for the case of HomePlug AV, to enable coexistence with HomePlug 1.0.1.

11.4.1   HomePlug AV Coexistence Modes

HomePlug 1.0.1 coexistence is an operating mode for HomePlug AV STAs. The CCo of each AVLN determines the mode of operation of STAs associated with it. There are three modes in which an AVLN can operate:

(1) *Fully Hybrid Mode:* The AVLN operates in Fully Hybrid Mode if interfering HomePlug 1.0.1 stations are detected. In this mode, all CP and CFP transmissions use Hybrid Mode delimiters (refer to Section 4.2.2) to coordinate medium access with HomePlug 1.0.1 stations.

(2) *Shared CSMA Hybrid Mode:* The AVLN operates in Shared CSMA Hybrid Mode if no interfering HomePlug 1.0.1 stations are detected. In Coordinated

Mode, the CCo may also specify a Shared CSMA Hybrid Mode if a Coordinating CCo in the INL specified a Shared CSMA or Fully Hybrid Mode. In this mode, only the Shared CSMA allocations use hybrid delimiters to coordinate medium sharing.

(3) *AV-only Mode:* The AVLN operates in AV-only Mode if no interfering Home-Plug 1.0.1 stations are detected. In Coordinated Mode, this further requires that none of the Coordinating CCos detects HomePlug 1.0.1 or HomePlug 1.1 stations. In this mode, all CP and CFP transmissions use AV-only delimiters.

Regardless of the AVLN's operating mode, all Beacon MPDUs always use a Hybrid delimiter structure. This choice simplifies the Neighbor Network operations in Coordinated Mode.

11.4.2 Detection and Reporting of Active HomePlug 1.0.1

Coexistence requires HomePlug AV STAs to detect active HomePlug 1.0.1 STAs on the medium, and to change the coexistence modes in response to their presence or absence. AV STAs continuously monitor the medium for HomePlug 1.0.1 transmissions, during the AVLN's CSMA allocations as well as during their CFP allocations (i.e., CFP allocations for which the STA is either the source or destination of the Global Link).

All AV stations that detected a HomePlug 1.0.1 transmission report it to the CCo by setting the HomePlug 1.0.1 Detect Flag (HP10DF) or HomePlug 1.1 Detect Flag (HP11DF), respectively, in the SOF and/or RTS delimiter of all subsequent transmissions they make for a duration of 1 s. AV stations also maintain statistics on the number of HomePlug 1.0.1/1.1 transmissions that were detected and report them to the CCo using the **CC_HP1_DET.CNF** message when explicitly requested by the CCo (using **CC_HP1_DET.REQ**). Stations may also send an unsolicited **CC_HP1_DET.CNF** message if they detect several HomePlug 1.0.1 transmissions and the network mode is either not changed to Fully Hybrid Mode or if none of the ongoing transmissions reports the detection of HomePlug 1.0.1 stations by setting the HP10DF.

11.4.3 HomePlug 1.0.1/1.1 Coexistence Mode Changes

The CCo of each AVLN determines the HomePlug 1.0.1 coexistence mode of the AVLN. The Hybrid Mode (HM) field in the Beacon is used by the CCo to indicate the operation mode of the AVLN.

On start-up, HomePlug AV STAs enter Fully Hybrid Mode and listen to the medium for existing activity. If a Beacon from an existing AVLN is detected and the station becomes a part of that AVLN, it changes its coexistence mode to that of the AVLN. The CCo and other stations in the AVLN exchange information to determine the operating mode of the AVLN.

When a CCo in AV-only Mode determines the presence of HomePlug 1.0.1 activity, it changes the network mode to Fully Hybrid Mode. To allow the AVLN to revert to AV-only Mode when all HomePlug 1.0.1 nodes become inactive, the CCo maintains a Fully Hybrid Mode Timer that is set to at least 1 s when it changes to Fully Hybrid Mode. This timer will be continuously reset when new HomePlug 1.0.1 activity is detected/reported while in Fully Hybrid Mode. If the Fully Hybrid Mode Timer expires and no HomePlug 1.1 activity is detected/reported, the CCo changes to AV-only Mode.

If the CCo in shared CSMA Mode determines the presence of HomePlug 1.0.1 activity, it changes the operating mode to Fully Hybrid and set the Fully Hybrid Mode timer to at least 1 s.

11.4.4 HomePlug 1.0.1-Compatible Frame Lengths

HomePlug AV Stations use hybrid MPDUs when coexisting with HomePlug 1.0.1 Stations during the Fully Hybrid and Shared CSMA Hybrid operating modes. Hybrid MPDUs use hybrid delimiters consisting of a Hybrid Preamble, the HomePlug 1.0.1 Frame Control, and HomePlug AV Frame Control(s). HomePlug AV Stations manipulate the fields of HomePlug 1.0.1 Frame Control to keep them synchronized and to control their access to the medium. This enables HomePlug AV stations to keep HomePlug 1.0.1 from access the medium during CFP allocation.

The CSMA channel access used by HomePlug 1.0.1 stations is the same as that used by HomePlug AV. During CSMA, HomePlug 1.0.1 Frame Control fields are manipulated by HomePlug AV transmitter to enable HomePlug 1.0.1 stations to maintain proper Virtual Carrier Sense. This enables HomePlug 1.0.1 Stations to determine the start of Priority Resolution Slots, thus enabling them to fairly content with HomePlug AV.

11.5 PROXY NETWORKING

Every STA in the AVLN must be able to communicate directly or indirectly with the CCo. Proxy networking provides the necessary functionality to enable stations that cannot directly communicate with the CCo to become part of the AVLN. It should be noted that Proxy networking does not enable stations in the AVLN that cannot directly communicate with each other to exchange data. The latter is accomplished by using the repeating and routing functionality which was not in HomePlug AV standard but was later added in IEEE 1901 standard (refer to Section 14.2.6).

A Hidden STA (HSTA) is a STA that cannot communicate directly with the AVLN's CCo and, as a result, must communicate indirectly, relaying messages via a proxy. By definition, an HSTA cannot hear the Beacons transmitted by the CCo; it can, however, infer the existence of the AVLN from Discover Beacons transmitted by other STAs or Proxy Beacons transmitted by Proxy Coordinators (PCos). The HSTA uses information from these Beacons to learn the identity of the AVLN and to identify a Proxy STA (PSTA), see below, or existing PCo to relay messages to/from the CCo.

Once the CCo receives a message and becomes aware of the HSTA, it appoints a STA (possibly the PSTA) as a PCo. From that point on, all messages will be relayed between the HSTA and the CCo via the PCo.

A Proxy Network (PxN) is a network established by the CCo when it appoints a PCo to support one or more HSTAs. A Proxy Network is always associated with an existing AVLN and is a wholly contained part of the AVLN. It consists of a PCo (appointed by the CCo) and one or more HSTAs. PxNs are strictly the concern of the AVLN to which they belong. Multiple networks that are neighbors may each independently support PxNs.

All operations of the HSTAs are controlled by the CCo with the PCo serving as a relay. Since the HSTA must obtain information contained in the CCo's Beacon, the PCo transmits a Proxy Beacon (refer to Section 11.5.4) once every Beacon Period. In Uncoordinated and Coordinated Modes, the CCo specifies TDMA allocations in which the Proxy Beacons need to be transmitted. During CSMA-only mode, the Proxy Coordinator should transmit the Proxy Beacon using CSMA/CA at CAP3 as soon as the Central Beacon from CCo is received.

Figure 11.5 shows an example of Network 1 containing a PxN. Within the PxN, STA B is the PCo, and STAs D and E are HSTAs.

Proxy Networking is an optional feature. Proxy Networking includes STAs functioning as Hidden Stations, Proxy Stations, Proxy Coordinators, and Central Coordinators. All these features are optional. In the remainder of this section, the terms Hidden Station, Proxy Station, Proxy Coordinator, and Central Coordinator refer to STAs that support the optional Proxy networking feature. Each STA advertises in its Discover Beacon whether it supports Proxy Networking.

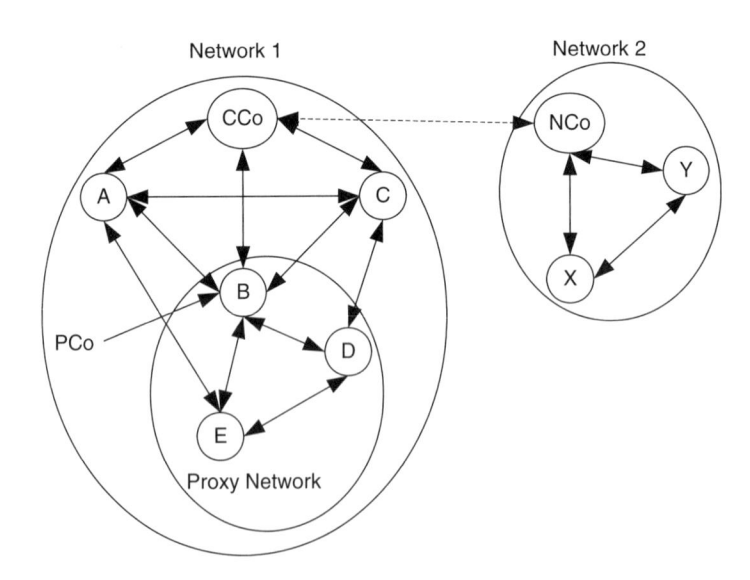

FIGURE 11.5 Proxy Network created by Network 1.

11.5.1 Identification of Hidden Stations

If the STA cannot hear a CCo's Beacon, but can determine the existence of the AVLN from Discover Beacons or Proxy Beacons, it knows it is an HSTA. If the HSTA hears a Proxy Beacon, it requests that PCo to relay its messages to the CCo. If the HSTA does not hear a Proxy Beacon, it selects a STA from among those from which it has heard Discover Beacons, and ask that STA to serve as a PSTA and relay its initial message to the CCo.

When the CCo receives the first relayed message from a NewHSTA, it recognizes that there is a new HSTA that wishes to associate with the AVLN. The method by which the CCo receives the association request message from the HSTA is described in Section 11.5.2. The accurate decryption and interpretation of the association request message from the HSTA informs the CCo of the presence of an HSTA in the network.

Before responding to the HSTA's association request message, the CCo appoints a PCo to support the HSTA as described in Section 11.5.3.

11.5.2 Association of Hidden Station

The association of a new STA which is in range with the CCo is described in Section 10.3.2. When the new STA (or more appropriately a New Hidden STA (NewHSTA)) is out of range of the CCo, it performs the association described below and shown in Figure 11.6.

Note: The fundamental association messaging is identical for both STAs and HSTAs. The only difference for a NewHSTA is the encapsulation of the messages within **CC_RELAY.REQ/IND** messages and the insertion of the **CP_PROXY_APPOINT.xxx** messaging between the **CC_ASSOC.REQ** and the **CC_ASSOC.CNF** messages.

The new STA encapsulates the **CC_ASSOC.REQ** MME within a **CC_RELAY.REQ** MME and send it to the PCo or the PSTA. If the HSTA does not know the CCo's TEI or MAC Address, it sets the FDA and FTEI field to the broadcast MAC Address and TEI, respectively. Otherwise, it sets the FDA and FTEI field to the MAC address and TEI of the CCo.

The PCo or the PSTA decapsulates the **CC_ASSOC.REQ** MME, re-encapsulate it inside a **CC_RELAY.IND** MME. The values of the OSA and OTEI fields inside the **CC_RELAY.IND** MME are set to the MAC address of the NewHSTA and the TEI used by the NewHSTA (which is **0x00**). The PCo sends the **CC_RELAY.IND** MME to the CCo. Normally the PCo would use the FDA and FTEI fields of the **CC_RELAY.REQ** MME to address the **CC_RELAY.IND** MME but in this case, the NewHSTA could not supply them. The PCo can identify the destination by virtue of the fact the OTEI field is **0x00** (new STA) and recognize that the only STA that can be addressed when the STEI = **0x00** is the CCo.

The CCo extracts the Payload field of the **CC_RELAY.IND** MME and processes the association request.

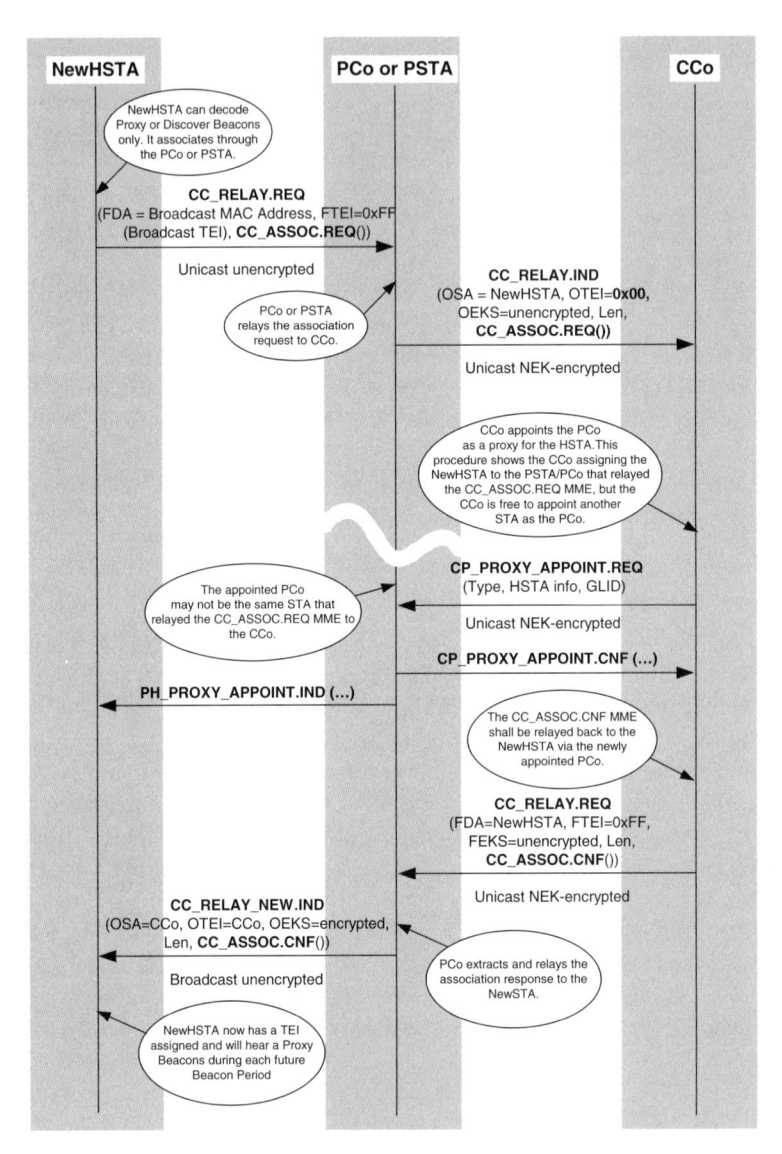

FIGURE 11.6 HSTA association.

If it determines to accept the association request, the CCo appoints a PCo, as described in Section 11.5.3. The messaging for PCo appointment occurs before the CCo replies to the association request.

After appointing a PCo, the CCo creates a **CC_ASSOC.CNF** MME and encapsulates it inside a **CC_RELAY.REQ** MME. The FDA and FTEI fields inside the **CC_RELAY.REQ** MME are set based on the OSA and OTEI fields inside the **CC_RELAY.IND** MME. The CCo sends the **CC_RELAY.REQ** MME to the newly appointed PCo.

If the association request is rejected, the **CC_ASSOC.CNF** MME is encapsulated as just described and sent to the PSTA that relayed the original **CC_ASSOC.REQ** MME.

If the association request is accepted, the **CC_ASSOC.CNF** MME contains the TEI assigned to the new STA and the lease time.

The PCo or the PSTA extracts the **CC_ASSOC.CNF** MME from the **CC_RELAY.REQMME**, encapsulates it inside a **CC_RELAY.IND** MME and sends it to the NewHSTA. The FDA and FTEI fields inside the **CC_RELAY.REQ** MME provide the required addressing information to send the **CC_RELAY.IND** MME.

11.5.3 Instantiation of Proxy Network

A PxN is established when the CCo appoints a PCo to support one or more HSTAs. The PxN consists of the PCo and the HSTA(s) that it supports.

A PxN is established when:

- The CCo learns of the presence of an HSTA through relay of a CC_ASSOC.REQ MME from a PSTA. There are two sub-cases:
  o A STA that cannot hear the CCo joins the AVLN for the first time.
  o A STA that is already authenticated becomes unable to hear the CCo because of changing channel conditions.
- The CCo decides to change the PCo because:
  o The PCo STA is leaving (or has left) the AVLN, or
  o The CCo determines that another STA is better suited to being the PCo for this particular PxN.

The CCo may also choose to reassign HSTAs from one PxN to another PxN. The CCo should attempt to minimize the number of PxNs that it creates. This may require it to reconfigure PxNs (e.g., combine two or more PxNs into a single PxN) as the Discovery Process progresses.

11.5.4 Proxy Beacons

A Proxy Beacon is a special type of Beacon that is transmitted by a PCo once every Beacon Period. It provides timing and schedule information for the HSTAs in the PxN. It contains the TEI of the transmitting STA (the PCo) and the Network ID of the network. It also contains any other BENTRIES found in the current Beacon, with the exception that the PCo may optionally omit some BENTRIES.

11.5.5 Provisioning the NMK to Hidden Stations

If the HSTA and the UIS are within range of each other, the provisioning of NMK to a HSTA is the same as described in Section 10.4.2. The HSTA may receive the broadcast message a second time, relayed from its PCo; it ignores this duplicate message.

If the HSTA and the UIS are not in rage of each other, the PCo's Relay functions will support NMK provisioning automatically.

All Encryption Key management methods rely on Encrypted Payloads, so they may be relayed freely and securely without intermediate knowledge of any encryption keys. Furthermore, all of the NMK provisioning methods begin with a broadcast CSMA message, which the PCo will encapsulate and relay to its HSTAs, as described in Section 11.5.7.

If the HSTA elects to respond to the broadcast message, the HSTA encapsulates its response (the encrypted payload MME) in a **CC_RELAY.REQ** MME and sends it to the PCo which relays it to the UIS. The UIS processes the MME encapsulated in the **CC_RELAY.IND** MME it receives as if it had been received directly except that it encapsulates its reply in a **CC_RELAY.REQ** MME that it sends to the PCo. The HSTA and the UIS continue encapsulating their messages in the **CC_RELAY.REQ** MMEs for the duration of the protocol run. The PCo acts as a relay for these messages.

11.5.6 Provisioning NEK for Hidden Stations (Authenticating the HSTA)

There is no difference in the process for an HSTA joining the AVLN than for any other STA except that the messages in the protocol run (Section 10.4.3) is encapsulated in **CC_RELAY.xxx** MMEs and relayed via the PCo.

Once the HSTA has been authenticated, the CCo sends a **CP_PROXY_APPOINT.REQ** (ReqType = Update) to tell the PCo that the HSTA is authenticated.

An HSTA may communicate with any other STA in the AVLN (subject to powerline channel characteristics) once the HSTA is associated and been authenticated by the CCo via the PCo.

11.5.7 Exchange of MMEs Through a PCo

Only MMEs are relayed through a PCo. Data transmission is not permitted unless the endpoints are in direct communication. STAs and HSTAs refrain from attempting to establish Connections with STAs or HSTAs from which they are hidden even if a PCo is available to support the relaying of MMEs.

The HSTA should be aware of which STAs in the AVLN it can directly communicate with and which ones it is hidden from. All unicast MMEs that the HSTA wants to send to the CCo or to another STA hidden from it should be encapsulated in a **CC_RELAY.REQ** MME and sent to the PCo. The PCo then extracts the Payload field of the **CC_RELAY.REQ** MME and encapsulates it inside a **CC_RELAY.IND** MME and sends it to the final destination STA.

STAs that want to communicate with an HSTA that is also hidden from them must know the identity of the PCo that supports this HSTA. MMEs from the CCo or another STA hidden from the HSTA are encapsulated in a **CC_RELAY.REQ** MME and sent to the PCo responsible for the particular HSTA. Fields in the **CC_RELAY.REQ** MME are used to identify the HSTA. The PCo extracts the payload MME and relays it to the HSTA in a **CC_RELAY.IND** MME.

When a PCo receives a broadcast MME from other STAs in the AVLN, it encapsulates that MME in a **CC_RELAY.IND** MME and sends it to all HSTAs that are under its control.

When a PCo receives a broadcast MME from a HSTA that is under its control, it encapsulates that MME in a **CC_RELAY.IND** MME and broadcasts it to all STAs in the AVLN.

All **CC_RELAY.REQ** MMEs are encrypted with the NEK unless the transmitting STA is not yet part of the Network. STAs that are not yet part of the network are permitted to send **CC_RELAY.REQ** MMEs unencrypted.

The PCo should encrypt each **CC_RELAY.IND** MME using the NEK unless it is directed to a station that has not yet been authenticated into the AVLN. The PCo broadcasts any unencrypted **CC_RELAY.IND** MME that contains a **CM_ENCRYPTED_PAYLOAD.IND**. This is necessary so that the UIS can monitor the Protocol Run for an MITM attack. The HSTA should not accept a **CC_RELAY.IND** MME that contains a **CM_ENCRYPTED_PAYLOAD.IND** message unless it is broadcast.

11.5.8 Transitioning from Being a STA to Being an HSTA

It is possible that an existing STA may no longer be able to decode the CCo's Beacons reliably. This might occur because channel conditions have changed significantly. It might also occur after a CCo handover if the existing STA is no longer within range of the new CCo. When this happens, the existing STA will re-associate with the CCo through a PSTA (or existing PCo). The procedure is the same as the association procedure for a HSTA (Section 11.5.2), except the ReqType field in the **CC_ASSOC.REQ** message will indicate that it is a renewal request. Since the message was delivered to the CCo in a **CC_RELAY.IND** MME, the CCo assigns a PCo to support the new HSTA and then processes the TEI renewal request normally.

11.5.9 Transitioning from Being an HSTA to Being a STA

It is possible that an existing HSTA may now be able to decode the CCo's Beacons reliably. This might occur because channel conditions have changed significantly. It might also occur after a CCo handover if the existing HSTA is within range of the new CCo. When this happens, the existing HSTA re-associate with the CCo directly as a regular STA. The procedure is the same as the association procedure for a STA except that the ReqType field in the **CC_ASSOC.REQ** message indicates that it is a renewal request.

Since the message was received directly from a STA that the CCo had previously listed as hidden, the CCo removes the STA from the PCo's list of supported HSTAs using the **CP_PROXY_APPOINT.REQ** message with a ReqType of "Delete." If the PCo now has no more HSTAs assigned to it, the CCo shuts down the PxN, as described in Section 11.5.11.

11.5.10 Recovering from the Loss of a PCo

It is possible for the PCo to drop out of the network without warning, either due to equipment failure or because the user unknowingly unplugged the STA that was serving as PCo. In this case, the HSTA will observe that it is no longer receiving Proxy Beacons from its PCo. When its schedule expires, it selects another PSTA (or PCo) and re-associates with the CCo using the procedure described in Section 11.5.8. The CCo notices that the TEI of the relay device is different than the TEI of the HSTA's assigned PCo and assigns the HSTA to a new PCo.

11.5.11 Proxy Network Shutdown

Once established, the PxN will continue until the PCo is instructed by the CCo to halt PCo functions. Instruction for this occurs via the **CP_PROXY_APPOINT.REQ** message with a ReqType = Shutdown.

The CCo shutdowns the PxN when all of the HSTAs assigned to the PCo have:

- disassociated,
- had their TEI leases expire,
- been transferred to another PCo, or
- transitioned from being an HSTA to a STA

11.6 SUMMARY

This chapter provided an overview the methodologies used in HomePlug AV to enable four key features of HomePlug AV, namely Channel Estimation, Bridging (used for example in PLC to Ethernet bridges), HomePlug 1.0.1 coexistence (to ensure seamless deployment of HomePlug AV with the large installed base of HomePlug 1.0.1 devices), and support for Proxy Networking (to communicate with hidden stations).

Chapter 12 examines the techniques used in HomePlug AV to ensure proper operation when there are two or more overlapping HomePlug Neighbor Networks.

12

NEIGHBOR NETWORKS

12.1 INTRODUCTION

A HomePlug AV network may need to operate in the presence of neighbor networks of other compatible Powerline Communication (PLC) devices. In this chapter, we describe the operation of HomePlug AV when there are such neighbor networks.

A Home AV Logical Network (AVLN) operates in one of the following three modes based on the Central Coordinator (CCo) capability and neighbor network detection:

- CSMA-only Mode;
- Uncoordinated Mode; and
- Coordinated Mode.

These modes of operation and the transition between the modes are described in the following sections.

12.1.1 CSMA-Only Mode

All stations (STAs) support Carrier Sense Multiple Access (CSMA)-only Mode. The Beacon Period structure and channel access mechanism in CSMA-only Mode are described in Section 8.3.1.

HomePlug AV and IEEE 1901: A Handbook for PLC Designers and Users, First Edition.
Haniph A. Latchman, Srinivas Katar, Larry Yonge, and Sherman Gavette.
© 2013 by The Institute of Electrical and Electronics Engineers, Inc. Published 2013 by John Wiley & Sons, Inc.

If a CCo operating in CSMA-only Mode can reliably detect Central or Proxy Beacons from one or more AVLN operating in Coordinated Mode, it adjusts its schedule and perform Passive Coordination with *one and only one* of those AVLNs. Passive Coordination provided by CCo operating in CSMA-only Mode enables Level-2 CCos to support Time Division Multiple Access (TDMA) even in the presence of neighboring AVLNs operating in CSMA-only Mode. Section 12.4 provides details on Passive Coordination.

12.1.2 Uncoordinated Mode

Uncoordinated Mode is only supported by Level-1 and Level-2 CCos. The Beacon Period structure and Channel Access Mechanism in Uncoordinated Mode are described in Section 8.3.2.

A CCo operating in Uncoordinated Mode generates its own timing and transmits its periodic Beacon independently of other networks. In Uncoordinated Mode, Quality of Service (QoS) can be guaranteed by allocating dedicated Contention Free allocations to applications that require QoS.

12.1.3 Coordinated Mode

Coordinated Mode is only supported by a Level-2 CCo. Section 8.3.3 provides details about the Beacon Period Structure Coordinated Mode. Coordinated sharing of bandwidth between neighboring networks is described in Section 12.3.

12.2 TRANSITION BETWEEN NEIGHBOR NETWORK OPERATING MODES

A new CCo may establish an AVLN in CSMA-only Mode irrespective of the presence of any neighboring networks. This option simplifies the power-on procedure.

Optionally, a new CCo may use the knowledge of the detected neighboring networks to determine the Network Mode in which the AVLN needs to be established. If this option is used, the CCo can immediately transition from CSMA-only Mode before transmitting any Beacons, based on the transitions described below.

A network operating in CSMA-only Mode transitions from CSMA-only Mode as follows:

- CSMA-only Mode to Uncoordinated Mode transitions
    - A Level-1 or Level-2 AVLN operating in CSMA-only Mode transitions to Uncoordinated Mode if it cannot detect any neighboring AVLNs.

TRANSITION BETWEEN NEIGHBOR NETWORK OPERATING MODES   **235**

- CSMA-only Mode to Coordinated Mode transitions
  - A Level-2 AVLN operating in CSMA-only Mode transitions to Coordinated Mode if it can successfully coordinate with neighboring Level-2 or higher AVLNs.
  - A Level-2 AVLN operating in CSMA-only Mode transitions to Coordinated Mode if only neighbor networks in CSMA-only Mode are detected.

A network operating in CSMA-only Mode transitions from Uncoordinated Mode as follows:

- Uncoordinated Mode to Coordinated Mode transition
  - A Level-2 AVLN operating in Uncoordinated Mode transitions to Coordinated Mode if it detects Beacons from neighboring Level-0 or Level-1 AVLN.
  - A Level-2 AVLN operating in Uncoordinated Mode transitions to Coordinated Mode if it can successfully coordinate with neighboring Level-2 AVLN(s).
- Uncoordinated Mode to CSMA-only Mode transition
  - A Level-1 AVLN operating in Uncoordinated Mode transitions to CSMA-only Mode if it detects any HomePlug AV in-home network.
  - A Level-2 AVLN operating in Uncoordinated Mode transitions to CSMA-only Mode if it cannot successfully coordinate with neighboring Level-2 AVLN(s).

A network operating in CSMA-Only Mode transitions from Coordinated Mode as follows:

- Coordinated Mode to Uncoordinated Mode transition
  - If all the networks in the Level-2 CCo's Interfering Network List (INL) have shut down through the procedure described in Section 12.3.8 and no neighboring Level-0 or Level-1 AVLNs are detected, the CCo transitions to Uncoordinated Mode.
  - If the Level-2 CCo can no longer decode any other Beacons from Coordinating Level-2 AVLNs for at least 10 Beacon Periods in a row, the CCo assumes that all Coordinating Level-2 AVLNs have been powered off. In this case, the CCo transitions to Uncoordinated Mode if no neighboring Level-0 or Level-1 AVLNs are detected.
  - If there are no Level-2 AVLNs present, and if Beacons from neighboring Level-0/Level-1 neighboring AVLNs cannot be detected for at least 10 s, the CCo assumes that all neighboring Level-0/Level-1 AVLNs have been powered off and transitions to Uncoordinated Mode.
- Coordinated Mode to CSMA-only Mode transition
  - If a Level-2 CCo cannot successfully coordinate with neighboring Level-2 AVLNs, it transitions to CSMA-only Mode.

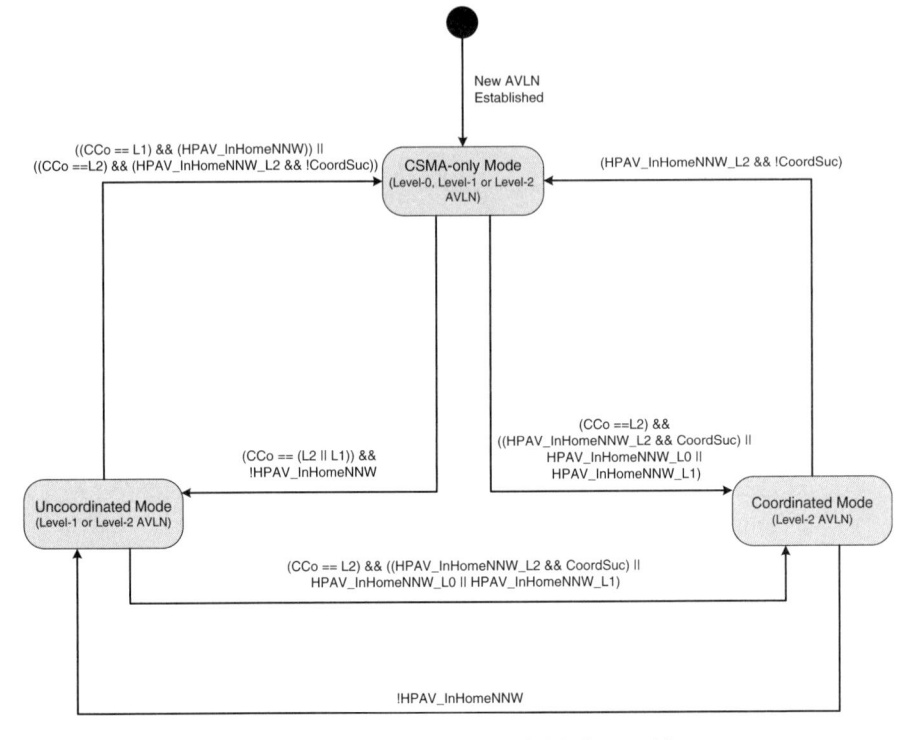

FIGURE 12.1 Neighbor Network Mode transitions.

Figure 12.1 shows the Neighbor Network Mode Transitions. Several concise terms were used in this figure to simplify the conditions for various transitions. The interpretation of these terms is as follows:

- HPAV_InHomeNNW = HomePlug AV in-home Neighbor Network.
- HPAV_InHomeNNW_L0 = Level-0 HomePlug AV in-home Neighbor Network.
- HPAV_InHomeNNW_L1 = Level-1 HomePlug AV in-home Neighbor Network.
- HPAV_InHomeNNW_L2 = Level-2 HomePlug AV in-home Neighbor Network.
- L1 = Level-1 CCo.
- L2 = Level-2 CCo.
- CoordSuc = Coordination with Neighboring Level-2 AVLN is successful.

12.3 COORDINATED MODE

Coordinated mode is only supported by Level-2 CCos. A Level-2 CCo typically operates in Coordinated Mode if it detects one or more Level-0, Level-1, or Level-2 CCos.

It is important to note that Level-2 CCo can operate in Coordinated Mode even if it detects only Level-0 or Level-1 AVLNs. In this case, the Beacon Region in Coordinated Mode will only have one Beacon Slot and the Level-2 CCo provides fair share of CSMA Region to the neighboring Level-0 or Level-1 AVLNs that are passively coordinating (refer to Section 12.4).

The following section provides details about neighbor network coordination between various Level-2 CCos operating in Coordinated Mode.

12.3.1 Interfering Network List

Each Level-2 CCo maintains an Interfering Network List. The INL of a CCo (or of a network) contains the list of all Level-2 networks that interfere with the network controlled by the CCo.

It is assumed that if two Level-2 CCos can detect each other's Beacon transmissions, the two networks interfere with each other. Furthermore, it is assumed that all STAs in the two networks also interfere with each other.

Note: Two interfering Level-2 networks might not coordinate with each other if they are in different groups (refer to Section 12.3.2). In this case, the two interfering networks might not be in each other's INL, even though the two networks interfere with each other.

12.3.2 Group of Networks

A group of networks is defined as a collection of one or more Level-2 networks that have the same system timing (that is, the Beacon Periods of these networks align with each other). It is possible to have two or more groups of networks in the vicinity of each other.

Network coordination between interfering networks that are in the same group is mandatory. Network coordination between interfering networks that are in different groups is optional.

12.3.3 Determining a Compatible Schedule

If the new CCo establishes a new network in Uncoordinated Mode, initially it specifies a Beacon Region with one Beacon Slot, a random Contention Free allocation for the transmission of Discover Beacons and CSMA allocations for the remainder of the Beacon Period.

Alternatively, if a new Level-2 CCo joins an existing group of networks in Coordinated Mode, the schedule of its Beacon must be compatible with the schedules of the existing networks in its INL. The rules to determine a compatible schedule are given in this section. First, the new CCo finds out the combined effect of the schedules of all the networks in its INL, called the INL allocation, using the procedure described in Section 12.3.3.1.

Once the INL allocation is computed, the rules used by a new CCo to set the Region Types (RTs) of the Regions BENTRY are as follows. Initially, the new CCo does not specify any Reserved Regions.

- If the INL allocation is a Beacon Region and the first entry, the new CCo specifies a Beacon Region. However, if it is not the first entry, the new CCo specifies a Protected Region.
- Otherwise, if the INL allocation is a Protected Region or a Reserved Region, the new CCo specifies a Stayout Region.
- Otherwise, the new CCo specifies a CSMA Region in all other intervals.

Once a network is established in Coordinated Mode, the rules used by an existing CCo to set the subsequent Region Types of the Regions BENTRY are as follows:

- If the INL allocation is a Beacon Region and if it is the first entry, the existing CCo specifies a Beacon Region. However, if it is not the first entry, the existing CCo specifies a Protected Region.
- If the INL allocation is a Protected Region or a Reserved Region, the existing CCo specifies a Stayout Region.
- If the INL allocation is a CSMA Region, the existing CCo specifies a CSMA Region. The existing CCo may propose to use this time interval in the future.
- If the INL allocation is a Stayout Region, the existing CCo may specify a CSMA Region or a Reserved Region. The existing CCo may propose to use this time interval in the future.

12.3.3.1 Computing the INL Allocation The CCo decodes the Beacons of all the networks in its INL and compute the combined effect of their allocations. This is called the INL allocation. For example, if one neighbor network in the INL specifies a Reserved Region and another neighbor specifies a CSMA or Stayout Region, the resultant INL allocation is a Reserved Region, because a Reserved Region "outweighs" both CSMA and Stayout Regions.

Figure 12.2 shows the algorithm that a CCo uses to compute its INL allocation. In the algorithm, it is assumed that numeric values are assigned to BEACON, PROTECTED, RESERVED, CSMA, and STAYOUT and the values are such that BEACON > PROTECTED > RESERVED > CSMA > STAYOUT to make the flowchart simpler. It has nothing to do with the RT field defined in the Regions BENTRY.

The inputs of the algorithm TYPE[n][i] and ENDTIME[n][i] are obtained from the RT and RET fields of the Region's message of all neighbor Beacons, where "n" represents which neighbor network, and "i" represents which schedule for that neighbor network. The entries TYPE[n][i] and ENDTIME[n][i] for each network are shifted, if necessary, to account for any difference in system timing.

Table 12.1 summarizes the rules in determining the INL allocation. The CCo concerned has two interfering networks in its INL:

- Network A; and
- Network B

The Regions BENTRY of networks A and B are shown in the first and second columns of Table 12.1. The resultant INL allocation is shown in the third column.

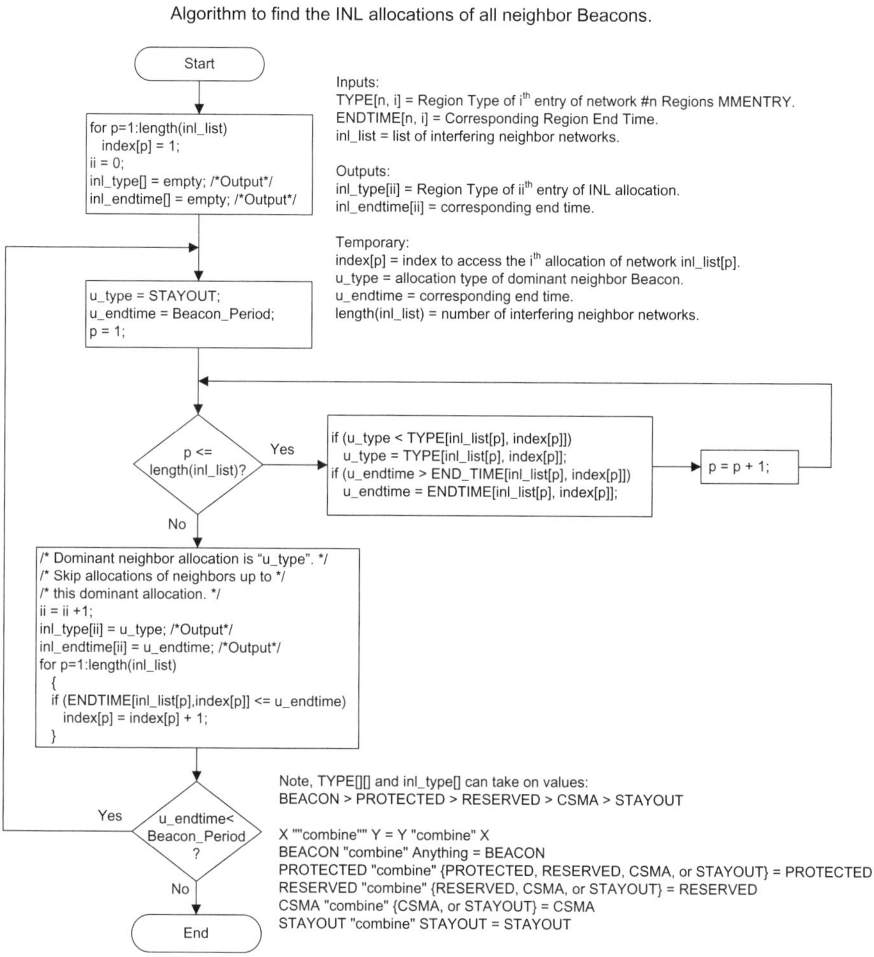

COORDINATED MODE

Algorithm to find the INL allocations of all neighbor Beacons.

Inputs:
TYPE[n, i] = Region Type of ith entry of network #n Regions MMENTRY.
ENDTIME[n, i] = Corresponding Region End Time.
inl_list = list of interfering neighbor networks.

Outputs:
inl_type[ii] = Region Type of ii^{th} entry of INL allocation.
inl_endtime[ii] = corresponding end time.

Temporary:
index[p] = index to access the ith allocation of network inl_list[p].
u_type = allocation type of dominant neighbor Beacon.
u_endtime = corresponding end time.
length(inl_list) = number of interfering neighbor networks.

```
for p=1:length(inl_list)
   index[p] = 1;
ii = 0;
inl_type[] = empty; /*Output*/
inl_endtime[] = empty; /*Output*/

u_type = STAYOUT;
u_endtime = Beacon_Period;
p = 1;
```

while p <= length(inl_list)?
   if (u_type < TYPE[inl_list[p], index[p]])
      u_type = TYPE[inl_list[p], index[p]];
   if (u_endtime > END_TIME[inl_list[p], index[p]])
      u_endtime = ENDTIME[inl_list[p], index[p]];
   p = p + 1;

```
/* Dominant neighbor allocation is "u_type". */
/* Skip allocations of neighbors up to */
/* this dominant allocation. */
ii = ii +1;
inl_type[ii] = u_type; /*Output*/
inl_endtime[ii] = u_endtime; /*Output*/
for p=1:length(inl_list)
{
   if (ENDTIME[inl_list[p],index[p]) <= u_endtime)
      index[p] = index[p] + 1;
}
```

u_endtime < Beacon_Period?

Note, TYPE[][] and inl_type[] can take on values:
BEACON > PROTECTED > RESERVED > CSMA > STAYOUT

X ""combine"" Y = Y "combine" X
BEACON "combine" Anything = BEACON
PROTECTED "combine" {PROTECTED, RESERVED, CSMA, or STAYOUT} = PROTECTED
RESERVED "combine" {RESERVED, CSMA, or STAYOUT} = RESERVED
CSMA "combine" {CSMA, or STAYOUT} = CSMA
STAYOUT "combine" STAYOUT = STAYOUT

FIGURE 12.2 Flowchart for computing INL allocation.

12.3.4 Communication between Neighboring CCos

Either MultiNetwork Broadcast, as in Section 6.9.1, or unicast MAC Protocol Data Units (MPDUs) are used for communication with a neighbor CCo. No association is required.

When unicast MPDUs are used for communication with a neighbor network, the Short Network IDentifier (SNID) and Destination Terminal Equipment Identifier (DTEI) of the neighbor CCo and an Source Terminal Equipment Identifier (STEI) of **0x00** are used. The Physical Layer (PHY) transmit clock for the unicast PHY Protocol Data Unit (PPDU) is based on the current estimate of the PHY Clock frequency of the neighbor CCo. These unicast MPDUs are limited to one segment and contain MAC Management Messages. The PHY Block Body is unencrypted and EKS is set to **0b1111**.

TABLE 12.1 Rules for Computing INL Allocation

Region Type of Neighbor A	Region Type of Neighbor B	INL Allocation of Neighbors A and B
BEACON	BEACON, PROTECTED, RESERVED, CSMA, or STAYOUT	BEACON
PROTECTED	BEACON	BEACON
PROTECTED	PROTECTED, RESERVED, CSMA, or STAYOUT	PROTECTED
RESERVED	BEACON	BEACON
RESERVED	PROTECTED	PROTECTED
RESERVED	RESERVED, CSMA, or STAYOUT	RESERVED
CSMA	BEACON	BEACON
CSMA	PROTECTED	PROTECTED
CSMA	RESERVED	RESERVED
CSMA	CSMA or STAYOUT	CSMA
STAYOUT	BEACON	BEACON
STAYOUT	PROTECTED	PROTECTED
STAYOUT	RESERVED	RESERVED
STAYOUT	CSMA	CSMA
STAYOUT	STAYOUT	STAYOUT

12.3.5 Neighbor Network Instantiation

If a CCo has determined that it will create a Neighbor Network, the CCo examines the Beacon Region of the group it chooses to join. On the basis of the number of Beacon Slots and the occupancy of the Beacon Slots in the Beacon Region, the CCo performs the following:

- If there are fewer than eight Beacon Slots in the Beacon Region, or if the CCo can find a vacant Beacon Slot to use, the CCo tries to create a network in Coordinated Mode. The neighbor network instantiation is completed.
- Otherwise, if the Beacon Region has eight Beacon Slots already and the CCo cannot find a vacant Beacon Slot to use, the behavior is to instantiate a network in Uncoordinated Mode.

12.3.5.1 Procedure to Establish a New Network in Coordinated Mode To establish a new network in Coordinated Mode, the (new) CCo chooses a SNID, find a vacant Beacon Slot in the Beacon Region to use, and specify a Beacon Period structure that is compatible (refer to Section 12.3.3 where the "compatible" rules are defined) with all the networks that are in its INL. Key steps include:

1. *Choosing a SNID.* The new CCo first exchanges the **NN_INL.REQ** and **NN_INL.CNF** messages with the neighbor coordinators (NCos) in its INL. This step allows the new CCo to ascertain the SNID and NID of the interfering

networks of an NCo. The new CCo then randomly chooses a SNID value which is not being used by any of its NCo, or by any interfering networks of its NCos.

2. *Finding a Vacant Beacon Slot.* The new CCo decodes all Central, Discover, and Proxy Beacons that can be reliably detected to determine what Beacon Slots are available. The NumSlots field in the Beacon Variant Fields and the SlotID and SlotUsage fields in the Beacon payload field provide information about the Beacon Region structure of a network. The new CCo finds a common Beacon Slot that is indicated as "free" in the SlotUsage field of all Beacons that are detected, including Central, Proxy, and Discover Beacons of any network.

    If a vacant Beacon Slot cannot be found, the new CCo proposes to use a new Beacon Slot (subject to the maximum limit of 8).

3. *Specifying a Compatible Beacon Period Structure* After the new CCo sends an **NN_NEW_NET.REQ** message to request to establish a new neighbor network, the existing NCo returns in the **NN_NEW_NET.CNF** message a Beacon Period structure (or schedule) that the NCo is going to use. The new CCo computes a compatible Beacon Period structure based on the schedules returned by all the NCos.

The procedures for establishing a new network in Coordinated Mode are summarized in Figure 12.3. In Figure 12.3, it is assumed that the new CCo can decode Beacons from two existing CCos (NCo 1 and NCo 2). In Case 1, both NCo #1 and NCo #2 accept the new CCo's request to set up a new AVLN. In Case 2, NCo #2 rejects the request.

12.3.5.1.1 New Network Instantiation Amid Two Groups of Networks A new CCo may detect two or more Groups of networks in the vicinity of each other (see Figure 12.4 for an example). In this example, two existing networks (NCo #1 and NCo #2) cannot detect each other's Beacons, and there is a fixed time offset between their Beacon Period boundaries.

When a new STA (CCo) wants to start a new network, the new CCo can detect and decode the Beacons from both NCo #1 and NCo #2. Since the timings of the two existing networks are different, the new CCo acquires only one of the two timings.

In the example, the new CCo chooses the same timing as NCo #1. The new CCo exchanges the **NN_NEW_NET.REQ/CNF/IND** messages with NCo #1 to establish a new neighbor network in Coordinated Mode. The new network and the network of NCo #2 may cause interference to each other because they do not coordinate with each other.

The new CCo may optionally exchange the **NN_NEW_NET.REQ/CNF/IND** messages with NCo #2, in addition to NCo #1. The Offset field in the **NN_NEW_NET.REQ** message sent to NCo #2 is set to a nonzero value to indicate that the new CCo has a different system timing than NCo #2. If NCo #2 accepts the request of the new CCo, all three networks coordinate with each other. The new CCo specifies a Protected Region in the same interval where NCo #2 has specified a Beacon Region.

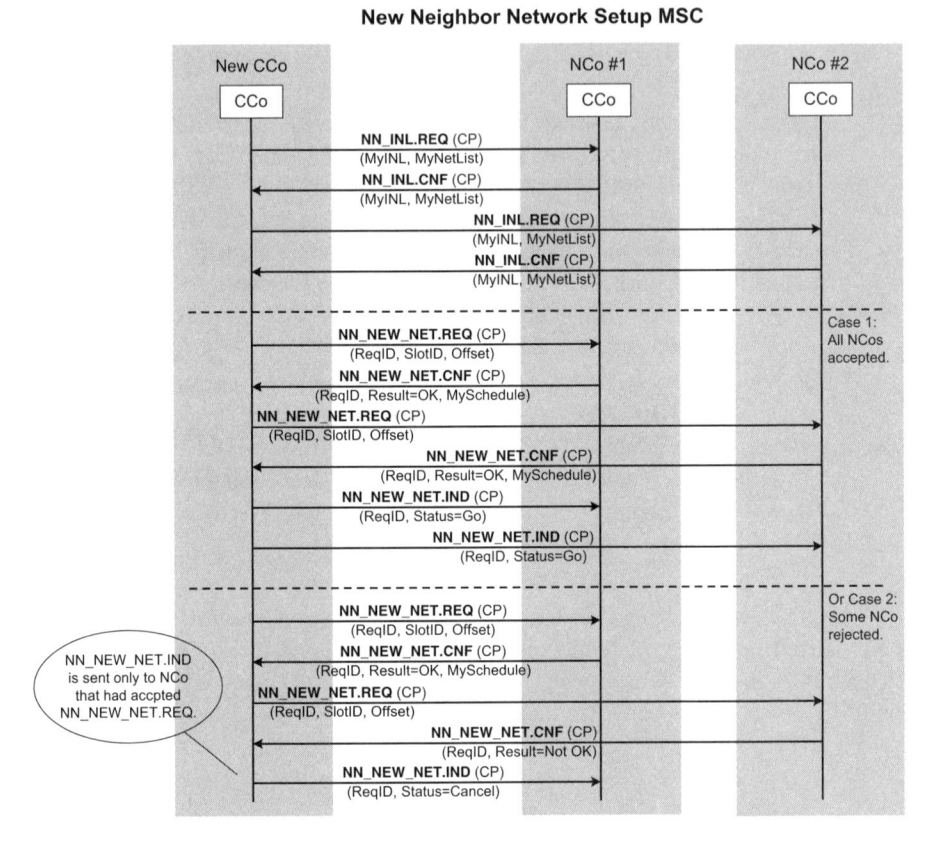

FIGURE 12.3 MSC to set up a new network in Coordinated Mode.

12.3.5.2 Changing the Number of Beacon Slots When a CCo in a group of networks determines that the number of Beacon Slots needs to be changed, the Change NumSlots BENTRY is added to the Central Beacon with NewNumSlots set to the new proposed value for NumSlots. The NumSlot Change Countdown (NSCCD) field should be set to a value large enough to ensure the BENTRY will propagate to and from the furthest CCo in the group. A value of at least 16 is recommended.

All CCos in a group of networks that detect a Change NumSlots BENTRY in the Central Beacon of another CCo in the group add the BENTRY to their Central Beacon. The value for NewNumSlots is set to be the larger of the highest SlotID of CCos it directly detected in the group or the value of NewNumSlots detected in the BENTRY of the other Central Beacons. NSCCD must be set to assure the value is identical for all Central Beacons in a Beacon Period of the group.

Any CCo in a group of networks that detects a higher value for NewNumSlots or NSCCD in the BENTRY of another Central Beacon in the Group should change the corresponding field in their BENTRY to match the larger value.

COORDINATED MODE

New CCo detects two Groups of Networks and decides to join one of them.

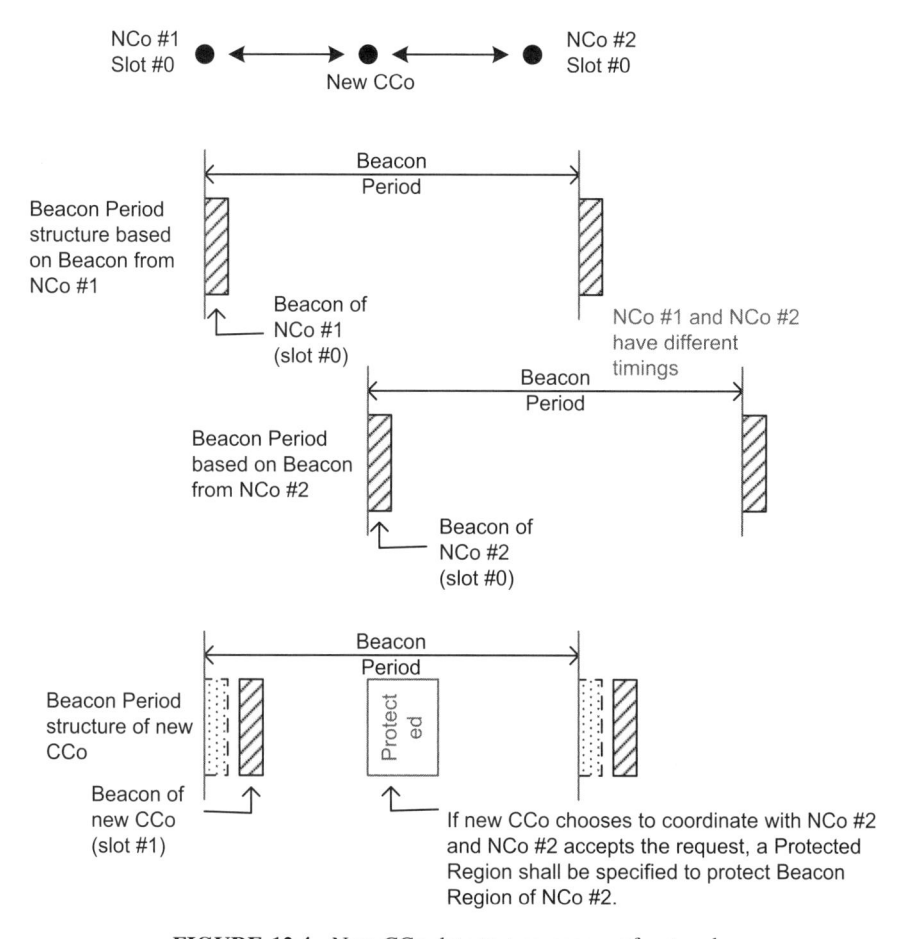

FIGURE 12.4 New CCo detects two groups of networks.

The change becomes effective in the Beacon Period following the Beacon Period where NSCCD reaches 1. When the change becomes effective, the NumSlots field in the Beacon MPDU Payload is set to the most recent value of NewNumSlots and the BENTRY be removed.

This process ensures that NumSlots for all networks in the group will be updated at the same time to the minimum value appropriate for the group.

When no change NumSlots BENTRY is present in any Central Beacon in the group, NumSlots is set in the next Beacon Period to the largest value heard in any other Central Beacon in the group.

When the change NumSlots BENTRY is present in a Beacon, Coordinating CCos of a different group may adjust their Protected Region based on the value of

NewNumSlots when NSCCD = 1. When the change NumSlots BENTRY is not present, the NumSlots field in the Beacon MPDU payload may be used.

12.3.5.3 Setting the Value of SlotUsage Field
A CCo in Coordinated Mode sets the SlotUsage field (refer to Section 5.2.2.3) according to the following rules:

- The bit corresponding to the Beacon Slot where the CCo transmits its own Beacon is set to **0b1**.
- The bit corresponding to the Beacon Slot where Beacons of a network are regularly detected is set to **0b1**.
- Remaining bits are set to **0b0**.

Discover and Proxy Beacons sets the SlotUsage field according to the following rules:

- The bit corresponding to the Beacon Slot where Beacons of a network are regularly detected a reset to **0b1**.
- Remaining bits are set to **0b0**.

Note: The SlotUsage Field of Beacon transmissions of STAs in an AVLN might not match the SlotUsage Field of the AVLN's CCo. A STA in the AVLN will set the SlotUsage field to **0b0** for the SlotID of the Central Beacon for its AVLN if it is a Hidden STA. A CCo will consider a Beacon Slot available if the bit is set to **0b0** in all Beacon transmissions that can be reliably detected including Discover and Proxy Beacons.

The HomePlug AV specification [1] gives several examples that are quite instructive on the operation of Neighbor Network Coordination.

12.3.6 Procedure to Share Bandwidth in Coordinated Mode

This section describes the procedure for sharing bandwidth between neighboring AVLNs operating in Coordinated Mode.

1. The (source) CCo that requests to share new bandwidth with the NCos in its INL first determine new time interval(s) that it desires to reserve. Section 12.3.3 describes the rules that an existing CCo uses to choose a compatible schedule.
2. The CCo sends the **NN_ADD_ALLOC.REQ** message to each of the NCos in its INL. The message contains the additional time interval(s) that the source CCo is requesting.
3. If the bandwidth request is accepted, the NCo replies with the **NN_ADD_ALLOC.CNF** message with a successful result code. The NCo also change the Region's message of its Beacon to reflect the changes in the schedule. Otherwise, the **NN_ADD_ALLOC.CNF** message with an unsuccessful result code is returned.

COORDINATED MODE

4. When the CCo receives responses from all the NCos in the INL, it sends the **NN_ADD_ALLOC.IND** message to the NCos in the INL. If all the NCos have replied with a successful result code in the **NN_ADD_ALLOC.CNF** message, the status field of the **NN_ADD_ALLOC.IND** message is set to "Go" to confirm that the CCo is going to reserve the time interval.

5. If one or more NCos have replied with an unsuccessful result code in the **NN_ADD_ALLOC.CNF** message, the status field of the **NN_ADD_ALLOC.IND** message is set to "Cancel" to inform the NCos that the request has been cancelled. In this case, the CCo sends the **NN_ADD_ALLOC.IND** message only to those NCos that have replied with a successful result code in the **NN_ADD_ALLOC.CNF** message. On receiving the **NN_ADD_ALLOC.IND** message with a "Cancel" status field, the NCo changes the Region's message of its Beacon to the original value.

Figure 12.5 shows an example. The CCo is operating in Coordinated Mode with two other CCos (NCo #1 and NCo #2). In Case 1, both NCo #1 and NCo #2 accept the CCo's request for additional bandwidth. In Case 2, NCo #2 rejects the request.

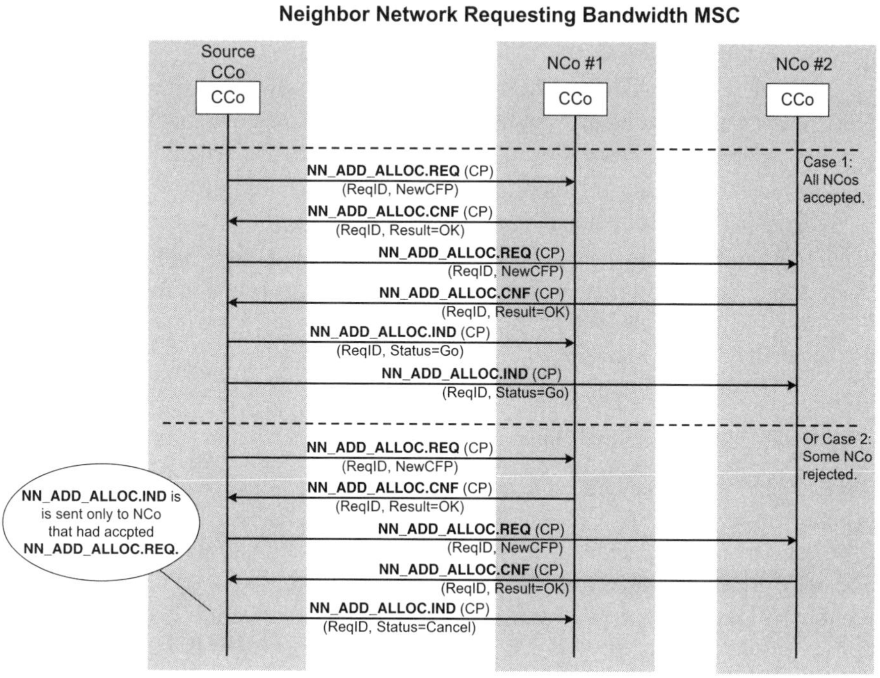

FIGURE 12.5 MSC to request additional bandwidth in Coordinated Mode.

12.3.7 Bandwidth Scheduling Rules

AVLNs operating in Coordinated Mode use **NN_ADD_ALLOC.REQ** messages to request Bandwidth from Coordinating AVLNs in the INL. The following rules should be used for sharing the bandwidth between Coordinating AVLNs:

1. The scheduling policy is on first-come, first-served basis when shared CSMA Region is available. Thus, an AVLN (CCo) may obtain any amount of bandwidth from the shared CSMA Region as far as the duration of MinCSMARegion is not violated.
2. When a CCo requires more bandwidth for a new Connection or to support an existing Connection, then:
3. If sufficient bandwidth is available in Shared CSMA Region (subjected to the MinCSMARegion restriction), the CCo can send **NN_ADD_ALLOC.REQ** asking for time allocation within the shared CSMA Regions of the Neighbor CCos.
4. If insufficient bandwidth is available in shared CSMA Region, bandwidth occupancy of the AVLN is compared with the bandwidth quota.
5. The CCo should not send **NN_ADD_ALLOC.REQ** to neighboring CCos if its AVLN is occupying more bandwidth than its bandwidth quota.
6. If the CCo is not currently using its bandwidth quota, it can send **NN_ADD_ALLOC.REQ** asking for time allocation within the Reserved Regions of the Neighbor CCos.
7. When an **NN_ADD_ALLOC.REQ** is received by a CCo that is currently using more than its share of the bandwidth quota, it should reconfigure and/or terminate existing Connection(s) to accommodate the request.
8. AVLNs operating in Coordinated Mode should reserve only the minimum duration of time required to support the on-going Connections.
9. A CCo that is releasing a reserved time interval sends the **NN_REL_ALLOC.REQ** message to each NCo in its INL. The message specifies the time interval that is being released by the CCo. Each NCo replies with an **NN_REL_ALLOC.CNF** message.

12.3.8 Procedure to Shut Down an AVLN

Figure 12.6 shows the procedure for shutting down an AVLN. The CCo that is shutting down its AVLN sends the **NN_REL_NET.IND** message to each NCo in its INL. The message specifies the Beacon Slot being used and the locations of the Reserved Regions that have been reserved by the CCo.

In addition, if the last Beacon Slot in the Beacon Region has been unoccupied for a period longer than **IDLE_BEACON_SLOT_TIMEOUT**, a CCo may request to reduce the number of Beacon Slots by using the process described in Section 12.3.5.2.

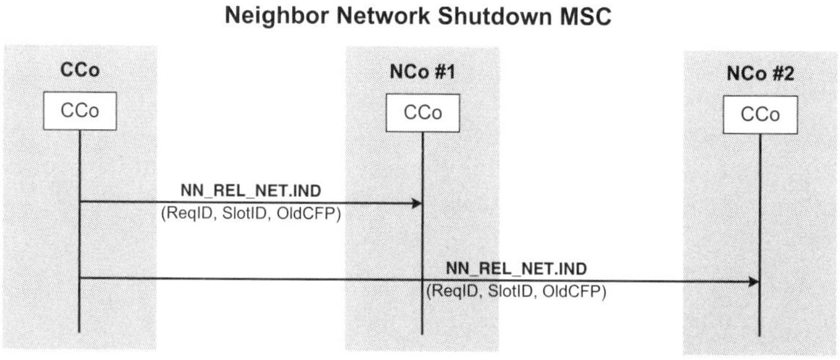

Neighbor Network Shutdown MSC

FIGURE 12.6 MSC to shut down an AVLN in Coordinated Mode.

When an AVLN operating in Coordinated or Uncoordinated Mode shuts down, it uses AC Line Sync Countdown BENTRY with the Reason Code = **0b00** (AVLN Shut Down or leaving Group) to indicate the number of Beacon Periods in which it is going to stop transmitting the Beacons. This information provided by the departing AVLN will enable other Coordinating CCos that are tracking the AC Line cycle to recover appropriately (refer to Section 12.3.9).

12.3.9 AC Line Cycle Synchronization in Coordinated Mode

In a group of Coordinating AVLNs, the CCo in the smallest occupied Beacon Slot number tracks the AC Line Cycle and should provide all four of the Beacon Transmit Offset values (BTO[0] through BTO[3]) to other coordinating CCos. This CCo also sets the AC Line Cycle Synchronization Status (ACLSS) field to the Beacon Slot number it is currently occupying in the Central Beacon to indicate that it is tracking the AC line cycle. All other CCos in the group tracks a CCo that is reliably detected. The CCo that is tracked should have a smaller Beacon Slot number.

To ensure that the group of AVLN's operations does not get disrupted when a Coordinating CCo shuts down, the departing CCo uses the AC Line Sync Countdown BENTRY to indicate the Beacon Period in which it will stop transmitting the Beacon. CCo's tracking the AC line cycle information provided by the departing CCo either chooses a different CCo to track (if one or more is detected) or start tracking the AC line cycle.

If a new AVLN joining a group of Coordinating CCos determines that a Beacon Slot number smaller than the Beacon Slot of the CCo that is currently tracking the AC line cycle is not used, it may request that Beacon Slot only if the Beacon from the CCo that is currently tracking the AC line cycle is reliably detected. In such cases, when the new AVLN becomes part of the group and starts transmitting the Beacons, it will initially track the AC line cycle information provided by the CCo transmitting Beacon in a larger Beacon Slot number and is currently tracking the AC line cycle. Once a CCo that is tracking the AC line cycle determines the presence of Beacon in a smaller Beacon Slot number, it hands over AC line cycle synchronization by using the AC Line Sync Countdown BENTRY.

If a CCo in a group determines that a Beacon Slot number smaller than the current Beacon Slot of the CCo is not used, it may use the Beacon Relocation BENTRY with Relocation Type set to SlotID to move to the smaller Beacon Slot number. A CCo in a group may also use the Beacon Relocation BENTRY with Relocation Type set to SlotID to move to a different available Beacon Slot number as may be necessary to assure reliable Beacon reception for its AVLN.

12.4 PASSIVE COORDINATION IN CSMA-ONLY MODE

Level-0 and Level-1 CCos do not support TDMA in the presence of neighbor networks. Level-2 CCos support Coordinated Mode, which enables neighboring Level-2 CCos to support TDMA. To further enhance the ability of Level-2 CCos in supporting TDMA in the presence of Level-0 and Level-1 CCos in CSMA-only Mode, a mandatory Passive Coordination procedure is defined for Level-0 and Level-1 CCos operating in CSMA-only Mode.

When a Level-1 or Level-0 CCo reliably detects Central or Proxy Beacons from one or more Level-2 CCos operating in Coordinated Mode, it tracks the Beacon Period Start Time of one and only one such Level-2 CCo. Further, it processes the Regions BENTRY of the Level-2 AVLN it is tracking and provide a Local Regions Schedule with CSMA Region during the interval where the neighboring AVLN has a CSMA Region and Stayout Region in the remainder of the Beacon Period. All STAs in the CSMA-only Mode AVLN refrain from transmitting (using Persistent and Nonpersistent Schedule BENTRIES) during the Stayout Region. This procedure is referred to as Passive Coordination.

When a Level-2 CCo operating in Coordinated Mode detects Central Beacons from one or more Level-0 or Level-1 CCos, it should provide a larger minimum CSMA Region during the Beacon Period to share the medium fairly with Level-0/Level-1 AVLNs. The duration of the minimum CSMA Region will be based on the sharing policy adopted by the service providers.

12.5 NEIGHBOR NETWORK BANDWIDTH SHARING POLICY

HomePlug Policy Creation and Management Working Group (PCMWG) is responsible for the definition of bandwidth sharing policies between neighboring network. PCMWG mandated the following policies;

- When one AVLN operating in Coordinated Mode detects one other CSMA AVLN, it must increase the MinCSMARegion to at least half of the Available Beacon Period (i.e., Beacon Period minus the duration of Beacon Region).
- When one AVLN operating in Coordinated Mode detects two other CSMA AVLN, it must increase the MinCSMARegion to at least two thirds of the Available Beacon Period.

- When two AVLN operating in Coordinated Mode detects three other CSMA AVLN, it must increase the MinCSMARegion to at least three fifths of the Available Beacon Period.

In general, if there are NAVLNs operating in Coordinated Mode and M AVLNs operating in CSMA-only Mode, the MinCSMARegion is determined as follows:

$$\text{MinCSMARegion} = \max\ (1.5\,\text{ms},\ \text{Available BeaconPeriod}^{*}M/(M+N))$$

12.6 SUMMARY

This chapter discussed bandwidth sharing when HomePlug AV is operating in the presence of Neighbor Networks, including the specific cases of CSMA-only, Uncoordinated and Coordinated Modes.

Chapter 13 presents a selection of important MAC management messages, divided based on functionality, as used in HomePlug AV.

13

MANAGEMENT MESSAGES

13.1 INTRODUCTION

Management messages are used by the MAC Layer for network management and provisioning of Quality of Service (QoS). Management messages also enable higher layers to configure HomePlug AV stations as well as to obtain performance statistics. The format of Management Messages is based on the standard Ethernet frame format, with a unique Ethertype assigned to HomePlug. The Ethernet format enables HomePlug AV management message to be exchanged with higher layers across an Ethernet network.

13.2 MANAGEMENT MESSAGE FORMAT

The general format of Management Message is shown in Figure 13.1. Details of various fields in the management message are provided in the following sections.

13.2.1 Original Destination Address (ODA)

Original Destination Address (ODA) is the 6-Octet MAC Address of the destination station(s) of the MAC Service Data Unit (MSDU). The format of this field is same as described in the IEEE Standard 802–2001. The prefix "Original" is used to

---

HomePlug AV and IEEE 1901: A Handbook for PLC Designers and Users, First Edition.
Haniph A. Latchman, Srinivas Katar, Larry Yonge, and Sherman Gavette.
© 2013 by The Institute of Electrical and Electronics Engineers, Inc. Published 2013 by John Wiley & Sons, Inc.

MANAGEMENT MESSAGE FORMAT

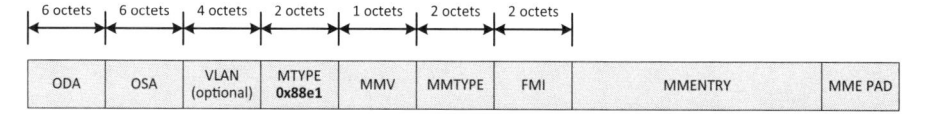

6 octets	6 octets	4 octets	2 octets	1 octets	2 octets	2 octets	
ODA	OSA	VLAN (optional)	MTYPE 0x88e1	MMV	MMTYPE	FMI	MMENTRY	MME PAD

FIGURE 13.1 Management Message format.

emphasize that the ODA can be a bridged station (e.g., an Ethernet station that is bridged through a powerline station).

13.2.2 Original Source Address (OSA)

Original Source Address is the 6-Octet MAC Address of the station that is the source of the MSDU. The format of this field is same as described in the IEEE Standard 802–2001. The prefix "Original" is used to emphasize that the OSA can be a bridged station.

13.2.3 VLAN Tag

VLAN Tag is a 4-Octet VLAN Tag field that is formatted as described in IEEE 802.1Q. This field is optional and is not necessarily present on all management messages.

13.2.4 MTYPE

MTYPE is set to the IEEE-assigned Ethertype value of **0x88e1**. This unique identifier is assigned by IEEE to HomePlug and enables HomePlug AV management messages to be distinguished from other Ethernet frames.

13.2.5 Management Message Version (MMV)

Management Message Version (MMV) is a 1-octet field that indicates the version of the Management Message. All HomePlug AV management messages have this field is set to **0x01**.

13.2.6 Management Message Type (MMTYPE)

Management Message Type (MMTYPE) is a 2-octet field that indicates the type of Management Message that follows.

The two LSBs of MMTYPE indicate that the message is a Request, Confirm, Indication, or Response. In the HomePlug AV specification, Request messages always end in REQ. The response (if any) to a Request message is always a Confirmation message, which ends in CNF. Indication messages always end in IND. The response (if any) to an Indication message is always a Response message, which ends in "RSP."

The three MSBs of the MMTYPE indicate the category to which the Management Message belongs. The categories of management messages are as follows:

- *Station—Central Coordinator:* Messages belonging to this category are exchanged between Station (STA) and Central Coordinator (CCo). A prefix "CC" is used in naming management messages exchanged between STA and CCo.
- *Proxy Coordinator:* Messages belonging to this category are exchanged with the Proxy Coordinator as part of handling hidden stations. A prefix "CP" is used for naming messages exchanged between the CCo and Proxy Coordinator. A prefix "PH" is used for naming messages exchanged between Proxy Coordinator and hidden station.
- *Central Coordinator–Central Coordinator:* Messages belonging to this category are exchanged between neighboring CCos. A prefix "NN" is used for naming messages belonging to this category.
- *Station–Station:* Messages belonging to this category are exchanged between two stations. A prefix "CM" is used for naming messages belonging to this category.
- *Manufacturer Specific:* These messages are defined by the AV chip manufacturers for exchanging manufacturer dependent control information between the host layer and the MAC. These messages are not transmitted over the powerline.
- *Vendor Specific:* These messages are defined by either the AV chip manufacturer or AV product vendor for exchanging chip or product implementation dependent control information across the host interface and/or over the powerline (i.e., between stations).

Sections 13.3–13.8 provides a summary of management messages types belonging to each of these categories.

13.2.7 Fragment Management Information

HomePlug AV restricts the maximum size of management messages transmitted using multinetwork broadcasting to 502 octets (refer to Section 6.9.1). For transmissions to STAs associated with the same HomePlug AV Local Network (AVLN), the maximum size of the management message is limited to 1518 octets (including VLAN Tag). The later restriction enables all HomePlug AV management messages to be transmitted across standard Ethernet interfaces.

In instances where the amount of management information that needs to be exchanged exceeds these limits, it is fragmented and transmitted using multiple management messages. Fragment management information contains the necessary information for receiver to obtain the complete management information individual

MANAGEMENT MESSAGE FORMAT

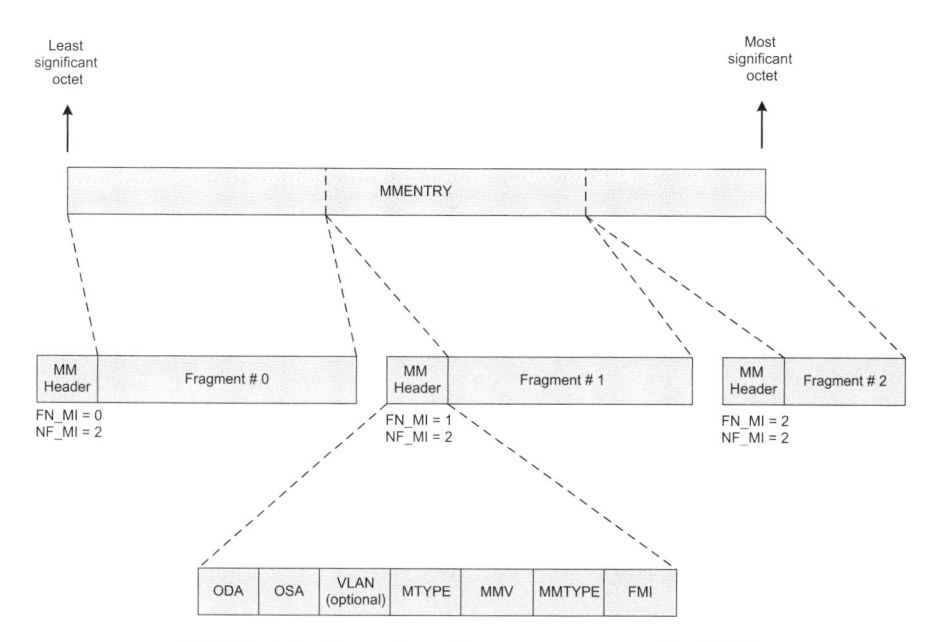

FIGURE 13.2 Illustration of fragmentation of a MMENTRY.

management messages. Fragment management information consists of the following fields:

- *Number of Fragments (NF_MI):* Number of fragments field is a 4-bit field that indicates the number of management messages into which the management information is fragmented. NF_MI field remains constant across all management messages that carry fragments of the same management information.
- *Fragment Number (FN_MI):* Fragment number is a 4-bit field indicates the fragment number of the management information contained within the management message.
- *FMSN:* The FMSN field is an 8-bit field that initialized to zero and incremented by one when management information has to be fragmented at the transmitter, regardless of the destination address or the type/version of the management message. FMSN field remains constant across all management messages that carry fragments of the same management information.

For fragmentation purposes, the MMENTRY is treated as an octet stream. Each management message carrying fragmented MMENTRY contains the ODA, OSA, MTYPE, MMV, MMTYPE, and FMI fields followed by a fragment of the MMENTRY. Figure 13.2 shows fragmentation of a MMENTRY into three management messages. The receiver uses the {ODA, OSA, MMV, MMTYPE, FMSN} to uniquely identify fragments belonging to the same management information.

13.2.8 Management Message Entry Data (MME)

The format of Management Message Entry Data (message) depends on the MMTYPE with which it is associated. The nominal priority settings for the Management Messages is PLID = **0x02**.

13.2.9 MMEPAD

All HomePlug AV management messages are required to be at least 60 octets long to enable them to be compatible with Ethernet frame format. MME PAD is a variable-length field that is added to management messages whose length, excluding the MME-PAD (i.e., from ODA to MMENTRY), is less than 60 octets. When MME PAD is present, its length is chosen to be the smallest possible value to ensure that the Management Message length, including the MME PAD (i.e., from ODA to MME PAD), is equal to 60 octets.

13.3 STATION–CENTRAL COORDINATION (CCO)

The set of management messages exchanged between a station and Central Coordinator and their usage is shown in Table 13.1.

TABLE 13.1 Station–Central Coordinator Messages

MMTYPE	Usage
CC_CCO_APPOINT.REQ	**CC_CCO_APPOINT.REQ** message is used to appoint a STA in the AVLN as a CCo. This message can also be used to un-appoint an existing CCo from being a user-appointed CCo
CC_CCO_APPOINT.CNF	**CC_CCO_APPOINT.CNF** message is sent in response to a received **CC_CCO_APPOINT.REQ** message to indicate the result of the request
CC_BACKUP_APPOINT.REQ	**CC_BACKUP_APPOINT.REQ** message is sent by the CCo to a STA to request the STA to become a Backup CCo. This message can also be used to release an existing Backup CCo from its duty as a Backup CCo
CC_BACKUP_APPOINT.CNF	**CC_BACKUP_APPOINT.CNF** message is sent by a STA to the CCo in response to a received **CC_BACKUP_APPOINT.REQ** message to indicate the result of the request
CC_LINK_INFO.REQ	**CC_LINK_INFO.REQ** message is sent by a STA to the CCo to request the CSPEC and BLE information of all active Global Links in the AVLN

TABLE 13.1 (*Continued*)

MMTYPE	Usage
CC_LINK_INFO.CNF	**CC_LINK_INFO.CNF** message is sent by the CCo in response to a received **CC_LINK_INFO.REQ** message. The message contains the CSPEC with CM-to-CCo QoS and MAC parameters and BLE information of all active Global Link(s) in the AVLN
CC_LINK_INFO.IND	**CC_LINK_INFO.IND** message is sent by a CCo to either a new CCo (during soft handover) or a Backup CCo (as part of CCo failure recovery) to provide the CSPEC with CM-to-CCo QOS and MAC parameters, and BLE information of the Global Link(s) that are active within the AVLN
CC_LINK_INFO.RSP	**CC_LINK_INFO.RSP** message is sent by the new CCo or Backup CCo to the current CCo to confirm the reception of the **CC_LINK_INFO.IND** message
CC_HANDOVER.REQ	**CC_HANDOVER.REQ** message is sent by the current CCo to another STA in the network to request the STA to become the new CCo
CC_HANDOVER.CNF	**CC_HANDOVER.CNF** message is sent in response to a received **CC_HANDOVER.REQ** message to indicate the result of the request
CC_HANDOVER_INFO.IND	**CC_HANDOVER_INFO.IND** message is sent by the current CCo to the new CCo during the handover process. This message is also sent by the current CCo to the Backup CCo to enable recovery from CCo failure
CC_HANDOVER_INFO.RSP	**CC_HANDOVER_INFO.RSP** message is sent by the new CCo or Backup CCo to the current CCo to confirm the reception of the **CC_HANDOVER_INFO.IND** messages
CC_DISCOVER_LIST.REQ	**CC_DISCOVER_LIST.REQ** message is sent by a STA to request the Discovered STA List and Discovered Network List of another STA Although this message is typically sent by the CCo to a STA in the AVLN, any STA in the AVLN can send this message to another STA in the AVLN and obtain the corresponding **CC_DISCOVER_LIST.CNF**
CC_DISCOVER_LIST.CNF	**CC_DISCOVER_LIST.CNF** message is sent by a STA in response to a received **CC_DISCOVER_LIST.REQ** message to report its Discovered STA List and Discovered Network List.

(*continued*)

TABLE 13.1 (*Continued*)

MMTYPE	Usage
CC_DISCOVER_LIST.IND	**CC_DISCOVER_LIST.IND** message is sent by a STA to the CCo in an unsolicited manner to report its Discovered STA List and Discovered Network List whenever the STA detects a new network
CC_LINK_NEW.REQ	**CC_LINK_NEW.REQ** message is sent by the initiating STA to the CCo to request connection setup in the CFP
CC_LINK_NEW.CNF	CCo sends the **CC_LINK_NEW.CNF** message to the initiating STA and terminating STA(s) of a Connection to confirm the completion of establishment of the Global Links associated with the Connection
CC_LINK_MOD.REQ	**CC_LINK_MOD.REQ** message is sent by either the initiating STA or the terminating station of a Connection to the CCo to request modification of Global Link(s)
CC_LINK_MOD.CNF	**CC_LINK_MOD.CNF** message is sent by the CCo to the STAs involved in a Connection to notify them that the reconfiguration of the CFP Link(s) has been completed successfully or failed
CC_LINK_SQZ.REQ	**CC_LINK_SQZ.REQ** message is set from the CCo to a STA to request it to Squeeze/De-Squeeze an existing connection
CC_LINK_SQZ.CNF	**CC_LINK_SQZ.CNF** message is set from the STA to the CCo in response to the corresponding **CC_LINK_SQZ.REQ** and indicates whether the request was successful or failed
CC_LINK_REL.REQ	**CC_LINK_REL.REQ** message is sent by a STA to the CCo to request release of the Global Links associated with a Connection
CC_LINK_REL.IND	**CC_LINK_REL.IND** message is sent by the CCo to the initiating STA and terminal station (s) of a Connection to indicate release of the Global Links associated with a Connection. The message is generated in response to the corresponding **CC_LINK_REL.REQ**. The CCo may also generate this message in an unsolicited manner when an existing Connection is terminated due to insufficient bandwidth, violation of the CSPEC, or at the request of another station within the AVLN.
CC_DETECT_REPORT.REQ	**CC_DETECT_REPORT.REQ** message is sent by the CCo to request a STA to perform the detect-and-report procedure

TABLE 13.1 (*Continued*)

MMTYPE	Usage
CC_DETECT_REPORT.CNF	**CC_DETECT_REPORT.CNF** message is sent by a STA to report to the CCo the results of the detect-and-report procedure
CC_WHO_RU.REQ	**CC_WHO_RU.REQ** is used to request the CCo to provide Human Friendly Identify (HFID) of the AVLN
CC_WHO_RU.CNF	**CC_WHO_RU.CNF** is sent by the CCo in response to the **CC_WHO_RU.REQ** and contains the Human Friendly Identifier of the AVLN
CC_ASSOC.REQ	**CC_ASSOC.REQ** is sent from a STA to the CCo to obtain anew TEI or extend its existing TEI lease
CC_ASSOC.CNF	**CC_ASSOC.CNF** is sent in response to the **CC_ASSOC.REQ** and contains the TEI assigned to the STA and the TEI lease duration
CC_LEAVE.REQ	**CC_LEAVE.REQ** message is sent by a station when it determines to leave the network.
CC_LEAVE.CNF	**CC_LEAVE.CNF** is sent by the CCo in response to a **CC_LEAVE.REQ** message
CC_LEAVE.IND	**CC_LEAVE.IND** is sent from the CCo to a STA to request it to leave the AVLN
CC_LEAVE.RSP	**CC_LEAVE.RSP** is sent in response to the corresponding **CC_LEAVE.IND** to indicate to the CCo that the STA is leaving the AVLN
CC_SET_TEI_MAP.REQ	**CC_SET_TEI_MAP.REQ** MME is sent to the CCo by an authenticated STA to request that the CCo send it a complete TEI_MAP of the AVLN
CC_SET_TEI_MAP.IND	**CC_SET_TEI_MAP.IND** MME is sent by the CCo to notify one or more STAs of any changes to the (TEI, MAC Address) mapping. This message is also sent in response to a **CC_SET_TEI_MAP.REQ**
CC_RELAY.REQ	**CC_RELAY.REQ** message is used to request to forward an unencrypted MME to a final STA. The TEI and MAC Address of the final STA are given as fields in the **CC_RELAY.REQ** message. Upon receiving this message, the PSTA extracts the Payload field, encapsulate it in a **CC_RELAY.IND** message and send it to the final destination STA
CC_RELAY.IND	**CC_RELAY.IND** message is used to forward an MME received through the corresponding **CC_RELAY.REQ**

(*continued*)

TABLE 13.1 (*Continued*)

MMTYPE	Usage
CC_BEACON_RELIABILITY.REQ	**CC_BEACON_RELIABILITY.REQ** is used by the CCo to request the detection reliability of Central Beacon from other station(s) within the AVLN
CC_BEACON_RELIABILITY.CNF	**CC_BEACON_RELIABILITY.CNF** is generated by a station in response to the corresponding **CC_BEACON_RELIABILITY.REQ**, and contains Central Beacon detection statistics. This message may also be generated in an unsolicited manner when a station observes poor Central Beacon detection
CC_DCPPC.IND	**CC_DCPPC.IND** message is sent by a station to the CCo to indicate that the station uses a different receive PHY clock correction during the CP than this network, identified by the SNID. The **CC_DCPPC.IND** message is also used to indicate when a station changes from using a different PHY Receive Clock Correction to using the correct PHY Receive Clock Correction for the network
CC_DCPPC.RSP	**CC_DCPPC.RSP** message is sent by the CCo in response to the corresponding **CC_DCPPC.IND** message
CC_HP1_DET.REQ	**CC_HP1_DET.REQ** message is a request for the CCo to the station(s) to provide statistics on the detected HomePlug 1.0.1 and HomePlug 1.1 transmissions
CC_HP1_DET.CNF	**CC_HP1_DET.CNF** message is generated in response to a corresponding **CC_HP1_DET.REQ** and contains the HomePlug 1.0.1 and HomePlug 1.1 detection statistics. This message may also be generated by AV stations in an unsolicited manner when HomePlug 1.0.1 and/or HomePlug 1.1 transmissions are detected
CC_BLE_UPDATE.IND	**CC_BLE_UPDATE.IND** message is sent from the STA that is the source of a Global Link to the CCo to provide the latest Bit-loading Estimates to the CCo. Reception of **CC_BLE_UPDATE.IND** causes the CCo to update the Bit-loading Estimates for the Global Link and the duration of CF allocation accordingly
CC_BCAST_REPEAT.IND	**CC_BCAST_REPEAT.IND** is sent by the CCo to provide a STA with the BMRAT, which is a list of STEIs for which the STA is responsible

TABLE 13.1 (*Continued*)

MMTYPE	Usage
	for retransmitting PBs for broadcast MPDUs received with a STEI in the BMRAT
CC_BCAST_REPEAT.RSP	**CC_BCAST_REPEAT.RSP** message is sent by the STA to confirm the reception of the **CC_BCAST_REPEAT.IND** message
CC_POWERSAVE.REQ	The **CC_POWERSAVE.REQ** message is sent by a STA to the CCo to request permission to go into Power Save mode
CC_POWERSAVE.CNF	The **CC_POWERSAVE.CNF** message is sent by the CCo in response to the corresponding **CC_POWERSAVE.REQ**. This message indicates whether the CCo accepted or rejected the request to enter Power Save mode. This message also includes the Power Save State Identifier and the list of all STAs (including CCo) in Power Save mode and their Power Save Schedules
CC_POWERSAVE_EXIT.REQ	The **CC_POWERSAVE_EXIT.REQ** message is sent by a STA to the CCo to indicate that it has exited the Power Save mode
CC_POWERSAVE_EXIT.CNF	The **CC_POWERSAVE_EXIT.CNF** message is sent by the CCo to confirm that it has received the corresponding **CC_POWERSAVE_EXIT.REQ** message. This message also includes the Power Save State Identifier and the list of all STAs (including the CCo) in Power Save mode and their Power Save Schedules
CC_POWERSAVE_LIST.REQ	The **CC_POWERSAVE_LIST.REQ** message is sent by a STA to the CCo to request the list of STAs in Power Save mode
CC_POWERSAVE_LIST.CNF	The **CC_POWERSAVE_LIST.CNF** message is sent by the CCo in response to the corresponding **CC_POWERSAVE_LIST.REQ** message. This message contains the Power Save State Identifier and the list of all STAs (including CCo) in Power Save mode and their Power Save Schedules
	Alternatively, the **CC_POWERSAVE_LIST.CNF** message is sent by the CCo as a unicast or broadcast message when the CCo is not able to indicate all changes to Power Save Schedules in the Power Save BENTRY due to the limited Beacon Size

TABLE 13.2 Proxy Coordinator Messages

MMTYPE	Usage
CP_PROXY_APPOINT.REQ	CP_PROXY_APPOINT.REQ message is sent by the CCo to a PSTA to promote it to PCo or to a PCo to update the PCo's information
CP_PROXY_APPOINT.CNF	CP_PROXY_APPOINT.CNF message is sent by a PSTA or PCo to the CCo in response to a received CP_PROXY_APPOINT.REQ message
PH_PROXY_APPOINT.IND	PH_PROXY_APPOINT.IND message is sent by a PSTA or PCo to an HSTA to indicate that the PCo is responsible for the HSTA
CP_PROXY_WAKE.REQ	CP_PROXY_WAKE.REQ may be sent by a PCo to request exit from Network Power Saving Mode when it detects transmission from HSTA

13.4 PROXY COORDINATOR (PCO) MESSAGES

Proxy Coordinator Messages are used for enabling hidden station to be part of an AVLN. The set of management messages exchanged with a Proxy Coordinator and their usage is shown in Table 13.2.

13.5 CENTRAL COORDINATOR–CENTRAL COORDINATOR

Central Coordinator of an AVLN exchanges messages with other Central Coordinators within in vicinity to facilitate neighbor network coordination. The set of management messages exchanged between neighboring CCos and their usage is shown in Table 13.3. All management messages between CCos of Neighbor Networks are unencrypted.

TABLE 13.3 Central Coordinator-Central Coordinator Messages

MMTYPE	Usage
NN_INL.REQ	NN_INL.REQ is used to request the Interfering Network List(INL) from neighboring networks CCo. NN_INL.REQ includes the INL of the CCo transmitting the request
NN_INL.CNF	NN_INL.CNF is generated in response to the Corresponding NN_INL.REQ and contains the INL of the CCo transmitting the confirmation

TABLE 13.3 (*Continued*)

MMTYPE	Usage
NN_NEW_NET.REQ	**NN_NEW_NET.REQ** message is sent by a new CCo to the CCos in its INL to request to set up a new network
NN_NEW_NET.CNF	**NN_NEW_NET.CNF** message is sent by a CCo in response to a received **NN_NEW_NET.REQ** message
NN_NEW_NET.IND	**NN_NEW_NET.IND** message is sent by the new CCo (i.e., CCo which sent the **NN_NEW_NET.REQ** message) to the CCo's in its INL to Indicate whether it has successfully formed a new network
	If all the CCos in the INL respond to the new CCo with **NN_NEW_NET.CNF** message indicating success, the new CCo will send **NN_NEW_NET.IND** message indicating that new network was successfully formed. Otherwise, it responds with **NN_NEW_NET.IND** message indicating the new network cannot be formed
NN_ADD_ALLOC.REQ	**NN_ADD_ALLOC.REQ** message is sent by a CCo to other CCo's in its INL to request to share additional bandwidth. The message contains the proposed schedules to be used by the CCo
NN_ADD_ALLOC.CNF	**NN_ADD_ALLOC.CNF** message is sent in response to a received **NN_ADD_ALLOC.REQ** message and indicates whether the additional bandwidth request was accepted or rejected
NN_ADD_ALLOC.IND	**NN_ADD_ALLOC.IND** message is sent by a CCo (which sent the **NN_ADD_ALLOC.REQ** message) to the CCo's in its INL to confirm whether the bandwidth request is successful or canceled
	If all CCos in the INL respond with **NN_ADD_ALLOC.CNF** message indicating success, the CCo will send an **NN_ADD_ALLOC.IND** message indicating that additional bandwidth was successfully allocated. Otherwise, it responds with **NN_ADD_ALLOC.IND** message indicating that additional bandwidth allocation was canceled
NN_REL_ALLOC.REQ	**NN_REL_ALLOC.REQ** message is sent by a CCo to the CCos of its INL to request to release part or all of its Reserved Regions
NN_REL_ALLOC.CNF	**NN_REL_ALLOC.CNF** message is sent by a CCo to another CCo to indicate successful reception of the corresponding **NN_REL_ALLOC.REQ** message
NN_REL_NET.IND	**NN_REL_NET.IND** message is sent by a CCo to the CCo's of its INL to release all its Reserved Regions and shut down the network

13.6 STATION–STATION

The set of management messages exchanged between two stations and their usage is shown in Table 13.4.

TABLE 13.4 Station–Station Coordinator Messages

MMTYPE	Usage
CM_UNASSOCIATED_STA.IND	CM_UNASSOCIATED_STA.IND message in transmitted by a STA that is not associated with any AVLN to indicate its presence to other stations in the vicinity
CM_ENCRYPTED_PAYLOAD.IND	CM_ENCRYPTED_PAYLOAD.IND message is used to transmit encrypted information within a management message. This message is used for authorization and authentication. This message may also be used by higher layers to exchange encrypted information
CM_ENCRYPTED_PAYLOAD.RSP	CM_ENCRYPTED_PAYLOAD.RSP message is only sent if the corresponding CM_ENCRYPTED_PAYLOAD.IND message failed. This message is never sent to indicate success
CM_SET_KEY.REQ	CM_SET_KEY.REQ message is used to provide Encryption Key. This message is usually embedded within the encrypted payload of a CM_ENCRYPTED_PAYLOAD.IND message
CM_SET_KEY.CNF	CM_SET_KEY.CNF message is sent in response to the corresponding CM_SET_KEY.REQ to indicate whether the request is successful
CM_GET_KEY.REQ	CM_GET_KEY.REQ message is used to request a key. This message is usually is embedded within the encrypted payload of a CM_ENCRYPTED_PAYLOAD.IND message
CM_GET_KEY.CNF	CM_GET_KEY.REQ message is transmitted in response to the corresponding CM_GET_KEY.REQ and contains the requested key. This message is usually is embedded within the encrypted payload of a CM_ENCRYPTED_PAYLOAD.IND message
CM_SC_JOIN.REQ	CM_SC_JOIN.REQ message used as part of simple connect (or push-button) based authorization. This message is transmitted by a station in SC-Join state to request joining or forming an AVLN

STATION–STATION

TABLE 13.4 (*Continued*)

MMTYPE	Usage
CM_SC_JOIN.CNF	CM_SC_JOIN.CNF message is transmitted by a STA if it accepts the corresponding CM_SC_JOIN.REQ message
CM_CHAN_EST.IND	CM_CHAN_EST.IND message is used send a new tone map to another station
CM_TM_UPDATE.IND	CM_TM_UPDATE.IND message is used to modify a previously communicated tone map
CM_AMP_MAP.REQ	CM_AMP_MAP.REQ message is used to send a new Amplitude Map to another station
CM_AMP_MAP.CNF	CM_AMP_MAP.CNF message is transmitted in response to the corresponding CM_AMP_MAP.REQ to indicate whether the new Amplitude Map was successfully applied by the station
CM_BRG_INFO.REQ	CM_BRG_INFO.REQ message is used to request Bridging Information from another station
CM_BRG_INFO.CNF	CM_BRG_INFO.CNF message contains the set of stations to which the current station is acting like a bridge. This message is transmitted in response to a CM_BRG_INFO.REQ. This message can also be transmitted by a station in an unsolicited manner (e.g., if the bridging information changes)
CM_CONN_NEW.REQ	CM_CONN_NEW.REQ message is a request from the station that is initiating the connection to the terminating station(s) to add a new connection.
CM_CONN_NEW.CNF	CM_CONN_NEW.CNF message is transmitted in response to the corresponding CM_CONN_NEW.REQ to indicate whether the connection request was accepted or not.
CM_CONN_REL.IND	CM_CONN_REL.IND message is used to indicate the release of a connection.
CM_CONN_REL.RSP	CM_CONN_REL.RSP message is transmitted in response to the corresponding CM_CONN_REL.IND message to indicate successful release of a connection
CM_CONN_MOD.REQ	CM_CONN_MOD.REQ message is used to request connection reconfiguration of an existing connection
CM_CONN_MOD.CNF	CM_CONN_MOD.CNF message is transmitted in response to the corresponding CM_CONN_MOD.REQ message to indicate whether the connection reconfiguration was successful or not

(*continued*)

TABLE 13.4 (*Continued*)

MMTYPE	Usage
CM_CONN_INFO.REQ	CM_CONN_INFO.REQ message is used to request information on ongoing connections that are either initiated or terminated at the STA.
CM_CONN_INFO.CNF	CM_CONN_INFO.CNF message is generated in response to the corresponding CM_CONN_INFO.REQ and contains information on ongoing connections at the station.
CM_STA_CAP.REQ	CM_STA_CAP.REQ message is used to request station capabilities of another station
CM_STA_CAP.CNF	CM_STA_CAP.CNF message is generated in response to the corresponding CM_STA_CAP.REQ and contains the station capabilities
CM_NW_INFO.REQ	CM_NW_INFO.REQ message is used to request the list of AVLNs to which the STA is a member and the information about the AVLN
CM_NW_INFO.CNF	CM_NW_INFO.CNF message is generated in response to the corresponding CM_NW_INFO.REQ and contains information about the AVLN(s) to which the station is a member
CM_GET_BEACON.REQ	CM_GET_BEACON.REQ message is a request to provide the Beacon Payload field of a recently received Central Beacon or Proxy Beacon (if station cannot hear the Central Beacon) of an AVLN to which the STA is a member. This message is primarily intended for use by the higher layers
CM_GET_BEACON.CNF	CM_GET_BEACON.CNF message is generated in response to the corresponding CM_GET_BEACON.REQ and contains the Beacon Payload information
CM_GET_HFID.REQ	CM_GET_HFID.REQ message is used to request Human Friendly Identifier of a STA or an AVLN
CM_GET_HFID.CNF	CM_GET_HFID.CNF message is generated in response to the corresponding CM_GET_HFID.REQ and contains the Human Friendly Identifier
CM_MME_ERROR.IND	CM_MME_ERROR.IND message is generated by a station upon the reception of management message that it does not support. This message may also be generated in response to the reception of a supported management message with invalid field(s)

TABLE 13.4 (*Continued*)

MMTYPE	Usage
CM_NW_STATS.REQ	CM_NW_STATS.REQ message is used to request to network statistics from another station
CM_NW_STATS.CNF	CM_NW_STATS.CNF message is generated in response to the corresponding CM_NW_STATS.REQ and contains the list of all associated and authenticated STAs in the AVLN and the physical layer data rates to all those stations
CM_LINK_STATS.REQ	CM_LINK_STATS.REQ is used to request statistics for a specific link (queue) at another station
CM_LINK_STATS.CNF	CM_LINK_STATS.CNF is generated in response to the corresponding CM_LINK_STATS.REQ and contains statistics (e.g., number of transmitted segments, number of dropped segments, and latency) associated with the link
CM_ROUTE_INFO.REQ	The CM_ROUTE_INFO.REQ message is a request to provide Distance Vector routing information
CM_ROUTE_INFO.CNF	CM_ROUTE_INFO.CNF is generated in response to the corresponding CM_ROUTE_INFO.REQ and provides the requesting STA information from the LRT
CM_ROUTE_INFO.IND	The CM_ROUTE_INFO.IND message has the same format as CM_ROUTE_INFO.CNF and is sent to other STAs in the AVLN when there is a significant change in the LRT information
CM_UNREACHABLE.IND	CM_UNREACHABLE.IND is generated when a STA becomes unreachable. It is sent if the STA was reachable directly but is no longer directly reachable, or in response to receipt of a CM_UNREACHABLE.IND MME from the STA that is the current Next TEI for the now unreachable STA
CM_EXTENDEDTONEMASK.REQ	The CM_EXTENDEDTONEMASK.REQ message is a request from the transmitting station to the receiving station to negotiate the Payload Tone Mask based on the Device Tone Masks that are supported by the transmitting station and receiving station. The CM_EXTENDEDTONEMASK.REQ indicates all carriers that are supported by the transmitting station

(*continued*)

TABLE 13.4 (*Continued*)

MMTYPE	Usage
CM_EXTENDEDTONEMASK.CNF	The **CM_EXTENDEDTONEMASK.CNF** message is generated by the receiver in response to the corresponding **CM_EXTENDEDTONEMASK.REQ**. The **CM_EXTENDEDTONEMASK.CNF** indicates all carriers that are supported by both the transmitting station and the receiving station
CM_SLAC_PARM.REQ	**CM_SLAC_PARM.REQ** is generated by the HLE of an PEV to request the parameters for the SLAC protocol from the EVSE(s)
CM_SLAC_PARM.CNF	**CM_SLAC_PARM.CNF** is generated by the HLE of a EVSE GP STA in response to the corresponding **CM_SLAC_PARM.REQ**
CM_START_ATTEN_CHAR.IND	HLE of an PEV transmits **CM_START_ATTEN_CHAR.IND** to indicate to the EVSEs that the SLAC Process is starting
CM_ATTEN_CHAR.IND	**CM_ATTEN_CHAR.IND** is transmitted by EVSEHLE to PEVHLE and contains the signal attenuation profile measured using the M-Sounds transmitted by the PEV
CM_ATTEN_CHAR.RSP	**CM_ATTEN_CHAR.RSP** is transmitted by the PEV in response to the corresponding **CM_ATTEN_CHAR.IND** from the EVSE
CM_MNBC_SOUND.IND	**CM_MNBC_SOUND.IND** is used as part of SLAC protocol to estimate the attenuation profile of a transmission from an PEV at EVSEs
CM_SLAC_MATCH.REQ	**CM_SLAC_MATCH.REQ** is used by the PEV to indicate to an EVSE that it is connected to it through a charging cordset
CM_SLAC_MATCH.CNF	**CM_SLAC_MATCH.CNF** is transmitted by the EVSE in response to **CM_SLAC_MATCH.REQ**. This message contains the NID and NMK for PEV to join the EVSE

13.7 MANUFACTURER-SPECIFIC MESSAGES

Manufacturer-specific messages are messages used by equipment manufacturers to implement manufacturer specific interfaces. The format and use of manufacturer-specific messages are manufacturer dependent. Manufacturer-specific messages are only exchanged between the MAC Layer and its host. Manufacturer-specific messages are never transmitted over the powerline.

13.8 VENDOR-SPECIFIC MESSAGES

Vendor-specific messages are used by implementers of AV STAs ("vendors") to enhance the functionality of the system when exchanged between STAs designed by the same vendor. The first three octets of the Vendor-Specific Management Message Entry contain IEEE-assigned Organizationally Unique Identifier (OUI) of the vendor. This enables each Vendor to distinguish their management messages from other vendor's messages. The format of the remaining field in the message is vendor dependent.

13.9 SUMMARY

MAC management messages (MMEs) are critical to the operation of HomePlug AV and this chapter provided an overview of several key MME structures used in HomePlug AV.

The remainder of the book is devoted to the discussion of several PLC technologies that are closely related to HomePlug AV. Namely, IEEE 1901, HomePlug GreenPHY, and HomePlug AV2. Chapter 14 provides an in-depth look at the IEEE 1901 PLC standard and shows its relationship to HomePlug AV.

14

IEEE 1901

14.1 INTRODUCTION

The IEEE 1901 standard was published in December 2010 after some five years of development under the sponsorship of the IEEE Communications Society (COMSOC). It features three technology areas or "clusters", namely the In-home cluster, the Access cluster, and the Coexistence cluster, and represents a standard of compromise between the Fast Fourier Transform (FFT) based Orthogonal Frequency Division Multiplexing (OFDM) PHY used in HomePlug and Wavelet based OFDM PHY used in Panasonic's HD-PLC devices. The standard specifies both PHYs as optional, with an Inter System Protocol (ISP) providing coexistence but not interoperability between the in-home FFT and Wavelet PHY realizations of IEEE 1901. In addition to enabling coexistence between these noninteroperable PHYs, the ISP is also designed to ensure coexistence between any in-home IEEE 1901 system and any IEEE 1901 Access or ITU-T G.hn standard based systems.

This chapter provides an overview of the key features of the IEEE 1901 standard, which relies heavily on the HomePlug AV specification, which is the main focus of this book. For the case of the IEEE 1901 FFT PHY, the chapter briefly summarizes the extensions of HomePlug AV that are included in IEEE 1901. A more detailed treatment is given for the case of the IEEE 1901 Wavelet PHY. The use of IEEE 1901 in powerline access networks has been very limited and is not covered in this book.

HomePlug AV and IEEE 1901: A Handbook for PLC Designers and Users, First Edition.
Haniph A. Latchman, Srinivas Katar, Larry Yonge, and Sherman Gavette.
© 2013 by The Institute of Electrical and Electronics Engineers, Inc. Published 2013 by John Wiley & Sons, Inc.

14.2 FFT

The IEEE 1901 FFT is an extension of HomePlug AV 1.1 specification. The extensions made to HomePlug AV in IEEE 1901 were done in a manner that preserves coexistence and interoperability between IEEE 1901 FFT stations and HomePlug AV stations. This enables IEEE 1901 FFT to build on the existing HomePlug AV 1.1 infrastructure while providing higher performance. The following sections provide an overview of the extensions to HomePlug AV in IEEE 1901.

14.2.1 30–50 MHz Frequency Band

HomePlug AV Physical layer operates in the 1.8–30 MHz frequency band. Furthermore, only 917 carriers are used in the 1.8–30 MHz band and the remaining carriers are masked (i.e., not used for transmitting data).

In IEEE 1901, the following two extensions were made to the frequency band:

- Frequency band was extended up to 50 MHz. The carrier spacing in the 30–50 MHz is the same as that in the 1.8–30 MHz band (i.e., 24.414 KHz).
- The masked carriers in the 1.8–30 MHz can also be used for transmitting data.

These two extensions enable IEEE 1901 FFT systems using the 1.8–50 MHz band to support up to 1974 carriers (i.e., $(50-1.8)^*1000/24.414$). These additional carriers along with a 16/18 code rate and a 1.6 μs Guard Interval enable IEEE 1901 systems to provide a peak PHY data rate of 500 Mbps ($1974^*12^*(16/18^*)/(40.96+1.6)$).

The 1901 FFT Physical layer defines two types of tone masks, namely a Broadcast tone Mask and an Extended tone Mask.

The Broadcast tone mask is used for frame control symbols and ROBO data symbols. All stations use the same Broadcast tone mask for coexistence and interoperability. The Broadcast tone mask uses 917 carriers in the 2–30 MHz band and is the same as the tone mask used by HomePlug AV devices.

The extended tone mask is used for the Priority Resolution Symbols, Preamble, and all types of data modulation. The extended tone mask is either the same as the Broadcast tone mask or it is a superset of the Broadcast tone mask. The Extended tone mask enables stations to use extended carriers (i.e., carriers in addition to the carriers used by Broadcast tone mask) for exchange of data, while maintaining coexistence and interoperability with stations that only support the Broadcast tone mask.

The use of an Extended tone mask that is different than the Broadcast tone mask requires negotiation between each pair of stations (STA1 and STA2). STAs indicate their ability to support Extended Carriers by including Extended Carriers Support BENTRY in their Discover Beacons and by using the Extended Carrier Support Flag (ECSF) in the Sound MPDU. This enables other STAs in the network that also support extended carriers to start negotiation of Extended tone masks.

The procedure used for negotiating the use of the Extended tone mask between two STAs (STA1 and STA2) is as follows:

- STA1 initiates a request for use of Extended tone mask by sending a CM_EXTENDEDTONEMASK.REQ message to STA2. The CM_EXTEN DEDTONEMASK.REQ message indicates all the carriers that can be supported by STA1.
- If STA2 respond with the CM_EXTENDEDTONEMASK.CNF message. The CM_EXTENDEDTONEMASK.CNF message indicates all the carriers that can be supported by both STA1 and STA2.
- On reception of the CM_EXTENDEDTONEMASK.CNF message, STA1 will reinitiate Channel adaptation with STA2 by transmitting sound MPDUs with Extended Carriers Used Flag set to **0b1**. This flag indicates to STA2 that STA1 is intending to use an Extended tone mask that includes all carriers that can be supported by both STA1 and STA2. STA2 responds with tone map(s) that include all carriers in the negotiated Extended tone mask.

14.2.2 Additional Guard Intervals

HomePlug AV Physical layer uses a 7.56 μs Guard Interval on the first two OFDM symbols of the PHY Protocol Data Unit (PPDU) payload. IEEE 1901 FFT allows the guard interval on the first two OFDM symbols to be either 7.56 μs or 19.56 μs. The 19.56 μs Guard Interval is optionally used in access networks to enable handling of large delay spreads. In-home networks are restricted to the 7.56 μs guard interval on the first two OFDM symbols, thus preserving compatibility with HomePlug AV. The selection of 7.56 μs versus 19.56 μs Guard Intervals in access networks is done as part of station configuration during deployment (i.e., stations are preconfigured with the desired Guard Interval).

HomePlug AV Physical layer limits the Guard Interval on the third and higher OFDM symbols in the PPDU payload to be either 5.56 μs or 7.56 μs. The receiver selects the Guard Interval to be used based on channel conditions and indicates it to the transmitter as part of channel adaptation. IEEE 1901 FFT extends the number of Guard Intervals by allowing Extended Smaller (i.e., smaller than 5.56 μs) Guard Intervals and Extended Larger (i.e., larger than 7.56 μs) Guard Intervals.

- Extended Smaller Guard Intervals enables support for {1.60, 3.92, 2.08, and 2.56} μs Guard Intervals. These enable IEEE 1901 stations to improve efficiency on channels with low delay spread channels.
- Extended Larger Guard Intervals enable support for {9.56, 11.56, 15.56, and 19.56} μs Guard Interval. These are primarily intended for use on high delay spread channels (e.g., in access networks).

Stations indicate their support for these Extended Guard Intervals by setting Extended Smaller Guard Interval Support Flag (ESGISF) and Extended Larger

Guard Interval Support Flag (ELGISF) in the Sound MPDU Frame Control. Receiver uses this information to determine the Tone Map Guard Interval based on the channel conditions and the Guard Interval capabilities of the transmitter.

14.2.3 4096 QAM

The highest modulation supported by HomePlug AV is 1024 Quadrature Amplitude Modulation (QAM). In IEEE 1901, support for 4096 QAM was added to enhance the performance on higher signal to noise ratio (SNR) channels.

Stations indicate their support for 4096 QAM by setting Extended Modulation Support (EMS) field in the Sound MPDU Frame Control. Receivers use this information to determine the modulation on each carrier of the Tone map based on channel conditions and receiver capabilities.

14.2.4 16/18 Code Rate

HomePlug AV supports FEC Code rates of $1/2$ and 16/21. In IEEE 1901, support for 16/18 code rate was added to enhance performance on higher SNR channels. 16/18 FEC Code uses a coding scheme similar to the Turbo Convolutional Encoder for the exiting rate $1/2$ and 16/21 codes; however, the puncturing pattern is modified to reduce the redundancy. Rate 16/18 Puncture pattern is shown in Table 14.1.

Stations indicate their support for 16/18 code rate by setting the Extended FEC Rate Support (EFRS) field in the Sound MPDU Frame Control. Receivers use this information to determine the FEC code rate for the Tone map based on channel conditions and receiver capabilities.

14.2.5 SNID Reuse

In HomePlug AV, the maximum number of simultaneous Home AV Logical Network (AVLNs) that can be supported is limited by the 4-bit Short Network Identifier (SNID). In particular, when the maximum number of AVLNs that an AVLN can directly hear is greater than 15, other AVLNs might end up using an SNID that is the same as its SNID. This can results in unreliable performance due to address conflicts in the Frame Control (i.e., stations in different network might end up having the same SNID and Terminal Equipment Identifier (TEI)).

In IEEE 1901, the maximum number of simultaneous AVLNs was increased to 64 by allowing up to four AVLNs to reuse (or share) the same SNID. Address conflicts are avoided by dividing the TEI range into four TEI subranges {[1–63], [64–127], [128–191], [192–254]}.

TABLE 14.1 Rate 16/18 Puncture Pattern

p	10000000 10000000
q	10000000 10000000

A Central Coordinator (CCo) that supports the SNID reuse indicates its capability using the Reuse SNID Flag (RSF) in its Beacon. Furthermore, the CCo chooses an SNID and TEI Subrange in a manner that is unique across all AVLNs in its vicinity. For example, if a neighboring CCo also supports SNID Reuse and one of the TEI subrange for that SNID is not used by any other neighboring AVLNs, the CCo can choose to reuse the same SNID and the unused TEI subrange. CCos that support SNID Reuse will restrict TEIs assigned to station in its AVLN to its TEI subrange.

The SNID Reuse procedure avoids addressing conflicts since the {SNID, TEI} Identifier is unique. Note that SNID Reuse reduces the number of stations that can be supported in each AVLN to 63 or 64. This should not be limiting for in-home deployments, where the maximum number of stations are significantly smaller than 64.

14.2.6 Repeating and Routing

In HomePlug AV, Proxy Networking feature is used to enable association and authentication of hidden stations. However, HomePlug AV does not provide any mechanism to repeat data MSDUs (i.e., repeating is limited to management message) and relies on higher layers to handle MSDU repeating.

In IEEE 1901, MAC layer repeating and routing mechanism was added for data MSDUs. This feature hides the complexity of powerline network topology from higher layers and provides connectivity between stations that cannot directly communicate.

14.2.6.1 Repeating and Routing of Unicast MSDUs Repeating of Unicast MSDUs is done at the MAC Frame level. A station acting as a repeater will first reassemble the received segments and extract the MAC Frame. MAC Frame that need to be repeated (based on ODA) are inserted into the appropriate transmit MAC Frame Stream, resegmented and transmitted as part of the MAC Protocol Data Unit (MPDU) payload.

Stations that support unicast repeating maintain routing information for identifying the next STA to deliver a MAC Frame based on the ODA for MAC addresses known to be on the AVLN or known to be bridged by other STAs in the AVLN. This information is determined from the following:

- TEI Map received from the CCo in the CC_SET_TEI_MAP.IND message,
- LBDAT received from other STAs in the CM_BRG_INFO.CNF messages, and
- Routing information received from other STAs in the CM_ROUTE_INFO. CNF/IND messages.

Stations maintain a Local Routing Table (LRT) that contains routing information for every associated STA in the AVLN. The LRT includes the TEI of the best next station through which to route (NTEI), the Route Data Rate (RDR) and Route Number of Hops (RNH) for every associated STA in the AVLN. Stations broadcast a **CM_ROUTE_INFO.IND** message containing routing information from its LRT on a periodic basis or when there is a significant change to LRT. An STA receiving a

CM_ROUTE_INFO.IND message computes a candidate RDR (RDR′) and RNH (RNH′) for each UDTEI entry in the LRT. RDR′ is computed from the RDR provided in the CM_ROUTE_INFO.IND MME and the average BLE of the transmit channel to the STA that sent the CM_ROUTE_INFO.IND MME as shown in Equation 14.1. RNH′ is computed by adding one to the value of RNH received in the CM_ROUTE_INFO.IND MME. Note that the RDR for an STA that can be communicated with directly is the average BLE of the transmit channel to that STA and the RNH is 0.

$$\text{RDR}' = \frac{1}{\frac{1}{\text{RDR}} + \frac{1}{\text{BLE}}} \quad (14.1)$$

An STA receiving a **CM_ROUTE_INFO.IND** MME updates the RDR/RNH/NTEI entry for each UDTEI in its LRT by selecting the best RDR/RNH pair of the existing entry in the LRT, the candidate RDR′/RNH′ pair computed from the received **CM_ROUTE_INFO.IND** MME and the RDR/RNH for direct communication with the STA. If a previously reachable destination becomes unreachable, stations use **CM_UNREACHABLE.IND** message to report this to other stations in the AVLN. This enables stations to update the route to the station in their LRT.

IEEE 1901 also supports multihop connections. This enables provisioning of end-to-end Quality of Service (QoS) in scenarios where retransmission is necessary due to the powerline topology.

14.2.6.2 Repeating and Routing of Broadcast/Multicast MSDUs Repeating of broadcast and multicast MSDUs is done at the PHY Block level. A station acting as a repeater will receive the PHY Block and retransmits them (i.e., segments are not reassembled into MSDUs). It should be noted that a station may be able to receive the same repeated PHY Blocks from multiple repeaters. In such cases, the Segment Sequence Numbers are used to filter out duplicate segment.

The CCo assigns STAs in the AVLN to repeat broadcast and multicast transmissions from certain source STAs. The CCo can determine optimum repeating assignment from the Topology Table. The CCo uses **CC_BCAST_REPEAT.IND** message to indicate the Source Terminal Equipment Identifier (STEIs) for which a station needs to act as a Broadcast/Multicast repeater. STAs maintain a Broadcast and Multicast Repeating Assignment Table (BMRAT) based on the **CC_BCAST_REPEAT.IND** message received from the CCo. The BMRAT contains a list of all TEIs for which the STA will retransmit PHY Blocks (PBs) when broadcast or multicast MPDUs are received with an STEI matching a TEI in the BMRAT. An STA assigned to repeat PBs received in broadcast and multicast MPDUs consults its BMRAT whenever a broadcast or multicast MPDU is received. If the originating STA for the MPDU (the STEI in the Frame Control) is in the BMRAT, the STA inserts the PBs into the reassembly buffer and will schedule any PBs that are not duplicates of ones already received to be broadcasted. These PBs are sent in an MPDU with the STEI copied from the incoming MPDU, the PPB set to **0xFF**, and the TEI of the repeating STA stored in the BLE field. Setting the PPB to **0xFF** enables other stations to determine the MPDU is retransmitted and to determine TEI of the station that retransmitted the MPDU using the 8-bits BLE field.

14.3 WAVELET

Wavelet OFDM modulation is resilient to selective frequency channel and narrowband noise, and provides efficient frequency utilization, partly because it does not require Guard Intervals (GIs) [30].

A wavelet OFDM frame may be transmitted either directly at baseband or by modulation to a bandpass carrier. In in-home and access applications, baseband is mandatory, while bandpass is optional. Baseband operation may be done at variable bandwidths being compatible with fixed frequency wideband operation, although it is fixed frequency wideband. Bandpass operation is frequency agile, operating at variable bandwidths over a wider range of frequencies. This agility allows channel allocation for QoS traffic classes.

14.3.1 Baseband PHY

The in-home baseband wavelet OFDM PHY places 512 evenly spaced carriers into the frequency band from DC to 31.25 MHz and uses the frequency band from 1.8–28 MHz, with notches used to avoid amateur radio frequency. Optional upper band above 30 MHz may be defined depending on country regulations. In high-speed mode, every carrier can be modulated with Pulse Amplitude Modulation, M-PAM (M: 2, 4, 8, 16, and optional 32) for its primary modulation. In diversity mode, every carrier can be modulated with 2PAM for the primary modulation, and further frequency diversity can be provided to improve the ability of the PHY to operate under adverse conditions. A variety of Foreword Error Correcting coding schemes (FEC) may be used, including Reed–Solomon encoder/decoder and convolution encoder/Viterbi decoder (Concatenated encoder/decoder), or Low Density Parity Check-Convolutional Code (LDPC-CC) encoder/decoder.

14.3.2 Bandpass PHY

The optional bandpass wavelet OFDM PHY is used for access networks and for indoor networks places 1024 evenly spaced carriers into any bands lying in the range 1.8–50 MHz. The tone mask for bandpass communication allows the use of more carriers since, for example, tones near DC do not need to be nulled. The adaptive coding, modulation, and diversity on used carriers is identical to the baseband case described in the previous paragraph. Details of Wavelet bandpass PHY were not included in this book. For additional details of the Wavelet bandpass PHY, refer to IEEE 1901 specification.

14.3.2.1 Wavelet MAC IEEE 1901 Wavelet systems use a CSMA/CA- and TDMA-based hybrid MAC layer. For additional details of the MAC layer that are specific to Wavelet PHY, refer to IEEE 1901 specification.

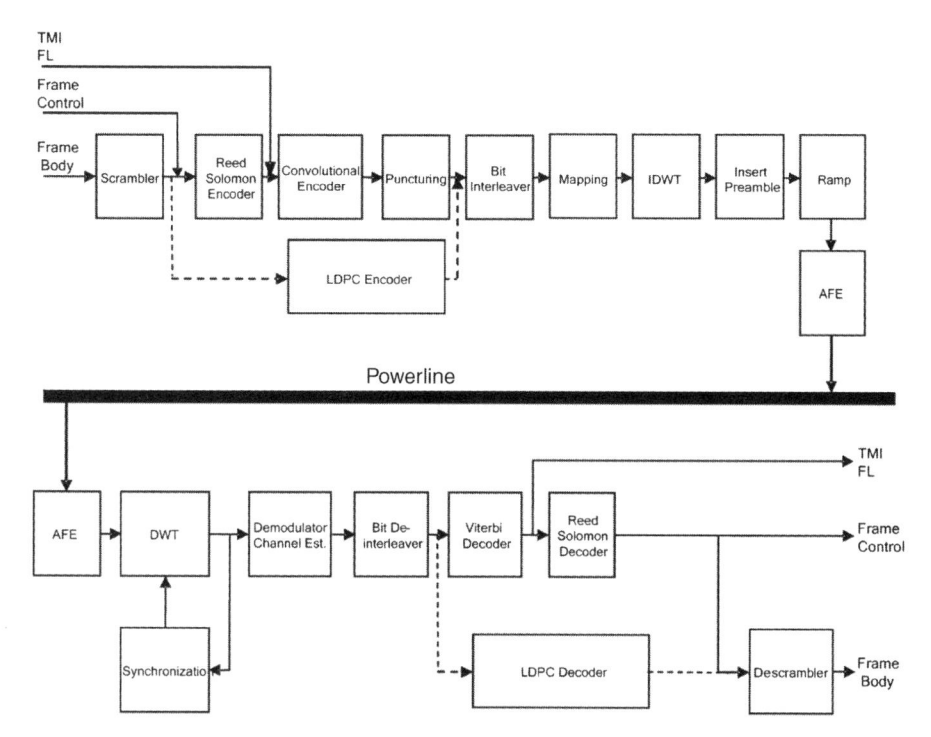

FIGURE 14.1 Wavelet transmitter and receiver.

14.3.3 Transceiver Block Diagram

The general block diagram of the transmitter and receiver for the wavelet PHY is shown in Figure 14.1. The transmitter's PHY receives its inputs from the MAC sublayer. The receiver's PHY provides its outputs to the MAC sublayer. The transmitter's PHY includes scrambler block, Reed–Solomon encoder block, convolutional encoder block, puncturing block, bit interleaver block, mapping block, inverse discrete wavelet transformer (IDWT) block, insert preamble block, ramp block, and AFE (analog front end) block. The Low Density Parity Check (LDPC) encoder block is optional. The scrambler block produces a scrambled data sequence from received data from the MAC sublayer. Then, these data are coded as concatenated code by using Reed–Solomon encoder block, convolutional block, and puncturing block or LDPC code by using the LDPC encoder block. The bit interleaver block interleaves data from the puncturing block. However, these data are coded using only the Reed–Solomon encoder block in the case of Reed–Solomon mode only. Then three blocks (convolutional, puncturing, and bit interleaver blocks) are not used in this mode. Evaluation data for the RCE frame is input into the mapping block directly. The mapping block produces a plurality of bit streams from the bit interleaver block so as to map the bit streams to signal point data of respective carriers for wavelet OFDM. The IDWT block modulates the respective carriers by wavelet waveforms that are orthogonal to each other

| Preamble | TMI (No-Payload flag : 1 bit TMI : 8 bits (or 5 bits) Tail : 6 bits) | Frame control (Info.: 34 bytes Tail : 6 bits) | FL (FL : 16 bits CRC : 8 bits Tail : 6 bits) | Frame body (Data : Variable Tail : 6 bits) | F-pad |

FIGURE 14.2 PPDU frame format.

based on the signal point data of the respective carriers mapped by the mapping block so as to produce time waveform data. The insert preamble block inserts the preamble that is generated (in advance) by the IDWT block. The ramp block performs ramp processing to the output composite waveform of the IDWT block. The Analog Front End (AFE) block converts the time waveform data into an analog time waveform signal. On the receiver side, opposite operations are done.

14.3.4 PPDU Format

Figure 14.2 shows the format for the PPDU including the wavelet OFDM PPDU preamble, Tone Map Information (TMI), Frame Control, Frame Length, Frame Body, and Pad bits.

The preamble length is from 11 to 17 wavelet OFDM symbols. The TMI and FL are each one wavelet OFDM symbol. And the Frame Control is 8 wavelet OFDM symbols. The TMI symbol consists of 1-bit No-Payload flag, 8-bit TMI (or 5-bit TMI), and 6-bit tail. The 5-bit TMI is used only when 1901 wavelet PHY communicates with a legacy HD-PLC node. The No-Payload flag means that only the preamble, TMI, and FC are transmitted. The TMI value of only Diversity-Orthogonal Frequency Division Multiplexing (OFDM) for Frame body (DOF) mode (not Advanced DOF: ADOF) is always zero. The Frame Control has 34 bytes of frame control information and a 6-bit tail. The FL symbol consists of a 16-bit FL, an 8-bit Cyclic Redundancy Code (CRC), and a 6-bit tail. The Frame Body is variable data and a 6-bit tail. All tail-bits are set to zero.

The FL indicates the number of symbols composed of Frame Body and F-pad. And it is also referenced as the common information with Frame Control among all STAs in the BSS. Since the FL is calculated with LENGTH_FB and tone map for transmission, it requires some delay for its calculation. The LENGTH_FB shows information bytes. Therefore, the FL is allocated between the Frame Control and the Frame Body to make implementation easy. If the FL value is **0xFFFE**, it indicates that the frame is transmitted with Postamble. (If errors are detected at the FL in receiving process in PHY layer, the FL value is set to **0xFFFF**.)

In the transmitter, the physical layer converts the Frame Control of the MPDU to the Frame Control of the PPDU.

In the receiver, the physical layer converts the Frame Control of the PPDU to the Frame Control of the MPDU.

In terms of modulation, the TMI and FL are transmitted with a diversity mode for them and convolutional code with the coding rate R = 1/2. Frame Control is transmitted with a diversity mode for it and concatenated code (R = 1/2, RS(50, 34)). Also, the FEC of Frame Body forms Reed–Solomon code (RS (255,239)), concatenated code (R is variable, RS(255,239)) or LDPC code (R is variable). In the case of the diversity mode in Frame Body, concatenated code (R = 1/2, RS(56, 40)), or LDPC code (R = 1/2) is used. The Frame Body and pad bits appended are transmitted at the data rate determined by the TMI field and may constitute multiple PAM wavelet OFDM symbols or the diversity mode for Frame Body. Only 2PAM is used in the diversity mode. The TMI, Frame Control, and Frame Length fields are required for decoding the Frame Body of the PPDU frame. TMI, Frame Control, FL, and Frame Body have a 6-bit tail at the end of each field. However, the PPDU frame format for the Request Channel Estimation (RCE) frame does not have the tail of the Frame Body and F-pad. Each of these fields is described in detail in Section 1.3.5.

LDPC-CC can be used instead of the concatenated code in the Frame Body. Then the 6-bit tail of the Frame Body is replaced with a tail of LDPC-CC. The tail and F-pad consist of all 0 bits in the case of LDPC-CC (refer to Table 14.3).

14.3.4.1 Overview of the PPDU Encoding/Decoding Process The following overview provides a brief summary of the PPDU encoders and decoding process to facilitate a basic understanding of the physical layer convergence procedure:

- Produce the PPDU Preamble field, composed of (2.5 + 11) wavelet OFDM symbols used for signal training.
- Produce the PPDU Header field from the TMI, Frame Control, and Frame Length fields by filling in the appropriate bit fields using appropriate coding rates and PAM modulation schemes.
- Calculate the frame length from tone map information.
- Initiate the scrambler with all ones as seed, generate a scrambling sequence, and XOR it with data.
- Encode the scrambled data with concatenated encoder with punctuations to achieve the desired coding rates.
- Perform an "interleaving" (reordering) of the encoded data of each symbol.
- Allocate the coded and interleaved data to each carrier according to tone map information and modulation scheme.
- Convert the carriers to the time domain at the appropriate clocking rate for the specified bandwidth using the inverse discrete wavelet transform (IDWT).
- If necessary, up-convert the time domain data to the specified carrier frequency for bandpass communication.

An illustration of the transmitted frame and its parts appears in Figure 14.2.

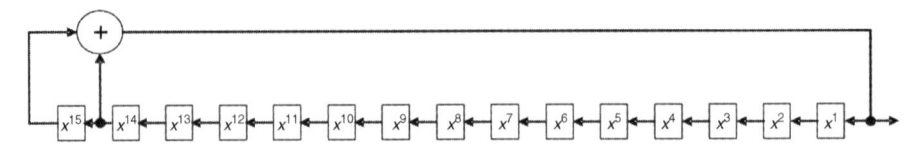

FIGURE 14.3 Generator.

14.3.4.2 Modulation-Dependent Parameters
A Request Channel Estimation frame (RCE frame) includes a wavelet OFDM symbol array for evaluation data, or in other words, the STA sends a CE request to obtain CINR properties and determine the parameters such as tone map carrier and modulation information.

If the destination address written in Frame Control matches the address of the receiver STA, it returns a RCE frame that includes these tone map parameters, after performing channel estimation and determining parameters such as tone map and FEC. With this operation, the tone maps of Source STA and Destination STA are shared.

The tone map includes Modulation Type, FEC Type, and Diversity Mode Flag.

14.3.5 PHY Encoder

All PHY encoder units are the same for the bandpass option as for baseband implementations, with the possible exception of different clock rates, as appropriate.

14.3.5.1 Generator for RCE Frame
Figure 14.3 shows the generator of evaluation data for the RCE frame. The evaluation data consist of the following generator polynomial (PN15 Code). The generator is initialized by all one in each RCE frame. The evaluation data is input into Mapping block directly without passing Scrambler, FEC Interleaver. Each carrier has 2PAM modulation in the order of increasing number of carriers.

14.3.5.2 Scrambler
Figure 14.4 shows the scrambler block. The data in the frame body is scrambled using the following generator polynomial (PN7 Code):

$$S(x) = x^7 + x^4 + 1$$

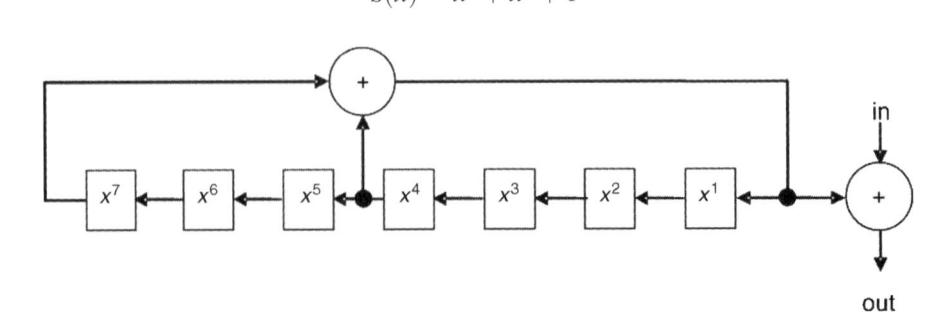

FIGURE 14.4 Scrambler.

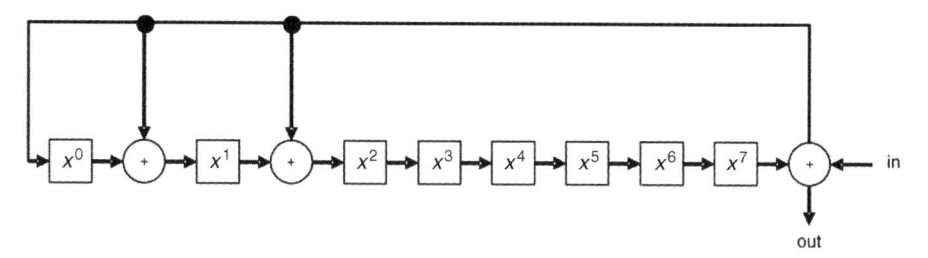

FIGURE 14.5 CRC encoder.

The bits in the scrambler are initialized to all ones at the start of processing for each frame. For each symbol, the first bit to be input into the data scrambler becomes the MSB of that symbol.

14.3.5.3 CRC Encoder for FL

Figure 14.5 shows the CRC encoder block. The 8-bit CRC bits are added to the 16-bit FL using the following generator polynomial.

$$G(x) = x^8 + x^2 + x + 1$$

The bits in the CRC encoder are initialized to all zeros at the start of processing for each frame. The MSB of the FL is the first bit at the input of the CRC encoder. After the input of the 16-bit FL at the encoder of CRC, the encoder outputs the 8-bit CRC bits in the order of x7, x6, ..., x0.

14.3.5.4 Concatenated Encoder

Reed–Solomon encoding is applied to the input data of the scrambler block, and then convolutional encoding is applied to the output of the Reed–Solomon encoder.

14.3.5.4.1 Reed–Solomon Encoder

The Reed–Solomon (n, k) encoder, shown in the following discussion, is applied to both the Frame Control block and the Frame Body block. For each symbol, the MSB is the first bit at the input/output of the Reed–Solomon encoder. For the last Reed–Solomon code of the Frame Body block, padding bits consist of all 0 bits at the input of the Reed–Solomon encoder.

Field generator polynomial	$P(x) = x^8 + x^4 + x^3 + x^2 + 1$ Field = GF $(2\,m)$, $m = 8$
Code generator polynomial	$G(x) = \{x - \alpha r\}\{x - \alpha(r+1)\}\{x - \alpha(r+2)\}, \ldots,$ $\{x - \alpha(2t + r - 1)\}$ where $t = 8$, $r = 0$
Code length (max.)	$n = 2m - 1 = 255$
The number of bits (max.)	$k = n - 2t = 239$
Coding rate	$(n, k) = (255, 239)$ for Frame Body (non-DOF) (56, 40) for DOF Frame Body (50, 34) for Frame Control

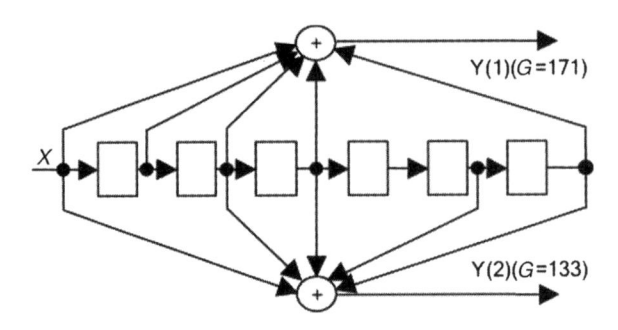

FIGURE 14.6 Convolutional encoder.

14.3.5.4.2 Convolutional Encoder/Puncturing The convolutional encoder, shown in the following discussion, is applied to the Tone Map Index (TMI), Frame Length, Frame Control block, and Frame Body block in the PPDU.

The coding rate of the encoder is 1/2, the constraint length is 7, and, as shown in Figure 14.6, the generator polynomials are 171 and 133 (in octal). The convolutional encoder is reset to zero state at the beginning of each Data field. In addition, a 6-bit tail is inserted at the end of each Data field for convergence to the zero state. Coded data is output in the order of Y(1), Y(2), ..., Y(1), Y(2).

The output of the 1/2 encoder can be punctured according to a puncture pattern, as shown in Table 14.2.

Puncture encoding thins out the output bits from the convolution encoder according to the puncture pattern.

The coding rate of the convolutional code is chosen on the basis of the channel estimation, but rate 1/2 is used in all other cases. In particular, for the Frame Control and Frame Body, it is used in concatenation with the Reed–Solomon code.

> Coding rate 1/2 to 7/8 for Frame Body (not DOF)
> 1/2 for TMI, Frame Control, FL, and DOF

14.3.5.5 Convolutional Codes Defined by Low-Density Parity-Check Polynomials (Optional) LDPC-CCs (low-density parity-check convolutional codes) are convolutional codes defined by low-density parity check polynomials. LDPC-CCs are

TABLE 14.2 Puncture Patterns

Coding Rate	Puncture Pattern (Y(1))	Puncture Pattern (Y(2))
2/3	10	11
3/4	101	110
4/5	1000	1111
5/6	10101	11010
6/7	100101	111010
7/8	1000101	1111010

optionally used in the Broadband over Power line (BPL) system as a high-performance Error Correction Code (ECC) technique, instead of the concatenated code (RS code + convolutional code).

14.3.5.5.1 Encoding Termination Termination via Zero-tailing leads to a stable LDPC-CC encoder state.

Zero-tailing can be done by the LDPC-CC encoder by adding m_z bits of information bits with a value set to zero, which is called as "Tail-bits." Note that the number of Tail-bits m_z depends on coding rate and information size. Since m_z bits of Tail-bits with a value set to zero are known at the receiver, Tail-bits are excluded from transmission bits but parity bits resulted from Tail-bit coding can be transmitted as termination. When the number of termination bits (transmission bits for tail-bit coding) is expressed as m_t, m_z with coding rate $R = (N-1)/N$ is expressed as follows:

$$q = 0: m_z = (N-1)m_t$$
$$q \neq 0: m_z = (N-1)m_t + (N-1) - q$$

where $I_s \bmod N - 1 = q$ and when information size expressed as I_s, the number of termination bits m_t is shown in Table 14.3.

14.3.5.6 FEC Type Field The Forward Error Correction (FEC) Type field shows the FEC Type parameter shown in Table 14.4 from a Modulation parameter that is determined by the received RCE frame.

14.3.5.7 Interleaver The convolutional (puncture) or the LDPC encoded data is interleaved by bit interleaver block and padded in the last symbol. However, the Reed–Solomon encoded data is not interleaved but is padded with scrambled data bits in the last symbol in the case of Reed–Solomon mode only. TMI and FL data are not interleaved.

TABLE 14.3 Number of Transmission Bits for Tail-Bit Coding

Coding Rate	Number of Transmission Bits for Tail-Bit Coding
1/2	$mt = 440$ bits
2/3	$mt = 540$ bits (when $I_s \leq 2040$ bits)
	$mt = 380$ bits (when $I_s > 2040$ bits)
3/4	$mt = 540$ bits (when $I_s \leq 2040$ bits)
	$mt = 380$ bits (when $I_s > 2040$ bits)
4/5	$mt = 680$ bits (when $I_s \leq 2040$ bits)
	$mt = 620$ bits (when 2040 bits $< I_s \leq 4080$ bits)
	$mt = 380$ bits (when $I_s > 4080$ bits)

TABLE 14.4 FEC Type Parameter

Value	Definition
0	Reed–Solomon only
1	LDPC-CC(1/2)
2	LDPC-CC(2/3)
3	LDPC-CC(3/4)
4	LDPC-CC(4/5)
5–7	(Reserved)
8	Reed–Solomon & Convolutional encoding (1/2)
9	Reed–Solomon & Convolutional encoding (2/3)
10	Reed–Solomon & Convolutional encoding (3/4)
11	Reed–Solomon & Convolutional encoding (4/5)
12	Reed–Solomon & Convolutional encoding (5/6)
13	Reed–Solomon & Convolutional encoding (6/7)
14	Reed–Solomon & Convolutional encoding (7/8)
15	(Reserved)

14.3.5.8 Wavelet Process Wavelet OFDM transmission requires several related processes, including mapping, symbol generation (baseband and bandpass), and Inverse Discrete Wavelet Transform (IDWT).

The order in which the carriers are mapped in the baseband case is from zero to $N-1$, where zero corresponds to the lowest frequency and $N-1$ corresponds to the highest frequency. In the bandpass case, the order in which the carriers are mapped from $-N$ to $N-1$, where $-N$ is the lowest negative frequency and $N-1$ is the highest positive frequency. In the latter case, index zero corresponds to DC before upconversion and corresponds to the carrier frequency after upconversion.

14.3.5.8.1 Mapping The mapping block creates maps of TMI (modulation method 2PAM), Frame Control (2PAM), FL (2PAM), and frame body (2PAM, 4PAM, 8PAM, 16PAM, and optional 32PAM). Diversity mapping is always used for TMI, FL, and Frame Control. Frame Body mapping is performed for DOF (Diversity-OFDM for the frame body), ADOF, and D2PAM (Double-2PAM) using the 2PAM modulation method on diversity mode, and 2PAM through 32PAM on high-speed mode. Table 14.5 shows the list of modulation methods for each information type.

Mapping is performed according to tone mask and tone map, which are set up. The tone mask specifies the carriers to be used. The tone map specifies a modulation type (2PAM, 4PAM, 8PAM, 16PAM, and optional 32PAM). The tone map permits the modulation type for each carrier to be changed according to the conditions.

For instance, carriers undergoing low carrier power to interference power plus noise power ratio (CINR) conditions can either use a low rate modulation type (e.g., 2PAM) or be masked. In addition, the carriers under high CINR conditions can use a high-rate modulation type (e.g., 32PAM). The tone map adapts the selection mechanism for modulation type for the frame body, and only the tone mask does

TABLE 14.5 Modulation Methods for Each Informational Type

Information Type	Bit(s)/Carrier	Modulation Type	Mode
TMI and FL	1	2PAM	Diversity
Frame Control	1	2PAM	
Frame Body	1	DOF, ADOF	
	1	D2PAM^a	High-speed
	1	2PAM	
	2	4PAM	
	3	8PAM	
	4	16PAM	
	5	32PAM	

^{a}D2PAM is Diversity technology, and it is done with the carrier pair in the high-speed mode. Refer to 14.3.5.8.1.2.5.

for preamble, TMI, FL, and Frame Control. Table 14.6 shows the applications of tone map/tone mask.

14.3.5.8.1.1 2PAM, 4PAM, 8PAM, 16PAM, AND OPTIONAL 32PAM MAPPING The data bit output from the bit interleaver block is mapped for PAM (2PAM, 4PAM, 8PAM, 16PAM, or optional 32PAM) in accordance with the tone map defining modulation types for each carrier. In 2PAM mapping, value zero is allocated to amplitude "−1", and value one to amplitude "+1". In 4PAM mapping, the initial data input to mapping block is used as MSB to create 2-bit data. In 8PAM mapping, the initial data input to mapping block is used as MSB to create 3-bit data. In 16PAM mapping, the initial data input to mapping block is used as MSB to create 4-bit data. In optional 32PAM mapping, the initial data input to mapping block is used as MSB to create 5-bit data.

14.3.5.8.1.2 DIVERSITY MODE There are three types of diversity modes, that is, Frame Control, TMI/FL, and Frame Body. In addition, there are two Frame Body diversity modes, DOF (Diversity-OFDM for Frame Body) and D2PAM (Double-2PAM).

14.3.5.8.1.2.1 Frame Control Diversity Mode In frame control diversity mode, data is transmitted repeatedly for a few times using different carriers. Provided the number of valid carriers is N_c, the number of frame control data transmitted is

TABLE 14.6 Applications of Tone Maps/Tone Masks

Tone MASK	Tone MAP
Applicable to all types	Applicable to the Frame Body field
	Not applicable to SYNCP, SYNCM, and TMI
	FL and Frame Control

	Carrier Number#												
		0–32	33	34	//	54	55–56	67	68	//	84	85–90	91	//	455	456

Symbol Number#													
0	MASK	in[80]	in[81]	//	in[101]	MASK	in[0]	in[1]	//	in[17]	MASK	in[18]	//	in[30]	in[31]
1	MASK	in[182]	in[183]	//	in[203]	MASK	in[102]	in[103]	//	in[119]	MASK	in[120]	//	in[132]	in[133]
2	MASK	in[284]	in[285]	//	in[305]	MASK	in[204]	in[205]	//	in[221]	MASK	in[222]	//	in[234]	in[235]
3	MASK	in[386]	in[387]	//	in[407]	MASK	in[306]	in[307]	//	in[323]	MASK	in[324]	//	in[336]	in[337]
4	MASK	in[488]	in[489]	//	in[509]	MASK	in[408]	in[409]	//	in[425]	MASK	in[426]	//	in[438]	in[439]
5	MASK	in[590]	in[591]	//	in[611]	MASK	in[510]	in[511]	//	in[527]	MASK	in[528]	//	in[540]	in[541]
6	MASK	in[692]	in[693]	//	in[713]	MASK	in[612]	in[613]	//	in[629]	MASK	in[630]	//	in[642]	in[643]
7	MASK	in[794]	in[795]	//	pad	MASK	in[714]	in[715]	//	in[731]	MASK	in[732]	//	in[744]	in[745]

FIGURE 14.7 Example of frame control transmitted data.

$N_b (= 816 (= 812$ databits $+ 4$ padding bits) and (fixed value)), the number of symbols used for frame control transmission is $L(=8$(fixed value)), and data to be input is in [i], the mth symbol data in frame control, $K_m[i]$, is expressed by the following equation. C offset depicts the offset value of the first input data.

$$K_m[i] = \text{in}[S \times m + mod(C_{\text{offset}} + i, S)] \quad 0 \leq i \leq N_c - 1, \quad 0 \leq m \leq L - 1$$

Here, $S = 102$ $(= N_b/L)$, $C_{\text{offset}} = 80$ (fixed value).
Figure 14.7 shows an example of frame control transmitted data.

14.3.5.8.1.2.2 TMI/FL Diversity Mode On TMI and FL diversity modes, frequency diversity gain can be achieved by simply repeating TMI or FL bit data for a few times within one symbol. The symbol data on TMI and FL modes equals the one gained by setting $L = 1$ to the equation for frame control diversity mode in 13.2.3.8.1.2.1.

For 8-bit TMI, S is 30 (fixed value) and C_{offset} is 8 (fixed value). For 5-bit TMI, S is 24 (fixed value) and C_{offset} is 2 (fixed value). Figure 14.8 is an example of TMI transmitted data.

For FL, S is 60 (fixed value) and C_{offset} is 38 (fixed value). Figure 14.9 is an example of FL transmitted data.

14.3.5.8.1.2.3 Frame Body Diversity Mode Two types of Frame Body diversity modes, DOF and D2PAM, are available.

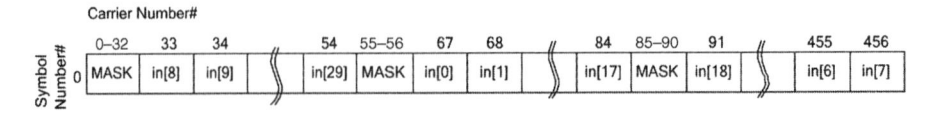

	Carrier Number#											
		0–32	33	34	//	54	55–56	67	68	//	84	85–90	91	//	455	456

Symbol Number#												
0	MASK	in[8]	in[9]	//	in[29]	MASK	in[0]	in[1]	//	in[17]	MASK	in[18]	//	in[6]	in[7]

FIGURE 14.8 Example of TMI transmitted data.

FIGURE 14.9 Example of FL transmitted data.

Symbol Number# \ Carrier Number#	0–32	33	34	...	54	55–56	67	68	...	84	85–90	91	...	455	456
0	MASK	in[38]	in[39]		in[59]	MASK	in[0]	in[1]		in[17]	MASK	in[18]		in[36]	in[37]

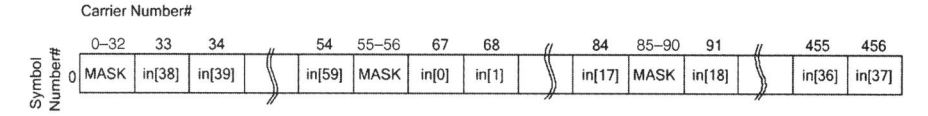

FIGURE 14.9 Example of FL transmitted data.

14.3.5.8.1.2.4 DOF Mode In DOF mode, data is transmitted repeatedly for a few times using different carriers. Data to be input is in[i], the mth symbol data for frame body, and $K_m[i]$ is expressed by the following equation. C_{offset} depicts the offset value of first input data. N_p depicts the number of symbols of the frame body.

$$K_m[i] = \text{in}[S \times m + \text{mod}(C_{\text{offset}} + i, S)] \quad \to 0 \leq i \leq N_c - 1, \quad 0 \leq m \leq N_p - 1$$

Here, $S = 84$ (fixed value) and $C_{\text{offset}} = 62$ (fixed value).

Figure 14.10 shows an example of frame body diversity transmitted data. A total of 84 bits of data between in[0] and in[83] are allocated to the first symbol (indicated by tone mask) of valid carriers in order starting from the low bit. After $S = 84$ bits, data allocation starts again at data in[0] so that the same data is transmitted repeatedly within a symbol. Similarly, for the second symbol, 84 bits of data transmission starts at in[84], 84 bits away from $S = 84$ and is repeated. Transmitting the same data at different frequencies allows for transmission performance improvement under the circumstances such as low SNR and narrow bandwidth interference. $S = 84$ is a fixed value in DOF mode. Modulation and FEC of this mode is 2PAM and concatenated code (code rate 1/2, RS(56,40)). 2PAM and LDPC-CC (code rate 1/2 to 4/5) are used in the case of ADOF, as an option.

14.3.5.8.1.2.5 D2PAM Mode In D2PAM mode, data bits are allocated to the pairs of adjacent carriers. A carrier pair includes $(2n - 1)$th and $(2n)$th carriers $1 \leq n \leq (M/2 - 1)$, where M is the number of carriers (0 to $M - 1$). Specifically, communication quality corresponding to carriers respectively is calculated and is compared with a threshold in the channel estimation process at the receiver side. Then a plurality of carriers corresponding to lower communication quality than the threshold of 2PAM are detected and extracted based on the comparison result.

Symbol Number# \ Carrier Number#	0–32	33	34	...	54	55–56	67	68	...	84	85–90	91	...	455	456
0	MASK	in[62]	in[63]		in[83]	MASK	in[0]	in[1]		in[17]	MASK	in[18]		in[0]	in[1]
1	MASK	in[146]	in[147]		in[167]	MASK	in[84]	in[85]		in[101]	MASK	in[102]		in[84]	in[85]
2	MASK	in[230]	in[231]		in[251]	MASK	in[168]	in[169]		in[185]	MASK	in[186]		in[168]	in[169]
3	MASK	in[314]	in[315]		in[335]	MASK	in[252]	in[253]		in[269]	MASK	in[270]		in[252]	in[253]

FIGURE 14.10 Example of frame body diversity transmitted data.

Finally, the same data is assigned to the pairs of adjacent carriers based on extracted results.

14.3.5.8.2 Baseband Symbol Generation The equations for IDWT block and Ramp block output signals are defined. In the baseband case, the number of used carriers is variable and is denoted by N_{used}. The N_{used} is 512 in the following figures in the baseband case. In all the formulas given in the following discussion, an overlap factor of 4 is assumed. Therefore the first sum in the formulas is built from 0 to 3.

14.3.5.8.2.1 RAMP PROCESS Preamble data is generated with well-known data (e.g., SYNCP) consisting of consecutive symbols and is used for carrier detection, synchronization, and equalization at the receiver side. The IDWT at the transmitter side modulates the preamble data (after phase rotation), generates a plurality of carriers, and outputs a composite waveform of time waveform of the plurality of carriers. As shown in Figure 14.11 , ramp processing is performed on the composite waveform with a predetermined delay (three symbols) from a reference position of the composite waveform in a manner such that the length of the composite waveform is shortened. More concretely, ramp processing is carried out by multiplying the composite waveform with a ramp function. The ramp function has zero value during the delay period and ramps up the composite waveform at the predetermined delay from the reference position wherein the reference position is defined as a starting position of the composite waveform. In particular, the starting position is initially the rising edge of the composite waveform.

14.3.5.8.2.2 SHORT PREAMBLE (OPTIONAL) Figure 14.12 shows a short preamble structure. In the following description, the ramp processing is performed on the composite waveform with a predetermined delay (three symbols) from the starting position of the composite waveform. SYNCP is modulated by all "+1" information. It is possible to use it in addition to the normal preamble. It is composed of SYNCP symbols, and its length is 2.5 symbols, including the ramp. The following is the equation for time domain waveform data, $S_{msp}[n]$. In the following equation, n is a sample position and *msp* is a short preamble symbol number. $w[n]$ is a ramp function and is adaptive only to the first frame symbol. $\theta(c)$ is a phase vector for peak power reduction and shows its value.

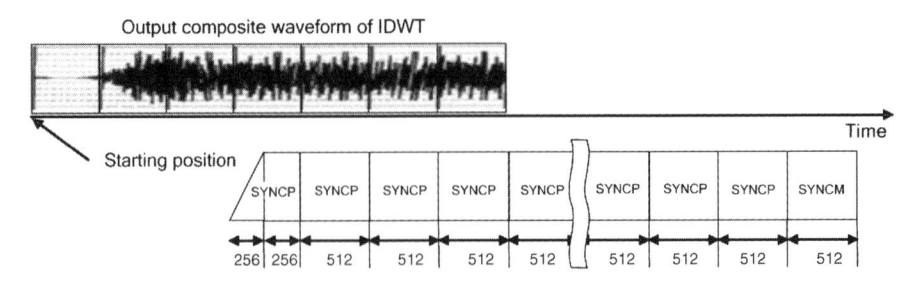

FIGURE 14.11 Image of the relation between ramp process and output wave of IDWT.

WAVELET

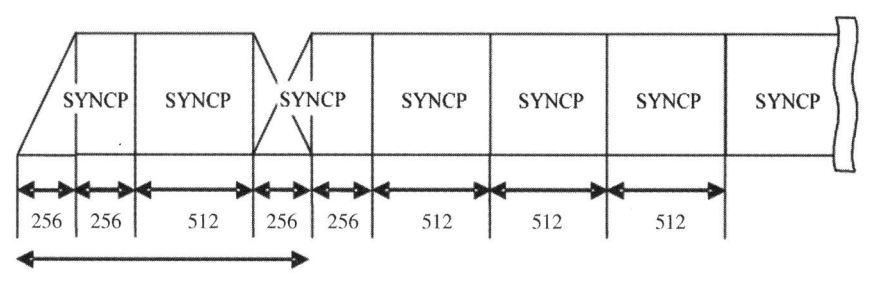

FIGURE 14.12 Short preamble signal.

$$S_{\text{msp}}[n] = \frac{w[n]}{8} \times \left\{ \sum_{k=0}^{3} \sum_{c=1}^{\text{floor}\{(N_{\text{used}}-1)/8\}} h(n+512k) \times \cos\left[\frac{\pi}{512} \times \left\{(n+512k) + \frac{512+1}{2}\right\}\right] \right.$$

$$\left. \times \left\{(8c-1) + \frac{1}{2}\right\} + \theta(8c-1)\right] \times \text{Tone\_MASK}(8c-1)$$

$$+ \sum_{k=0}^{3} \sum_{c=1}^{\text{floor}\{(N_{\text{used}}-1)/8\}} h(n+512k) \times \cos\left[\frac{\pi}{512} \times \left\{(n+512k) + \frac{512+1}{2}\right\}\right]$$

$$\times \left(8c + \frac{1}{2}\right) + \theta(8c)] \times \text{Tone\_MASK}(8c)\}$$

for $0 \leq n < 512$ and $0 \leq msp < 3$

$$w[n] = \begin{cases} \dfrac{n}{256} & (msp = 0 \text{ and } n < 256) \\ 1 & (msp = 0 \text{ and } n > 255) \\ 1 & (msp = 1) \\ \dfrac{255-n}{256} & (msp = 2 \text{ and } n < 256) \\ 0 & (msp = 2 \text{ and } n > 255) \end{cases}$$

14.3.5.8.2.3 PREAMBLE Figure 14.13 shows the preamble structure. In the following description, the ramp processing is performed on the composite waveform with a

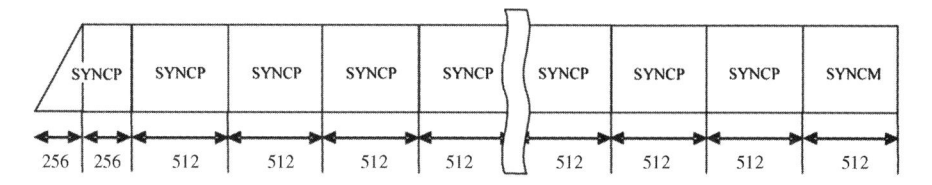

Preamble (11 symbols)

FIGURE 14.13 Preamble signal.

predetermined delay (three symbols) from the starting position of the composite waveform. SYNCP is modulated by all "+1" information. SYNCM is defined as the SYNCP multiplied by -1. The preamble consists of N_{SYNCP} SYNCP symbols (including lamp processing symbol) followed by one SYNCM symbol. The first symbol is ramp-processed by ramp block. And the preamble length is variable from 11 to 17 symbols (N_{SYNCP} is variable from 10 to 16).

The following equation is for time domain waveform data, $S_{mp}[n]$. "mp" is a preamble symbol number.

$$S_{amp}[n] = \frac{w[n]}{16} \times \sum_{k=0}^{3} \text{sign} \sum_{c=0}^{N_{used}-1} h(n+512k)$$
$$\times \cos\left[\frac{\pi}{512} \times \left\{(n+512k) + \frac{512+1}{2}\right\} \times \left(c + \frac{1}{2}\right) + \theta(c)\right] \times \text{Tone\_MASK}(c)$$

for $0 \leq n < 512$ and $0 \leq mp < N_{SYNCP} + 1$ {Tone_MASK $(2x+1)$ and Tone_MASK $2(x+1)$ are the same setting. (x:integer)}
where

$$w[n] = \begin{cases} \dfrac{n}{256} & (mp = 0 \text{ and } n < 256) \\ 1 & \text{otherwise} \end{cases}$$

$$\text{sign} = \begin{cases} -1 & (k = 0 \text{ and } mp = N_{SYNCP}) \\ 1 & \text{otherwise} \end{cases}$$

14.3.5.8.2.4 TMI SYMBOL The equation for the time domain waveform signal for TMI symbol, $S_{mt}[n]$, is shown as follows. In the definitions for this equation, "mt" is the TMI symbol number and $PAM_{mt}(c)$ is the signal position allocated to each carrier from the mapping block. The TMI symbol consists of one symbol ($mt = 0$).

$$S_{mt}[n] = \frac{1}{16} \times \sum_{k=0}^{3} \sum_{c=0}^{N_{used}-1} h(n+512k) \times PAM_{mt-k}(c)$$
$$\times \cos\left[\frac{\pi}{512} \times \left\{(n+512k) + \frac{512+1}{2}\right\} \times \left(c + \frac{1}{2}\right) + \theta(c)\right] \times \text{Tone\_MASK}(c)$$

for $0 \leq n < 512$ {Tone_MASK $(2x+1)$ and Tone_MASK $2(x+1)$ are the same setting. (x:integer)}
where

$$\text{mt-k} \leq -2 \quad PAM_{mt-k}(c) = 1$$
$$\text{mt-k} = -1 \quad PAM_{-1}(c) = -1$$
$$\text{mt-k} = 0 \quad PAM_0(c) = \text{PAM Results of TMI}$$

14.3.5.8.2.5 FRAME CONTROL The equation for the time domain waveform signal for frame control symbol, $S_{mf}[n]$, is shown as follows. In the definitions for this equation, "mf" is the frame control symbol number and $PAM_{mf\text{-}k}(c)$ is the signal position allocated to each carrier from the mapping block for the frame control symbol number (mf-k). The frame control symbol consists of eight symbols (mf = 0–7).

$$S_{mf}[n] = \frac{1}{16} \times \sum_{k=0}^{3} \sum_{c=0}^{N_{used}-1} h(n+512k) \times PAM_{mf\text{-}k}(c)$$

$$\times \cos\left[\frac{\pi}{512} \times \left\{(n+512k) + \frac{512+1}{2}\right\} \times \left(c + \frac{1}{2}\right) + \theta(c)\right] \times Tone\_MASK(c)$$

for $0 \leq n < 512$ {Tone_MASK (2x + 1) and Tone_MASK 2(x + 1) are the same setting. (x:integer)}
where

mf-k = −3 $PAM_{-3}(c) = 1$

mf-k = −2 $PAM_{-2}(c) = -1$

mf-k = −1 $PAM_{-1}(c)$ = PAM results of TMI

mf-k ≥ 0 $PAM_{mf\text{-}k}(c)$ = PAM results of Frame Control

14.3.5.8.2.6 FL SYMBOL The equation for the time domain waveform signal for frame length (FL) symbol, $S_{mfl}[n]$, is shown as follows. In the definitions for this equation, "mfl" is the FL symbol number and $PAM_{mf}(c)$ is the signal position allocated to each carrier from the mapping block. The FL symbol consists of one symbol (mfl = 0).

$$S_{mfl}[n] = \frac{1}{16} \times \sum_{k=0}^{3} \sum_{c=0}^{N_{used}-1} h(n+512k) \times PAM_{mfl\text{-}k}(c)$$

$$\times \cos\left[\frac{\pi}{512} \times \left\{(n+512k) + \frac{512+1}{2}\right\} \times \left(c + \frac{1}{2}\right) + \theta(c)\right] \times Tone\_MASK(c)$$

for $0 \leq n < 512$ {Tone_MASK (2x + 1) and Tone_MASK 2(x + 1) are the same setting. (x:integer)}

mfl-k = −3 $PAM_{-3}(c)$ = PAM results of Frame Control 5

mfl-k = −2 $PAM_{-2}(c)$ = PAM results of Frame Control 6

mfl-k = −1 $PAM_{-1}(c)$ = PAM results of Frame Control 7

mfl-k = 0 $PAM_{0}(c)$ = PAM results of FL

14.3.5.8.2.7 FRAME BODY The equation for the time domain waveform signal for the Frame Body, $S_{md}[n]$, is shown as follows. In the definitions for this equation, "md" is the Frame Body data symbol number and $PAM_{md-k}(c)$ is the signal position allocated to each carrier from the mapping block for the Frame Body data symbol number (md-k).

$$S_{md}[n] = \frac{1}{16} \times \sum_{k=0}^{3} \sum_{c=0}^{N_{used}-1} h(n+512k) \times PAM_{md-k}(c)$$

$$\times \cos\left[\frac{\pi}{512} \times \left\{(n+512k) + \frac{512+1}{2}\right\} \times \left(c + \frac{1}{2}\right) + \theta(c)\right] \times Tone\_MASK(c)$$

for $0 \leq n < 512$ {Tone_MASK $(2x+1)$ and Tone_MASK $2(x+1)$ are the same setting. (x:integer)}
    where

        md-k $= -3$ $PAM_{-3}(c) = $ PAM results of Frame Control 6

        md-k $= -2$ $PAM_{-2}(c) = $ PAM results of Frame Control 7

        md-k $= -1$ $PAM_{-1}(c) = $ PAM results of FL

        md-k ≥ 0 $PAM_{md-k}(c) = $ PAM results of Frame body

14.3.5.8.2.8 FRAME TERMINATION The equation for the time domain waveform signal for the frame termination, $S_{mft}[n]$, is shown as follows. In the definitions for this equation, "mft" is the frame termination symbol number and $PAM_{mft-k}(c)$ is the signal position allocated to each carrier from the mapping block for the frame body data symbol number and the frame termination symbol number. N is the last number of the frame body data symbol. The Frame Termination symbol consists of three symbols (mft $= 0$–2).

$$S_{mft}[n] = \frac{1}{16} \times \sum_{k=0}^{3} \sum_{c=0}^{N_{used}-1} h(n+512k) \times PAM_{mft-k}(c)$$

$$\times \cos\left[\frac{\pi}{512} \times \left\{(n+512k) + \frac{512+1}{2}\right\} \times \left(c + \frac{1}{2}\right) + \theta(c)\right] \times Tone\_MASK(c)$$

for $0 \leq n < 512$ {Tone_MASK $(2x+1)$ and Tone_MASK $2(x+1)$ are the same setting. (x:integer)}
    where

        mft-k $= -3$ $PAM_{-3}(c) = $ PAM results of Frame body $(N-2)$

        mft-k $= -2$ $PAM_{-2}(c) = $ PAM results of Frame body $(N-1)$

        mft-k $= -1$ $PAM_{-1}(c) = $ PAM results of Frame body N

        mft-k ≥ 0 $PAM_{mft-k}(c) = 0$

Preamble	TMI (No-Payload flag : 1 bit TMI : 8 bits (or 5 bits) Tail : 6 bits)	Frame Control (Info. : 34 bytes Tail : 6 bits)	FL (FL : 16 bits CRC : 8 bits Tail : 6 bits)	Frame Body (Data : Variable Tail : 6 bits)	F-pad	Postamble

FIGURE 14.14 PPDU frame format with postamble.

14.3.5.8.2.9 POSTAMBLE (OPTIONAL) Figure 14.14 shows the PPDU frame format with postamble. The postamble is the signal type that is almost the same as the preamble (using different phase vector), consisting of six symbols, and it is added to the end of the PPDU frame.

The equation for the time domain waveform signal for the postamble, $S_{mpo}[n]$, is shown as follows. In the definitions for this equation, "mpo" is the postamble symbol number and $PAM_{mpo\text{-}k}(c)$ is the signal position allocated to each carrier from the mapping block for the frame body and postamble symbol number. N is the last number of the Frame Body data symbol. The postamble symbol consists of six symbols (mpo = 0–5).

$$S_{mpo}[n] = \frac{w(n)}{16} \times \sum_{k=0}^{3} \sum_{c=0}^{N_{used}-1} h(n+512k) \times PAM_{mpo\text{-}k}(c)$$
$$\times \cos\left[\frac{\pi}{512} \times \left\{(n+512k) + \frac{512+1}{2}\right\} \times \left(c+\frac{1}{2}\right) + \theta_{mpo\text{-}k}(c)\right] \times Tone\_MASK(c)$$

for $0 \leq n < 512$ {Tone_MASK $(2x+1)$ and Tone_MASK $2(x+1)$ are the same setting. (x:integer)}
where

mpo-k = −3 $PAM_{-3}(c)$ = PAM results of Frame body $(N-2)$
mpo-k = −2 $PAM_{-2}(c)$ = PAM results of Frame body $(N-1)$
mpo-k = −1 $PAM_{-1}(c)$ = PAM results of Frame body N
mpo-k ≥ 0 $PAM_{mpo\text{-}k}(c) = 1$

mpo-k < 0 $\theta_{mpo\text{-}k}(c) = \theta(c)$
mpo-k ≥ 0 $\theta_{mpo\text{-}k}(c) = \theta(\text{mode}(c+34, 512))$

$$w[n] = \begin{cases} \dfrac{511-n}{256} & (mpo = 5 \text{ and } n > 255) \\ 1 & \text{otherwise} \end{cases}$$

14.3.5.8.2.10 PILOT SYMBOL (OPTIONAL) Figure 14.15 shows frame bodies using pilot symbols. We can use the same type of signal as for the preamble, consisting of nine symbols (eight SYNCP + SYNCM) and it is inserted every 128 symbols in the

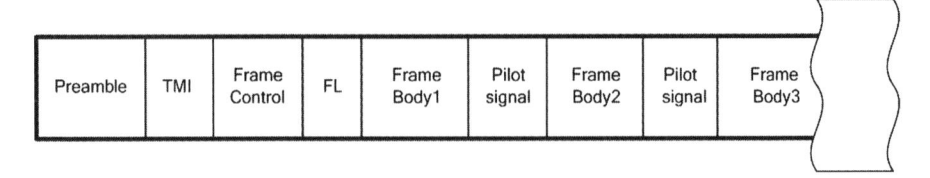

| Preamble | TMI | Frame Control | FL | Frame Body1 | Pilot signal | Frame Body2 | Pilot signal | Frame Body3 |

FIGURE 14.15 Frame bodies using pilot signals.

frame body. Each frame body consists of 128 symbols. By use of information (channel estimation result, equalizer information, error rate, retransmission rate, transmission rate, and so on) that is obtained on the basis of a received signal at a receiver side, it is determined whether the pilot symbols are inserted into the frame body of a transmission signal or not.

14.3.5.9 Major Specifications Table 14.7 lists the major specifications of the wavelet OFDM PHY. Basically, mandatory wavelet OFDM consists of 512-point IDWT with 61.03515625 kHz carrier spacing and bandwidth (1.8–28 MHz). This list includes extended major parameters of the wavelet OFDM PHY. Channels can be created above and below 30 MHz, allowing use of a band that is greater than 30 MHz. For instance, a transmission signal can be upconverted to the upper channel and downconverted again upon receipt. Baseband communication is mandatory and bandpass communication as optional can use both channels. The number of carriers is decided on carrier spacing and bandwidth. And the phase vector is constituted by repeating the phase group from carrier number 1 to 512 according to the number of carriers. Optional upper bands may be defined depending on country regulations and may include frequencies above or below 50 MHz.

These carrier spacings are realized with 512-point, 1024-point, and 2048-point IDWT. The ability to support symbol sizes of 16.384 μs and 32.768 μs is optional for

TABLE 14.7 Wavelet OFDM Major Specifications

Communication Method	Wavelet OFDM
Carrier spacing	61.03515625 kHz, 30.517578125 kHz (optional), and 15.2587890625 kHz (optional)
Symbol length	8.192 μs, 16.384 μs (optional), 32.768 μs (optional)
Primary modulation (per carrier)	32PAM (optional), 16 PAM to 2PAM
Frequency range used	1.8–28 MHz, 1.8–50 MHz: (optional)
Maximum PHY transmission rate (HAM notches and no FEC)	220 Mbps 420 Mbps (optional)
Forward Error Correction (FEC)	Reed–Solomon encoder/decoder, Convolutional encoder/Viterbi decoder, Reed–Solomon encoder/decoder and Convolutional encoder/Viterbi decoder or LDPC-CC encoder/decoder: (optional)
Diversity mode	Provided

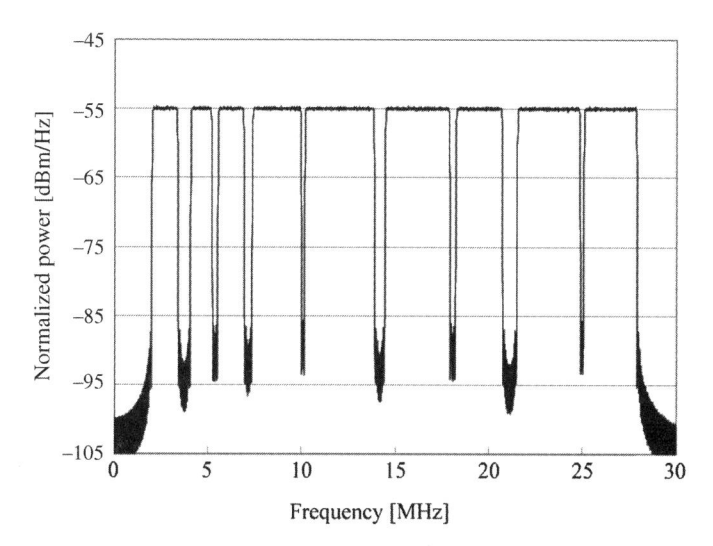

FIGURE 14.16 Example transmission spectrum of Wavelet OFDM with notches (up to 30 MHz).

stations using wavelet PHY. Stations that support these optional symbol sizes are capable of upbanding the corresponding baseband anywhere in the 2–30 MHz frequency range.

14.3.5.10 Notch and Power Control Controlling two or more carriers using the wavelet OFDM creates various power level bands of up to −35 dB, which significantly reduces interference to other systems (e.g., shortwave radio) using the same frequency bands. Thus, wavelet OFDM can adapt flexibly to the regulations of various countries or even to regulation changes, simply by changing the carriers that are left unused. Several notches are performed for amateur radio spectrums with mask processing in Figure 14.16. Power control values are met by at least a 1 dB step with an accuracy of ± 1 dB through −20 dB reduction and by at least a 2 dB step with an accuracy of ± 2 dB from more than −20 dB through −30 dB reduction. Power control applies to only the unmasked carriers per carrier. Using the default value and no power control express the same thing.

In addition, because the proposed MAC and PHY provide power-control functionality, notches can be created flexibly, and notch bandwidth can be controlled by independently controlling the power applied to each carrier.

If the notch is controlled dynamically, specific carriers may be off when the specific carriers have the same frequency of narrowband noise that is generated by other systems and that noise level is equal to or more than a predetermined value at the receiver side (Figure 14.17).

Certain carriers are always masked for protection of the amateur radio spectrum. These masked carriers are not output at any time. Each two carriers controlled by a set of Management Information Base (MIB) attributes approximately between 0 dB and −35 dB.

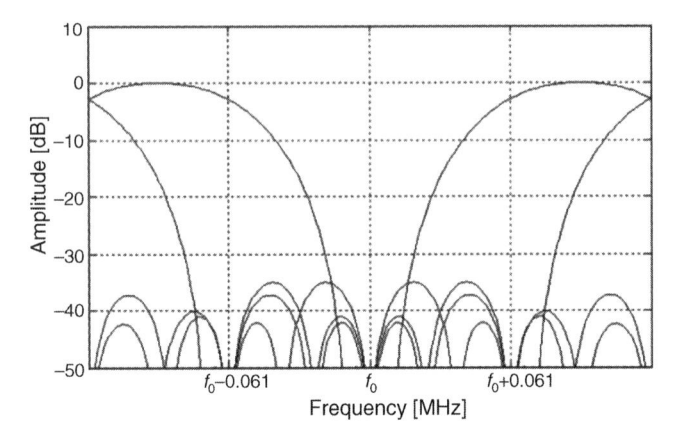

FIGURE 14.17 Notch frequency characteristics.

14.3.5.11 System Clock Frequency Tolerance The system clock is the electrical clock signal used by each device for its signal processing. The system clock frequency tolerance is ±25 ppm maximum.

14.4 COEXISTENCE

IEEE 1901 specification uses an Inter-System Protocol (ISP) for coexistence between IEEE 1901 access and in-home stations that use IEEE 1901 wavelet or IEEE 1901 FFT PHY. ISP can also be used for coexistence between IEEE 1901 and G.hn stations.

The ISP is a simple resource sharing mechanism that regulates access to the powerline medium. ISP uses coexistence signals (refer Section 14.4.1) that enable IEEE 1901 stations to signal their presence and also determine the presence of other IEEE 1901 systems and G.hn systems. Coexistence signals are transmitted in a periodic manner during the ISP windows assigned to various systems. Details of coexistence signaling scheme is presented in Section 14.4.2. IEEE 1901 systems share resources based on a resource allocation scheme described in Sections 14.4.3 and 14.4.4. ISP-related parameters and management messages are described in Sections 14.4.5 and 14.4.6, respectively.

14.4.1 Coexistence Signals

ISP signals consist of 16 consecutive OFDM symbols. Each OFDM symbol, formed by a set of all "one" BPSK data, is modulated onto the carrier waveforms using a 512-point Inverse Fast Fourier Transform (IFFT) at a 100 MHz Sampling frequency. The time domain symbols (that span 81.92 μs duration) are then multiplied by a window function $W(n)$ to reduce the out-of-band energy to be compliant with the transmit spectrum mask. The ISP protocol establishes the necessity for five different signal phases, all of them based on the previously defined OFDM symbol but using

COEXISTENCE

TABLE 14.8 ISP Signal Phases and their Use

ISP Signal	Use
Phase 1	Access
Phase 2	In-home Wavelet (IH-W) & Resync
Phase 3	In-home FFT OFDM (IH-O) & Resync
Phase 4	Access & FDM Interference
Phase 5	In-home G.hn (IH-G) & Resync

different phases at each carrier. Table 14.8 below shows the use of each of the five ISP signals.

Stations that belong to the same system (e.g., all stations belonging to IEEE 1901 FFT in-home system) transmit ISP signal at the same time. To avoid noncompliance with radiated emissions regulation, ISP signals are sent at 8 dB lower power than normal transmission. Lowering power on ISP signals also limits IEEE 1901 systems that are well isolated from detecting each other, thus enabling them to simultaneously use the channel (i.e., full spatial reuse without sharing the time on wire).

14.4.2 ISP Signaling Scheme

Coexistence signaling is carried out by the use of periodically repeating ISP windows that are used to convey information: system presence, resource requirements, and resynchronization requests. Each PLC system category is allocated a particular ISP window.

The ISP window is a region of time used by coexisting PLC devices exclusively for transmitting/detecting ISP signals. The ISP window is allotted periodically every T_{ISP}, a multiple of AC_CYCLE at a fixed offset, T_{off}, from a sync point. The period formed by four consecutive T_{ISP} is given by T_H.

A Sync Point is defined as a time located at 0, 60, 120, 180, 240, or 300 degrees relative to a zero-cross point of the AC main. Figure 14.18 shows the sync points relative to the AC line cycle phases. As there are two sync points per mains cycle and there may be up to three phases, then there may be up to six possible sync points. All coexisting systems synchronize to a common sync point.

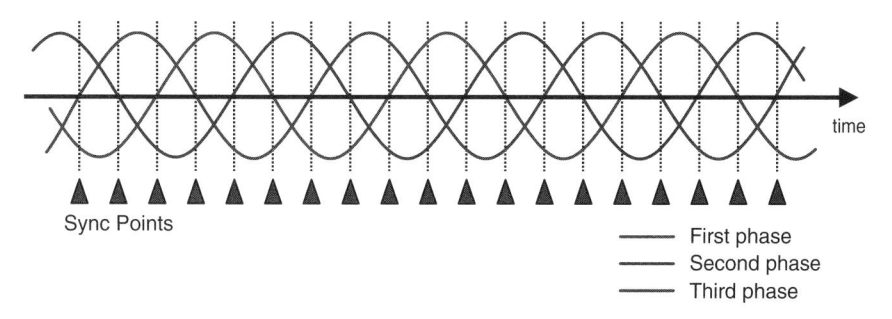

FIGURE 14.18 Sync points.

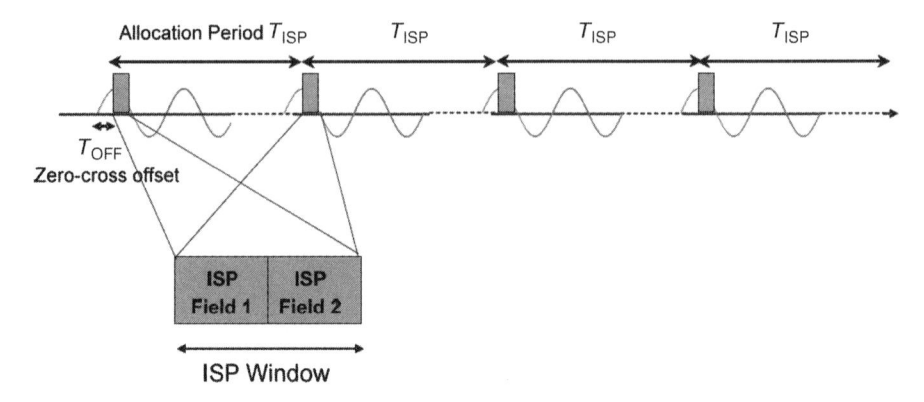

FIGURE 14.19 ISP time Window and ISP fields concept.

Each ISP window consists of two ISP fields. The ISP window and ISP fields are shown in Figure 14.19. Each coexisting system is allocated a particular ISP window. ACC uses an ISP window, IH-W uses the following ISP window, IH-O uses the next ISP window, and IH-G uses the next ISP window, as shown in Figure 14.20.

ISP signals are transmitted by all systems. All devices that belong to the same system transmit ISP signals simultaneously in the ISP window allocated to their system with a periodicity of $4 \times TISP$. Every device monitors the remaining ISP windows to assess the network status.

Successful operation of ISP requires all the systems to be coordinated with regard to the ISP windows. In some situations, some systems may become out-of-sync with other. In such cases, Resync the signals. This resync mechanism enables all systems to get back into coordination.

14.4.2.1 ISP Fields The ISP window contains two ISP fields. The devices of a coexisting system transmit ISP signals within these fields. The signals within each field are transmitted with a defined phase to convey a particular status or request. The combination of system category, field, and signal phase contributes to the network status.

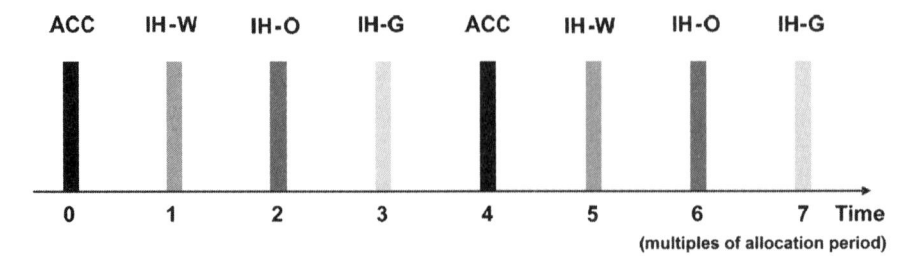

FIGURE 14.20 Periodicity of ISP windows.

COEXISTENCE 297

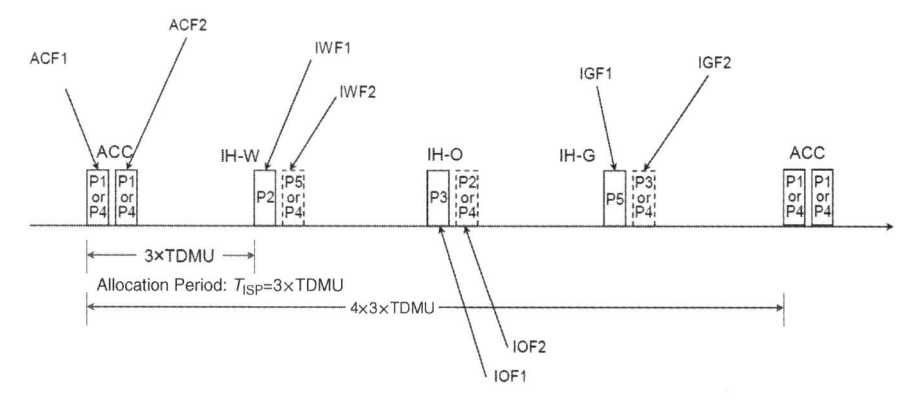

FIGURE 14.21 ISP fields.

Figure 14.21 shows the ISP fields. Solid lines denote fields where ISP signals are always transmitted if PLC devices belonging to the appropriate system category are present. Dashed lines denote fields where ISP signals may or may not be present. Note that ACF1 indicates Access Field 1, similarly IOF2 indicates in-home FFT OFDM field 2 and so on. Table 14.9 below shows the meaning of ISP window fields for various systems.

TABLE 14.9 Meaning of ISP Window Fields

System	Field 1	Field 2	Meaning for Network Status
Access (ACC)	Ph1	Ph1	Access TDM request partial bandwidth
	Ph1	Ph4	Access TDM request full bandwidth
	Ph4	Ph1	Access FDM below 10 MHz on all TDMSs
	Ph4	Ph4	Access FDM below 14 MHz on all TDMSs
In-home Wavelet (IH-W)	Ph2		1901 in-home wavelet system present
	Ph2	Ph5	1901 in-home wavelet system must start resync procedure
	Ph2	Ph4	1901 in-home wavelet interference threshold exceeded
In-home FFT OFDM (IH-O)	Ph3		1901 in-home OFDM system present
	Ph3	Ph2	1901 in-home OFDM system must start resync procedure
	Ph3	Ph4	1901 in-home OFDM interference threshold exceeded
In-home G.hn (IH-G)	Ph5		ITU-T G.hn system present
	Ph5	Ph3	ITU-T G.hn system must start resync procedure
	Ph5	Ph4	ITU-T G.hn interference threshold exceeded

14.4.2.2 Network Status By monitoring the signals transmitted within the ISP windows allocated to other systems, a coexisting system is able to determine the number and type of coexisting systems present on the line and their resource requirements. By monitoring the signals within its own ISP window (second field), a coexisting system is able to detect a resynchronization request from any of the other coexisting systems.

Information about the systems on the power line includes

- In-home—{Present, Resynchronization request from other system, and "interference threshold exceeded" to access system}.
- Access—{Present, TDM, or FDM, and Full or partial resource allocation requirement}.

This set of instantaneous information is termed the network status. Network status is used to determine the allocation of resources to each coexisting system.

14.4.3 Coexistence Resources

The ISP allows systems to coexist in the time domain (TDM) and/or the frequency domain (FDM).

14.4.3.1 FDM FDM may only be invoked by an access system. The overall FDM scheme consists of two frequency bands. The upper band is shared by in-home systems, and the lower band is reserved for the access system. One of the two frequency split points are signaled by the access system, denoting "full bandwidth" and "partial bandwidth" FDM modes. The "full bandwidth" split point is 14 MHz, and the "partial bandwidth" split point is 10 MHz. Whichever mode is chosen by the access system is dependent on its overall bandwidth requirements.

14.4.3.2 TDM TDM is used for coexisting between in-home systems and between coexisting in-home systems and an access system operating in TDM. An access system may indicate its TDM resource requirements as either "full bandwidth" or "partial bandwidth." This request impacts the TDM resource allocation between the coexisting systems.

The overall synchronization period for the in-home and access systems is T_H, and within TH, there are four ISP windows. The period between the starts of consecutive ISP windows is further divided into three TDM Units (TDMUs), so there are 12 TDMUs in each period T_H, labeled as TDMU#0 through TDMU#11. Each TDMU is further subdivided into eight TDM Slots (TDMS), labeled as TDMS#0 through TDMS#7.

Depending on the network status, each TDMS within a TDMU is allocated to a particular system category. The same TDMS is allocated to the same system category in each TDMU (e.g., "1901 access" could be allocated "TDMS#3" and "TDMS#4" in each TDMU).

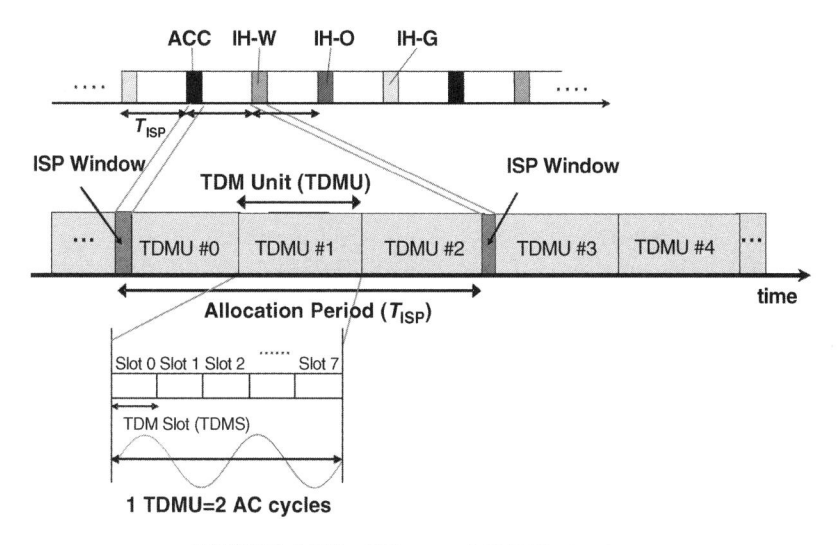

FIGURE 14.22 ISP general TDMA structure.

The overall ISP TDM structure is shown in Figure 14.22; there are four such structures in every TH. The structure of one TDMU is shown in Figure 14.23.

14.4.4 ISP Resource Allocation

The policy for resource sharing between access and a group of in-home systems is based on the following rule:

> When the access system requires 50% or more of the available resources and the group of in-home systems requires 50% or more of the available resources, the access system generally receive 50% of the resources and the group of in-home systems generally receive the remaining 50% of the resources. The access system or the group of in-home systems may use additional resources above 50% if not required by the other system.

TDM resources are allocated according to Table 14.10. The right-hand side of this table indicates the systems that are allocated to the corresponding TDMS within the

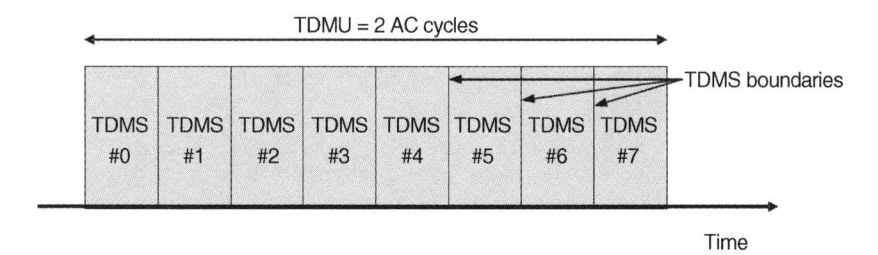

FIGURE 14.23 Structure of TDMA.

TABLE 14.10 ISP Parameter Specification

Index	ISP Field					TDM Slot Number							
	ACC	IH-W	IH-O	IH-G	BW	0	1	2	3	4	5	6	7
1	-	-	-	IH-G		IH-G	IH-G	IH-G	IH-G	IH-G	IH-G	IH-G	IH-G
2	-	IH-W	-	-		IH-W	IH-W	IH-W	IH-W	IH-W	IH-W	IH-W	IH-W
3	-	-	IH-O	-		IH-O	IH-O	IH-O	IH-O	IH-O	IH-O	IH-O	IH-O
4	-	IH-W	-	IH-G		IH-W	IH-W	IH-G	IH-W	IH-G	IH-G	IH-G	IH-G
5	-	IH-W	IH-O	-		IH-W	IH-W	IH-O	IH-W	IH-O	IH-W	IH-W	IH-O
6	-	-	IH-O	IH-G		IH-G	IH-W	IH-O	IH-O	IH-O	IH-G	IH-G	IH-G
7	-	IH-W	IH-O	IH-G		IH-W	IH-W	IH-O	IH-O	IH-O	IH-G	IH-G	IH-G
8	ACC	-	-	-	FB	ACC	ACC	ACC	ACC	ACC	ACC	ACC	ACC
9	ACC	-	-	IH-G	PB	IH-G	IH-G	IH-G	IH-G	ACC	ACC	IH-G	IH-G
10	ACC	-	-	IH-G	FB	IH-G	IH-G	IH-G	ACC	ACC	ACC	ACC	IH-G
11	ACC	IH-W	-	-	PB	IH-W	IH-W	IH-W	IH-W	ACC	ACC	IH-W	IH-W
12	ACC	IH-W	-	-	FB	IH-W	IH-W	IH-W	ACC	ACC	ACC	ACC	IH-W
13	ACC	-	IH-O	-	PB	IH-O	IH-O	IH-O	IH-O	ACC	ACC	IH-O	IH-O
14	ACC	-	IH-O	-	FB	IH-O	IH-O	IH-O	ACC	ACC	ACC	ACC	IH-O
15	ACC	IH-W	-	IH-G	PB	IH-W	IH-W	IH-G	IH-W	ACC	ACC	IH-G	IH-G
16	ACC	IH-W	-	IH-G	FB	IH-W	IH-W	IH-G	ACC	ACC	ACC	IH-W	IH-W
17	ACC	IH-W	IH-O	-	PB	IH-W	IH-W	IH-W	ACC	ACC	ACC	ACC	IH-O
18	ACC	IH-W	IH-O	-	FB	IH-W	IH-W	IH-O	ACC	ACC	ACC	IH-W	IH-O
19	ACC	-	IH-O	IH-G	PB	IH-G	IH-O	IH-O	ACC	ACC	ACC	IH-G	IH-G
20	ACC	-	IH-O	IH-G	FB	IH-G	IH-O	IH-O	ACC	ACC	ACC	ACC	IH-G
21	ACC	IH-W	IH-O	IH-G	PB	IH-W	IH-W	IH-O	IH-O	ACC	ACC	IH-G	IH-G
22	ACC	IH-W	IH-O	IH-G	FB	IH-W	IH-W	IH-O	ACC	ACC	ACC	ACC	IH-G

TABLE 14.11 ISP Parameter Specification

Parameter	Definition	Value
AC_CYCLE	AC mains cycle	~20 ms on 50 Hz
		~16.667 ms on 60 Hz
T_0	Sync point interval	$(1/6) \times$ AC_CYCLE
TDM_UNIT_LEN	Time length of TDM Unit	$2 \times$ AC_CYCLE
TDM_SLOT_LEN	Time length of TDM Slot	$(1/8) \times$ TDM_UNIT_LEN
T_{ISP}	Period of ISP windows	$3 \times$ TDM_UNIT_LEN
T_H	Synchronization period for in-home systems and for access system	$4 \times T_{ISP}$ ($24 \times$ AC_CYCLE)
T_{OFF}	Offset length from the Sync Point to the beginning of the ISP window	200 µs

TDMU. The left-hand side indicates the systems that are present on the power line, that is, the network status. The last column on the left-hand side indicates the bandwidth (BW) used by the access system, either full bandwidth (FB) or partial bandwidth (PB), as described in Table 14.9. Each system determines the network status and adjusts its use of resources accordingly.

14.4.5 ISP Parameters

Table 14.11 shows the ISP-related parameters and their definition.

14.4.6 Management Messages

IEEE 1901 systems (e.g., In-home FFT systems) operating in the same logical network exchange ISP-related information using management messages. These messages enable all stations within a logical network to be coordinated with respect to ISP. Additional details of these messages can be found in IEEE 1901 specification [2].

14.5 SUMMARY

This chapter provided a description of the IEEE 1901 PLC standard, first giving a brief description of the FFT PHY as extensions of HomePlug AV, and then describing the Wavelet PHY as well as the Inter-PHY System Protocol (ISP) for coexistence.

Chapter 15 gives an overview of the HomePlug GreenPHY specification, showing how HomePlug AV is used as a fundamental building block in the development of this Smart Energy communication specification.

15

HomePlug GREEN PHY

15.1 INTRODUCTION

HomePlug Green PHY is designed to enable Smart Grid applications using powerline as the communication medium. Since all appliances that consume electricity are already plugged into the powerline, Green PHY is a natural fit for such application. Smart Grid applications do not require high bandwidth, but they require robust and reliable communication. Further low cost and low power consumption is also a key requirement for Smart Grid. HomePlug Green PHY specifically designed to address these needs.

HomePlug Green PHY is designed to be fully interoperable with HomePlug AV, IEEE 1901, and HomePlug AV2. This enables Green PHY to fit seamlessly into the existing infrastructure and enables a single powerline network that addresses all communication within the home network. Green PHY also leverages the field proven and robust communication techniques used by HomePlug AV and IEEE 1901.

Green PHY provides significant simplification that enables it to be cost effective for Smart Grid applications [31].

15.2 PHYSICAL LAYER

Green PHY physical layer is same as HomePlug AV with the PHY Protocol Data Unit (PPDU) Payload restricted to STD-ROBO, HS-ROBO, or MINI-ROBO.

HomePlug AV and IEEE 1901: A Handbook for PLC Designers and Users, First Edition.
Haniph A. Latchman, Srinivas Katar, Larry Yonge, and Sherman Gavette.
© 2013 by The Institute of Electrical and Electronics Engineers, Inc. Published 2013 by John Wiley & Sons, Inc.

TABLE 15.1 ROBO Mode Parameters

ROBO Mode	Modulation	FEC Code Rate	Number of Copies (N_{copies})	PHY Rate (Default Tone Mask)	Payload Capacity
STD-ROBO_AV	QPSK	$1/2$	4	4.9226 Mbps	One PB520
HS-ROBO_AV	QPSK	$1/2$	2	9.8452 Mbps	One, two, or three PB520
MINI-ROBO_AV	QPSK	$1/2$	5	3.7716 Mbps	One PB136

ROBO modes are the most reliable modes for transmitting data. In these modes, data are modulated using Quadrature Pulse Shift Keying (QPSK), $1/2$ Turbo Code along with time/frequency-based repetition coding. This enables significant resilience to errors. Restricting the PPDU payload to low data rate ROBO modes also enable Green PHY stations to significantly reduce the cost and power consumption.

Table 15.1 summarizes the key parameters for various ROBO modes.

15.3 MAC LAYER

Green PHY MAC Layer is based on HomePlug AV and eliminates several MAC features to reduce the cost and complexity. The various simplifications to HomePlug AV MAC in Green PHY are as follows:

- Green PHY stations only support Level-0 Central Coordinator (CCo) functionality. Carrier Sense Multiple Access with Collision Avoidance (CSMA/CA) based channel access is used for exchanging data.
- Green PHY stations are not required to support connection-based services. This eliminates the need for supporting connection setup procedures, bandwidth management, arrival time stamps, and so on.
- Green PHY stations are not required to support channel sounding and exchange to channel specific tone maps. The transmitter decides which ROBO mode to select for various transmissions.
- Green PHY stations are not required to support Mac Protocol Data Unit (MPDU) bursting.
- Green PHY stations are not required to support bidirectional bursting.
- Green PHY stations are not required to support request SACK retransmission mechanism.
  Green PHY MAC also adopts the following features from IEEE 1901.
- Repeating and routing (refer to Section 14.2.6).
- Short Network Identifier (SNID) reuse (refer to Section 14.2.5).
- Inter System Protocol (ISP) (refer to Section 14.4).

Apart from these Green PHY MAC includes additional features that reduce their power consumption and enable fair bandwidth sharing with IEEE 1901 and other HomePlug technologies. Another key feature in Green PHY is the Plug-in Electric Vehicle (PEV)—Electric Vehicle Supply Equipment (EVSE, a.k.a. charging station) association protocol for secure electric vehicle charging and billing. These new features are described in the following sections.

15.3.1 Power Save

Power Save mode in Green PHY enables stations to reduce their average power consumption. In Power Save mode stations reduce their average power consumption by periodically transitioning between Awake and Sleep states. STAs in the Awake state can transmit and receive PPDUs over the powerline. In contrast, STA in Sleep state temporarily suspend transmission and reception of PPDUs over the powerline, thus reducing the station's average power consumption.

The periodic interval of time during which a station in Power Save mode is in the Awake state is referred to as Awake Window. Similarly, the periodic interval of time during which a STA in Power Save mode is in Sleep state is referred to as Sleep Window. The duration of Awake Window and Sleep Window can range from a few milliseconds to several Beacon Periods. The sum of Awake Window duration and Sleep Window duration is referred to as Power Save Period (PSP). Power Save Periods is restricted to 2^k multiples of Beacon Periods (i.e., 1 Beacon Period, 2 Beacon Periods, 4 Beacon Periods, etc.). The Power Save Schedule (PSS) for each STA in power save mode is indicated by the combination of the 4-bit Power Save Period and Awake Widow duration values. To communicate with a STA in Power Save mode, other STAs in the HomePlug AV Local Network (AVLN) need to know its PSS.

Central Coordination of the AVLN coordinates the Power Save mode for all STAs (including itself) in the AVLN. Before entering Power Save mode, STA send **CC_POWERSAVE.REQ** message to the CCo and includes in the message the requested Power Save Schedule. CCo responds to **CC_POWERSAVE.REQ** with a **CC_POWERSAVE.CNF**. **CC_POWERSAVE.CNF** indicates whether the CCo accepted the STA's request to enter Power Save mode. Reception of a **CC_POWERSAVE.CNF** with Result Code set to accept causes the STA to go into Power Save mode.

CCo provides a Beacon Period Count (BPCnt) to enable stations to synchronize the start of their Awake Window. BPCnt is a 12-bit field that is incremented by one in each Beacon Period. The Awake Window of a Station in Power Save. Awake Window of stations start in Beacon Periods where their PSP is an integral multiple of BPCnt. If there is a Green PHY Preferred (GPP) allocation with the Beacon Period, then the start time of Awake Window is the same as the start of GPP allocation. Otherwise, the start time of the Awake Window is the same as the start of the first CSMA allocation within the Beacon Periods. Since PSP is always a 2^k multiple of Beacon Period and Awake Window start time in a Beacon Period is aligned for all stations, there is a guaranteed overlap of Awake Windows for all stations in Power

MAC LAYER   305

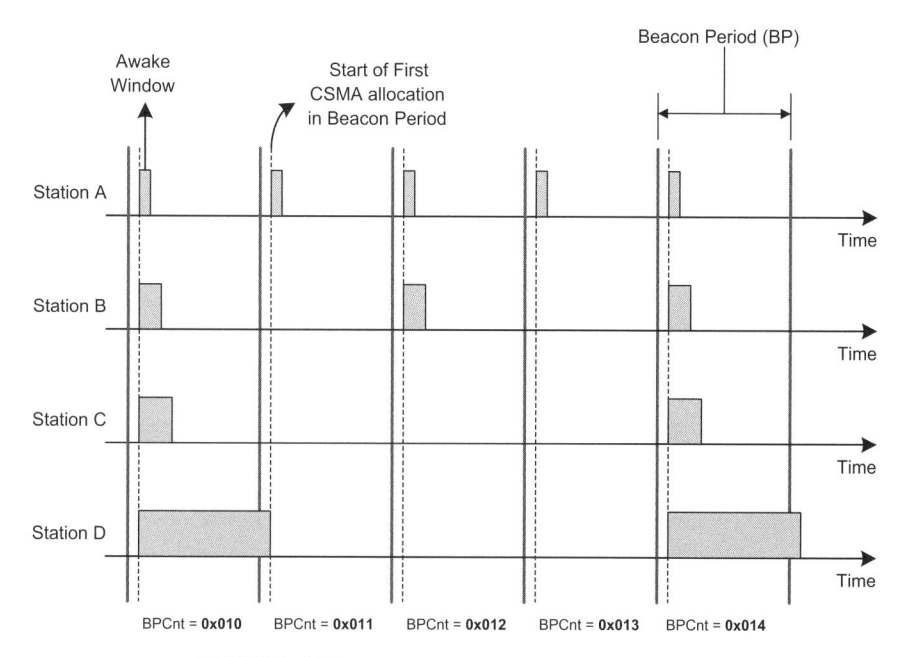

FIGURE 15.1 Examples of Power Save Schedules.

Save mode. This overlap interval can be used for transmission of information that needs to be received by all stations within the AVLN.

Figure 15.1 shows an example of Power Save Schedule of four stations {A, B, C, D}. In this example:

- Station A has PSP of 1 Beacon Period, and Awake Window duration of 2 ms. Station A is awake for 2 ms every Beacon Period, with Awake Window start time aligned with the start of the first CSMA allocation within each Beacon Period.
- Station B has PSP of 2 Beacon Periods and Awake Window duration of 4 ms. Station B's 4 ms Awake Window starts at the start of the first CSMA allocation in every Beacon Period where BPCnt is an integer multiple of 2 (e.g., BPCnt = **0x010**, **0x012**, and **0x014**).
- Station C has PSP of 4 Beacon Periods and Awake Window duration of 6 ms. Station C's 6 ms Awake Window starts at the start of the first CSMA allocation in every Beacon Period where BPCnt an integer multiple of 4 (e.g., BPCnt = **0x010** and **0x014**).
- Station D has PSP of 4 Beacon Periods and Awake Window duration of 1 Beacon Period. Station D's Awake Window starts at the start of the first CSMA allocation in every Beacon Period where BPCnt an integer multiple of 4 (e.g., BPCnt = **0x010** and **0x014**).

Note that in this example, stations are always awake at the same time once every 4 Beacon Periods for a duration of 2 ms. It should be noted that the precise start time of the Awake Window can vary as the Beacon Period varies (due to AC line cycle variations). STAs take such variations into account to ensure that STAs remain in the Awake state for the entire Awake Window.

A STA exiting Power Save mode notifies the CCo by sending the **CC_POWERSAVE_EXIT.REQ** message. The CCo acknowledges the reception of **CC_POWERSAVE_EXIT.REQ** with **CC_POWERSAVE_EXIT.CNF**.

15.3.1.1 Distribution of Power Save State Information All stations in the AVLN need to be aware of PSS of all other stations. This information is used to ensure that stations only transmit to other stations during the Awake Window.

CCo maintains a Power Save State Identifier (PSSI) and Power Save schedules of STAs in the AVLN. PSSI is used to identify the Power Save state of the AVLN. PSSI is incremented by one whenever a STA (including the CCo) in the AVLN enters or exits Power Save mode. All stations in the AVLN also maintain a local copy of the Power Save State Identifier and Power Save schedules of STAs in the AVLN. Stations update their local copy based on the information provided by the CCo.

CCo uses the Power Save BENTRY to announce recent changes in the Power Save Schedule of STAs (including the CCo) in the AVLN. Each {TEI[n], PSS[n]} in the Power Save BENTRY indicates a single Power Save Schedule change. These are in chronological order with the most recent schedule change indicated first. Note that because schedule changes are indicated (rather than simply the current schedule for STAs), multiple schedule changes for a single STA may be indicated. The CCo repeats the changes in Power Save Schedules in multiple Beacons to increase the probability of reception by all STAs in the AVLN. In some cases, CCo may not be able to indicate all changes to Power Save Schedules in the Power Save BENTRY due to the limited Beacon Size. In such case, CCo broadcasts the Power Save schedules of all stations in the AVLN using **CC_POWERSAVE_LIST.CNF**.

STAs update their Power Save schedules based on the Power Save BENTRY and **CC_POWERSAVE_LIST.CNF**. STAs compare their local PSSI with the PSSI in the Beacon. If they are the same, then STA's already has the current set of PSSs. If the STAs PSSI is different than the PSSI in the Beacon, then STAs can use the difference between them to determine the number of changes in the Power Save schedules. If the Beacon contains all the changes, then STA can update its local PSSI and Power Save Schedules. If Beacon does not contain all the changes in the Power Save Schedule, STAs may request the CCo using **CC_POWERSAVE_LIST.REQ** message.

For example, if the local PSSI is **0x22** and the PSSI in the Beacon is **0x26**, then the Power Save state of four STAs in the AVLN has changed. If the Beacon contains four recent changes (i.e., TEI[1], PSS[1], . . . ,TEI[4], PSS[4]), then the STA can update its local PSSI to **0x26** and the local Power Save Schedule is updated based on the four recent changes. In this example, if the Beacon contains only two recent changes (i.e., TEI[1], PSS[1], TEI[2], PSS[2]), then the STA may request the CCo for the Power Save List using **CC_POWERSAVE_LIST.REQ** message.

STAs may request CCo to provide the Power Save list any time using the **CC_POWERSAVE_LIST.REQ** message. Reception of a **CC_POWERSAVE_LIST.REQ** causes the CCo to respond with a **CC_POWERSAVE_LIST.CNF** message containing PSSI and the latest PSSs of all stations in the AVLN.

When a STA has a MPDU to transmit, it uses the scheduling information in the Beacon and the Power Save Schedule of the receiver(s) to determine when the MPDU has to be transmitted. When a STA sends MPDUs to a receiver and fails to get any acknowledgment (e.g., SACK, CTS, and RSOF) after multiple attempts, STAs should request the CCo to provide the Power Save List using **CC_POWERSAVE_LIST.REQ** to confirm that the receiver is not in Power Save mode. This enables stations to recover gracefully when their local copy of PSSs of other stations in the AVLN is not accurate.

15.3.1.2 CCo Power Save CCo of an AVLN can also operate in Power Save mode. The maximum Power Save Period of a CCo in Power Save mode should not be greater than the shortest Power Save Period currently being used by the STAs in the AVLN. CCo may handover its functionality to another station (e.g., backup CCo) in the network to reduce its power consumption.

CCo exits Power Save mode when CCo functionality is handed over to a new station. Further, CCo also force all STAs in the AVLN to exit the Power Save mode by setting the Stop Power Save Flag in the Beacon before handing over the CCo functionality. The CCo may also force stations to exit Power Save when the network state needs to be changed (i.e., changing SNID). Any STA that receives a Beacon with the Stop Power Save Flag set immediately exit the Power Save mode.

15.3.2 Bandwidth Sharing between Green PHY and HomePlug AV and IEEE

Green PHY stations are likely to operate in environments consisting of HomePlug AV, IEEE 1901, HomePlug AV2, and future HomePlug technologies. This makes it critical to have mechanisms that enable fair bandwidth sharing. In particular, Green PHY stations need to be guaranteed a minimum bandwidth to support their low data rate Smart Grid applications. Similarly, Green PHY stations should not consume excessive bandwidth there by starving high data rate applications (e.g., video streaming) supported by IEEE 1901 and other HomePlug technologies (e.g., HomePlug AV).

The following section provides mechanisms that enable bandwidth sharing between Green PHY, IEEE 1901, and other HomePlug technologies.

15.3.2.1 Green PHY Preferred Allocation Green PHY preferred allocations is a mechanism by which future versions of HomePlug CCos (e.g., HomePlug AV2) can provide Green PHY stations with a guaranteed $\sim 7\%$ time-on-wire each Beacon Period. When the CCo detects the presence of Green PHY stations, it reserves $\sim 7\%$ of time-on-wire for Green PHY stations and includes that in the Central Beacon. Green PHY stations use GPP allocation to obtain their minimum guaranteed bandwidth. Even when GPP allocation is provided by the CCo, Green PHY stations can transmit in other CSMA allocations within the Beacon Periods. However, such transmissions are subjected to the Distributed Bandwidth Control protocol.

15.3.2.2 Distributed Bandwidth Control Distributed Bandwidth Control (DBC) protocol enables Green PHY stations to limit the maximum time-on-wire consumed by their transmissions at Channel Access Priority 1, 2, and 3 (i.e., CAP1, CAP2, and CAP3) to ~7%. Each Green PHY stations execute the DBC protocol during GPP allocations as well as CSMA allocations.

Green PHY stations use the Pending PHY Blocks (PPB) in the start-of-frame (SOF) to indicate the Channel Access Priority used by their transmissions. PPB field is set to 0xDF to indicate transmissions at CAP0, otherwise it is set to 0xEF (i.e., for CAP1, CAP2, and CAP3). Note that these values of PPB are not used by HomePlug AV and IEEE 1901. Thus, these values also enable detection of transmissions from Green PHY.

Green PHY stations store the PPDU Type and the time (i.e., time stamp of local clock or NTB) associated with the last two Green PHY transmissions (including its own transmissions and transmissions from other Green PHY stations) with PPB field in SOF set to 0xEF. A Green PHY station with a pending MPDU only contend at CAP1, CAP2, or CAP3 if its pending PPDU and the PPDUs transmitted within the last 33.33 ms does not exceed the combinations shown in Table 15.2. This limits the maximum time-on-wire used by Green PHY stations at CAP1, CAP2, and CAP3 to ~7%.

Green PHY PPDUs transmitted at CAP0 do not use DBC and may transmit at any time during the CSMA allocation. It is important to note that packet belonging to Priority Link Identifier (PLID) queues with PLID = 0x01, 0x02, and 0x03 may be transmitted by contending at CAP0. This enables Green PHY stations to use up to 100% of time-on-wire, when the bandwidth is not used by non-Green PHY stations.

In certain scenarios, Green PHY stations may not get their fair share of bandwidth because of the presence of higher priority HPAV traffic (e.g., when traffic from Green PHY station is at CAP1 and AV station(s) is CAP2). In such cases, Green PHY stations can promote the priority at which PRS signaling is sent (e.g., send PRS signals at CAP3), to ensure that they get their fair share of the bandwidth.

TABLE 15.2 Allowed PPDU Combinations

Combination	PPDU Combination
1	One ROBO PPDU and one Mini ROBO PPDU
2	Three Mini ROBO PPDUs
3	Two Mini ROBO and One High Speed ROBO with 1 PB
4	One Mini ROBO and One High Speed ROBO with 2 PBs
5	Two High Speed ROBO with 1 PB
6	One Mini ROBO and One High Speed ROBO with 3 PBs
7	One ROBO and One High Speed ROBO with 1 PB
8	One High Speed ROBO with 1 PB and One High Speed ROBO with 2 PBs

Distributed Bandwidth Control protocol only limits the total bandwidth consumed by monitoring only SOF MPDUs (i.e., MPDUs carrying data or management messages). Beacon transmissions from Green PHY stations are ignored by the DBC protocol.

15.3.3 PEV-EVSE Association

Plug-in Electric Vehicles (PEVs) need a convenient means of charging at any location they are parked (e.g., at home, at work, and in public parking lots). The simplest user experience for electric vehicle owner would be to just plug the vehicle into the charging station (a.k.a. Electric Vehicle Supply Equipment) and let the electric vehicle and charging station interact to ensure secure charging and billing. To facilitate this, electric vehicle needs to uniquely identify the charging station with which it is "associated" (i.e., directly connected via the charging harness) and exchange information to facilitate secure charging and billing.

Figure 15.2 illustrates an example where a blue car gets plugged into an EVSE. Communication signals that are transmitted by the blue car can be received by multiple EVSE as communication signals could not only propagate through the charging cables but also due to RF coupling. This creates an ambiguity for associating the PEV with the EVSE to which it is directly connected. Green PHY defines a PEV–EVSE association procedure that is designed to eliminate this ambiguity and to provide a unique association between PEV and EVSE.

When a PEV transmits Green PHY signal over the charging cable, the EVSE to which it is connected will receive the signal with minimal attenuation. For the Green PHY signals to reach any other EVSE, the signal needs to either propagate through the electric wiring or through RF coupling, which would result in higher signal attenuation. PEV–EVSE association procedure in Green PHY exploits this asymmetry in signal attenuation to associate (or match) the PEV with the EVSE.

PEV–EVSE association procedure was developed by Green PHY technical working group with input from several automotive companies, ISO/IEC 15118 standards group and from Society of Automotive Engineers (SAE) members. Figure 15.3 illustrates the various wires in a charging harness. AC1 and AC2 are used for power supply. Control Pilot (CPLT) wire is used to carry a 1 kHz pulse width

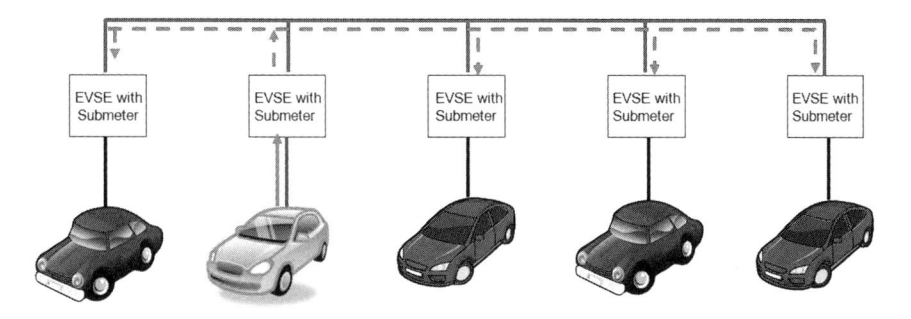

FIGURE 15.2 Illustration of PEV–EVSE association.

```
        AC1
<─────────────────>
        AC2
<─────────────────>
        GND
<─────────────────>
        CPLT
<─────────────────>
        PROX
<─────────────────>
```

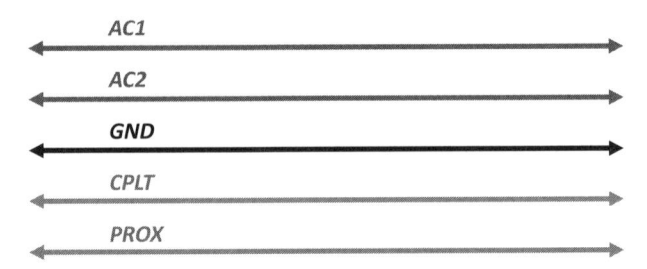

FIGURE 15.3 Charging harness.

modulated signal used by legacy equipment. Proximity Detect Line (PROX) is used to detect if vehicle is plugged into a charging station and is used to disable the vehicle from moving when it is plugged. Green PHY signaling between the PEV and EVSE is carried over using the Control Pilot (CPLT) and Ground wires in the Charging harness. The selection of control pilot wire (rather than AC1 or AC2 wires) enables Green PHY to be used in all regulatory domains (e.g., regulations in Japan do not allow powerline communications outside the home). Further, this also limits the Green PHY signal leakage from one EVSE to another, making the PEV–EVSE association procedure more robust. The transmit power of the Green PHY signal on the control pilot will also be lower than on normal powerline channels to limit interference with the pulse width modulated signal on control pilot. Lower transmit power level also increases security as the range of Green PHY signal propagation will be lower.

15.3.3.1 PEV–EVSE Association Procedure The PEV–EVSE association procedure is executed between the PEV and EVSEs to identify the PEV–EVSE association. EVSEs are all preconfigured to always be the CCo (i.e., master), while the PEV is preconfigured to never be a CCo. PEV initiates the association procedure after it determines that it is plugged into the EVSE. Note that the association procedure is executed when PEV and EVSE are part of different AVLNs (i.e., different logical network). Hence all messages exchanged during this procedure use Multinetwork Broadcasts at the MAC/PHY Layer even if the Ethernet Frame has a unicast destination addresses (for example, message sent from EVSE to a particular PEV).

Step-1: Signal Level Attenuation Characterization (SLAC) Parameters Exchange
  PEV sends a **CM_SLAC_PARM.REQ** to EVSEs to request the parameters associated with the SLAC protocol. EVSEs that receive this request will respond with a **CM_SLAC_PARM.CNF** that indicates parameters such as number of M-Sounds that need to be transmitted during the association procedure.

Step-2: Signal Attenuation Characterization PEV indicates the start of attenuation characterization by transmitting **CM_START_ATTEN_CHAR.IND** message. This message notified EVSEs that the PEV is starting the signal attenuation characterization.

PEV then transmits multiple **CM_MNBC_SOUND.IND** (or M-Sounds) messages. The number of M-Sounds transmitted will be based on the information provided by EVSEs in **CM_SLAC_PARM.CNF**. EVSEs measure the average attenuation from the received M-Sounds and reports the measurement results to PEV using **CM_ATTEN_CHAR.IND** message.

Step-3: Matching Decision PEV uses the **CM_ATTEN_CHAR.IND** messages received from EVSEs to make the matching decision. Typically, PEV selects that EVSE that has attenuation below a threshold as the matching (or associated) EVSE.

Step-4: Validation PEV may choose to validate the matching decision using non-Green PHY related techniques. This procedure is optional.

Step-5: Inform EVSE of Decision PEV indicates the matching decision to the EVSE using **CM_SLAC_MATCH.REQ** message.

Step-6: Matching Confirm by EVSE EVSE confirms the matching request by the PEV by sending a **CM_SLAC_MATCH.CNF** message that contains the NID and NMK of the EVSE.

Step-7: Network Join PEV uses the NID and NMK provided by the EVSE to join the EVSE's logical network. This step completes the PEV–EVSE association procedure at the MAC/PHY layer of Green PHY. Higher layers will then exchange the necessary information to enable secure charging and billing between the PEV and its associated EVSE.

15.4 SUMMARY

With the emerging emphasis on Smart Grids and Intelligent Energy Networking, this chapter explains how the HomePlug Green PHY Specification exploits HomePlug AV technologies and in particular the relatively lower bit rate but much more reliable ROBO modes.

Chapter 16 is devoted to a description of HomePlug AV2 which again builds on and in this case expands on HomePlug AV structures by the addition of (i) a larger bandwidth, (ii) higher order modulation, (iii) more efficient FEC codes, (iv) MIMO transceivers, (v) repeating, and (vi) power save mode. These augmentations of HomePlug AV result in HomePlug AV 2.0 (AV2) featuring a data rate of about 1.5 Gbps.

16

HomePlug AV2

16.1 INTRODUCTION

The HomePlug AV 2.0 Specification (AV2) adds new features to HomePlug AV that provide a significant increase in throughput and coverage performance. The HomePlug AV 2.0 Specification provides a significantly higher data rate of 1.5 Gbps compared to 200 Mbps for HomePlug AV. A notable performance point for AV2 is the coverage performance of ~90 Mbps User Datagram Protocol (UDP) network throughput (three equal UDP streams of 30 Mbps each) for 99% of networks with four or more devices assuming immediate repeating is implemented, based on field test measurements.

New physical layer features include Multiple Input Multiple Output (MIMO), wider frequency band, efficient notching and short (lower overhead) delimiter. New MAC Layer features include delayed acknowledgment, immediate repeating, and a power save mode. The following sections provide detail on each of these features.

16.2 MIMO

Most powerline wiring contains three individual wires: Line (or Phase), Neutral, and Ground (or Protective Earth). Most powerline modems such as HomePlug AV 1.1 modems use the Line-Neutral wire pair for communication. However, the third wire, Ground, may be used in combination with either Line or Neutral to support sending a

---

HomePlug AV and IEEE 1901: A Handbook for PLC Designers and Users, First Edition.
Haniph A. Latchman, Srinivas Katar, Larry Yonge, and Sherman Gavette.
© 2013 by The Institute of Electrical and Electronics Engineers, Inc. Published 2013 by John Wiley & Sons, Inc.

second, independent signal to increase capacity. Note that only two unique signals can be transmitted on three wires due to Kirchoff's voltage law: the signal third wire pair is the sum of the signals on the other two wire pairs.

On receive, there are up to four possible receive channels: Line-Neutral, Line-Ground, Neutral-Ground, and common mode. Note that the three wire pairs are subject to Kirchoff's voltage law, but in practice the third wire pair does provide additional diversity evidently because of non-ideal implementations of the coupling to the powerline wiring. Common mode is the signal between true earth ground and the three wire pairs and is generally only available in products that have a relatively large effective ground plane, such as a flat screen television.

Note that is some cases (e.g., older construction), and in some countries such as Japan, the Ground wire is not present. MIMO cannot be used in these cases.

2xN MIMO spatial multiplexing with Eigen beamforming is specified in Home-Plug AV. The "2" in "2xN MIMO" refers to two transmitters and "N" refers to the number of receivers, which can be 2, 3, or 4 receivers. The typical performance gain from MIMO compared to HomePlug AV 1.1 (SISO—Single-Input Single-Output) is 100% (two times) on all channels ranging from poor channels to good channels. Eigen beamforming is required in order to achieve a performance benefit from MIMO on poor channels.

Figure 16.1 shows a block diagrammatic representation for the physical layer of the HomePlug AV2 transmitter and receiver based on a 200 MHz Sampling Clock. On the transmitter side, the PHY layer receives its inputs from the Medium Access Control (MAC) Layer. Three separate processing chains are shown because of the different encoding for HomePlug 1.0.1 Frame Control (FC) data, HomePlug AV2 Frame Control data, and HomePlug AV2 Payload data. AV2 Frame Control data is processed by the AV2 Frame Control Encoder, which has a Turbo Convolutional Encoder and Frame Control Diversity Copier while the HomePlug AV2 payload data stream passes through a Scrambler, a Turbo Convolutional Encoder, and an Interleaver. The HomePlug 1.0.1 Frame Control data passes through a separate HomePlug 1.0.1 Frame Control Encoder. The outputs of the FC Encoders and Payload Encoder lead into a common MIMO Orthogonal Frequency Division Multiplexing (OFDM) Modulation structure, consisting of a MIMO Stream Parser that provides up to two independent data streams to two transmit paths which includes two Mappers, a phase shifter that applies a 90° phase shift to one of the two streams, a MIMO precoder to apply transmitter beamforming operations, two Inverse Fast Fourier Transform (IFFT) processors, Preamble, and Cyclic prefix insertion, and symbol Window and Overlap blocks, which eventually feeds the Analog Front End (AFE) module with one or two transmit ports that couple the signal to the powerline medium.

At the receiver, an AFE with N_{RX} = one, two, three, or four receive ports operates with individual Automatic Gain Control (AGC) modules and one or more time-synchronization modules to feed separate Frame Control and Payload data recovery circuits. Receivers plugged to power outlets which are connected to the three wires Line, Neutral, and Protective Earth might utilize up to three differential mode Rx ports and one common mode Rx port. The common mode signal is the voltage between the sum of signals on the three wires and the ground. The Frame Control

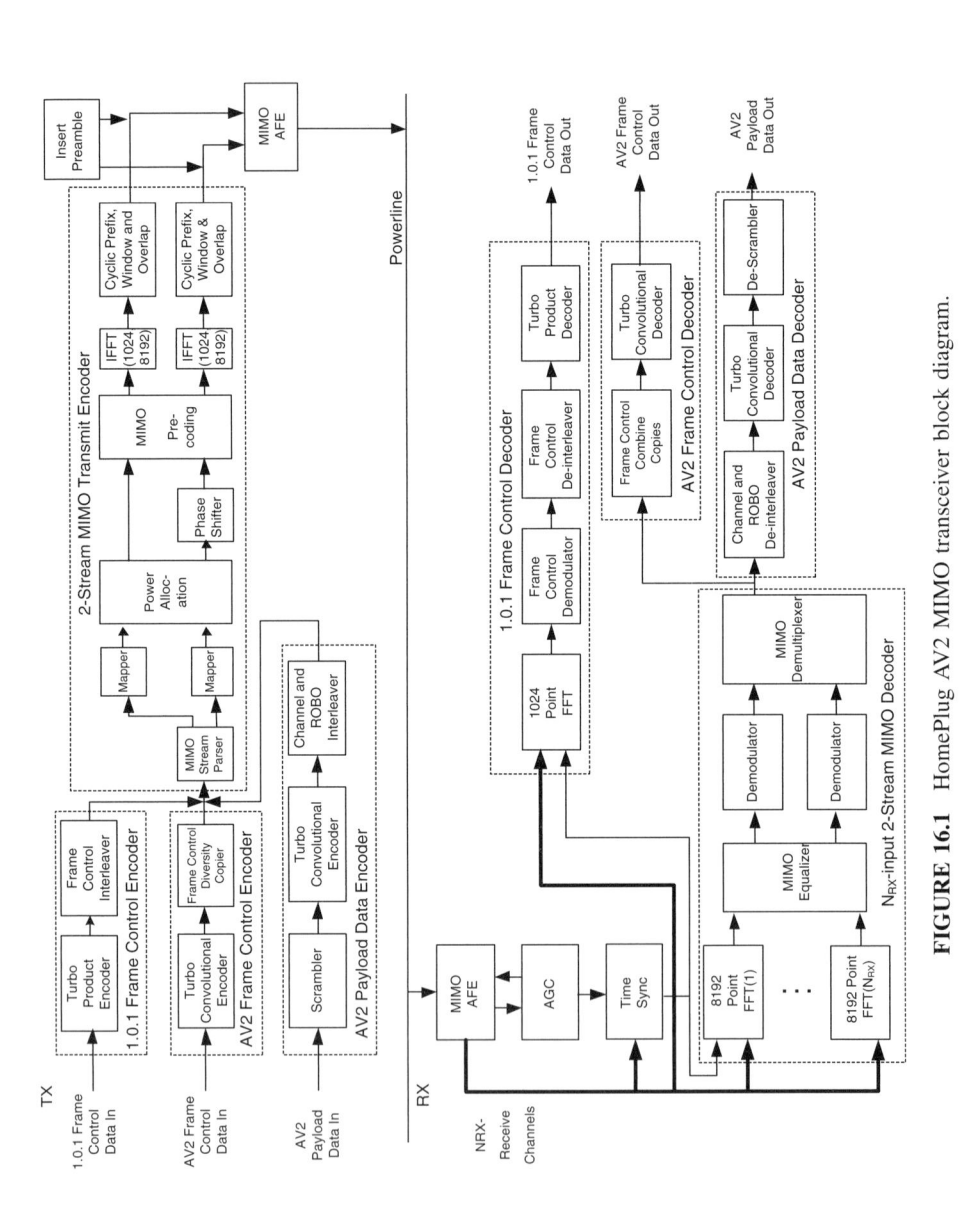

FIGURE 16.1 HomePlug AV2 MIMO transceiver block diagram.

data is recovered by processing the received signals through a 1024-point FFT (for HomePlug 1.0.1 delimiters) and multiple 8192-point FFTs, and through separate Frame Control Decoders for the HomePlug AV2/AV1.1 and HomePlug 1.0.1 Modes. The payload portion of the sampled time domain waveform, which contains only HomePlug AV2 formatted symbols, is processed through the multiple 8192-point FFT (one for each receive port), a MIMO Equalizer that receives N_{RX} signals, performs receive beamforming and recovers the two transmit streams, two Demodulators, a Demultiplexer to combine the two MIMO streams and a Channel De-interleaver followed by a Turbo Convolutional Decoder and a De-scrambler to recover the AV2 Payload data.

MIMO has an added benefit when communicating with legacy SISO devices or AV2 SISO devices. The two transmitters can use beamforming to improve the signal arriving at a SISO device. The two or more receivers can use Maximal Ratio Combining to improve performance when receiving from a SISO device.

16.3 EXTENDED FREQUENCY BAND

The HomePlug AV 1.1 Specification utilizes the frequency band from 1.8 to 30 MHz, and the IEEE 1901 Standard increase that to 1.8–50 MHz. HomePlug AV2 extends the frequency band even further to 1.8–86.13 MHz.

One of the challenges with the frequency band above 30 MHz is the regulations require a 25–30 dB reduction in transmit power spectral density (PSD) above 30 MHz. The powerline wiring radiation characteristics obviously do not change suddenly at 30 MHz. Rather, this is an artifact due to the way the regulations are specified for different frequency bands. In the United States, at least three factors in the FCC Part 15 regulations contribute to this step in PSD at 30 MHz: a reduction in the measurement distance from 30 to 3 m, an increase in the field strength from 30 to 100 μV/m and an increase in the measurement bandwidth from 9 to 120 kHz.

The performance gain provided by the 30–86.13 MHz band is generally quite high on medium to good channels due to the relatively wide channel bandwidth. However, this additional band does not provide much benefit on the poorest channels, for example, the worst 5% of connections, because of the low PSD level allowed at the transmitter. However, this band does contribute to the coverage performance in two ways. First, most powerline channels fall in the good to medium category, and the higher data rate provided on these channels enables them to reduce the time-on-wire for their traffic thus enabling more time-on-wire to be available for traffic on poorer channels. Also, when the higher frequency band is combined with immediate repeating, described in Section 16.6, the performance on the poorest paths can typically be improved dramatically by taking advantage of higher data connections through a repeater. This combination is a significant factor in achieving the 99% coverage performance mentioned in Section 16.1.

16.3.1 Power BackOff

Power backoff is a feature that enables higher performance on relatively good powerline channels. Practical implementations of the transmitter and receiver have limited dynamic range and thus the OFDM carriers above 30 MHz suffer distortion at the reduced PSD level. This is largely due to quantization noise of the digital-to-analog and analog-to-digital converters and the limited linearity of the transmit line driver. For example, a system may support 40 dB dynamic range for the band below 30 MHz, but only 10 dB is available for the band above 30 MHz with a 30 dB step in PSD.

To address this problem on good channels where the channel SNR can be quite high, the transmit PSD in the band below 30 MHz can be reduced so that the distortion to the OFDM carriers above 30 MHz can be reduced. For example, a 10 dB backoff in PSD below 30 MHz can increase the dynamic range of the carriers above 30 MHz by 10 dB (i.e., allows 20 dB of dynamic range).

16.4 EFFICIENT NOTCHING

In HomePlug AV 1.1, OFDM with windowing was specified to support 30 dB deep notches without the need for FIR or IIR filters. This feature provides for programmable notches as may be needed for different regulatory regions or to accommodate changes in regulations and to permanently notch the Amateur bands. In HomePlug AV 1.1, the default tone mask specifies 917 active carriers out of 1155 total carriers. A total of 238 OFDM carriers are permanently turned off out to create notches for the 10 Amateur bands in the 1.8–30 MHz frequency band.

However, the approach to deep notches used in HomePlug AV1.1 requires ~12% in overhead for additional guard interval needed to support the time domain window, which is 4.96 μs in duration, and 8% in overhead to support a guard band of 195 kHz (eight OFDM carriers) for each notch required to achieve the 30 dB notch depth.

The requirement to notch using windowing was removed from the HomePlug AV 2.0 Specification to allow for implementations to allow alternate implementation of the transmitter to reduce this overhead, such as adding fixed and/or programmable FIR or IIR filters, or a combination of windowing and filters. To support this, smaller guard intervals were added and protocol changes were made to support additional OFDM carriers when supported by a transmitter (e.g., when a small guard band can be used, some carriers in the guard band used in AV1.1 can be enabled). This feature will also help products that need to meet CENELEC EN50561-1 requirements (refer to Section 1.3.3.2) to minimize the performance loss due to the additional notch requirements for the aeronautical and broadcast bands.

16.5 SHORT DELIMITER AND DELAYED ACKNOWLEDGMENT

The Short Delimiter and Delayed Acknowledgment features were added to HomePlug AV2 to improve efficiency by reducing the overhead associated with transmitting payloads over the powerline channel. In HomePlug AV 1.1, this overhead results in relatively poor efficiency for TCP payloads. One goal that was achieved with these new features was TCP efficiency was improved to be relatively close to that of UDP.

In order to send a packet carrying payload data over a noisy channel, signaling is required for a receiver to detect the beginning of the packet and to estimate the channel so that the payload can be decoded, and additional signaling is needed to acknowledge the payload was received successfully. Inter-frame spaces are also required between the payload transmission and the acknowledgment for the processing time at the receiver to decode and check the payload for accurate reception and to encode the acknowledgment. This overhead is even more significant for TCP payload since the TCP acknowledgment payload must be transmitted in the reverse direction.

16.5.1 Short Delimiter

The delimiter specified in AV 1.1 contains the preamble and frame control symbols and is use for the beginning of data PPDUs as well as for immediate acknowledgments. The length of the AV 1.1 delimiter is 110.5 µs and can represent a significant amount of overhead for each channel access. A new single OFDM symbol delimiter is specified in AV2 to reduce the overhead associated with delimiters by reducing the length to 55.5 µs. Figure 16.2 shows that every fourth carrier in the first OFDM symbol is assigned as a preamble carrier, and the remaining carriers encode the Frame Control. The following OFDM symbols encode data the same as in AV1.1.

One of the limitations of the Short Delimiter is that it cannot be easily detected asynchronously, which is necessary for CSMA channel access. Thus, reception of the short delimiter requires that the receiver know the position in time where the Short Delimiter was transmitted. Thus, the use of the Short Delimiter is limited to Selective Acknowledgment of CSMA and TDMA Long MPDUs, Reverse Start-of-Frame, and TDMA Start-of-Frame.

Figure 16.3 shows a PPDU with the Short Delimiter. The sample points indicated are based on a 200 MHz clock. The guard interval for the Short Delimiter is 9.56 µs

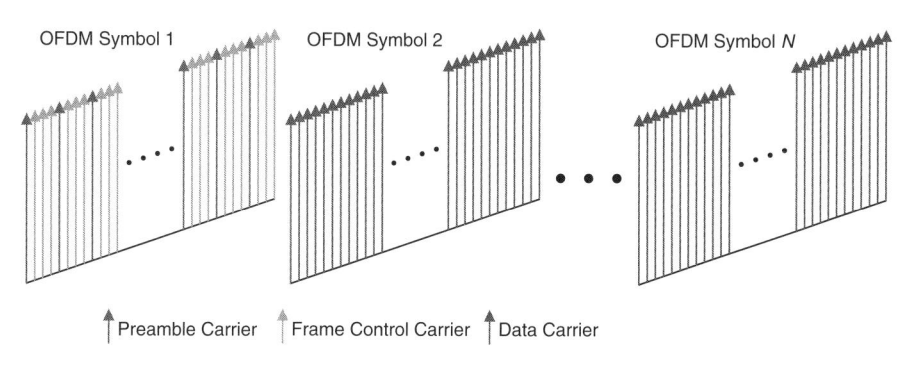

FIGURE 16.2 Short delimiter.

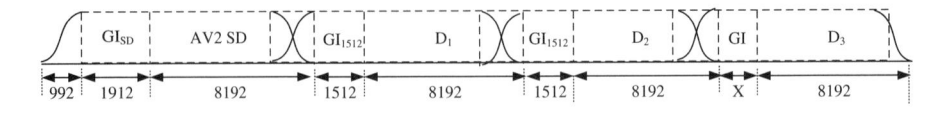

GI_SD	AV2 SD	GI_1512	D_1	GI_1512	D_2	GI	D_3
992	1912	1512	8192	1512	8192	X	8192

FIGURE 16.3 PPDU format with Short Delimiter.

to provide for sampling error by the receiver. The receiver can detect the sampling error from the Preamble Carriers, and since every fourth carrier is a Preamble carrier the receiver can detect and error up to one fourth of the OFDM symbol time, or ± 5.12 μs, and will use the correction for the sampling location of the data symbols as well as to correct for the sampling location for receiving future Short Delimiters from a particular transmitter.

The typical decoding process for receiving an AV1.1 PPDU is as follows:

1. Detect the presence of the Preamble signal.
2. Estimate the position of the Preamble signal.
3. Generate an estimate of the channel from the Preamble signal.
4. Sample the Frame Control symbol based on the position estimate from the Preamble, perform an FFT and decode the Frame Control using the estimate of the channel from the Preamble signal.
5. Estimate the channel from the Frame Control by re-encoding the Frame Control to determine the reference.
6. Perform an FFT and decode each payload symbol using the estimate of the channel from the Frame Control and the tone map based on the index indicated in the Frame Control data.

Decoding of the payload of an AV2 PPDU with the Short Delimiter is similar:

1. Perform an FFT of the Short Delimiter symbol based on the position estimate
2. Generate an estimate of channel from the Preamble carriers
3. Decode the Frame Control using the estimate of the channel from the Preamble carriers
4. Re-encoded the Frame Control and generate and estimate of the channel from the Preamble and Frame Control carriers
5. Perform an FFT and decode each payload symbol using the estimate of the channel from the Frame Control and the tone map based on the index indicated in the Frame Control data

Figure 16.4 demonstrates the efficiency improvement when the AV2.0 Short Delimiter is used for the acknowledgment of a CSMA Long MPDU compared to the AV1.1 delimiter. Not only is the length of the delimiter reduced from 110.5 to 55.5 μs, the Response Inter-Frame Space (RIFS) and Contention Inter-Frame Space

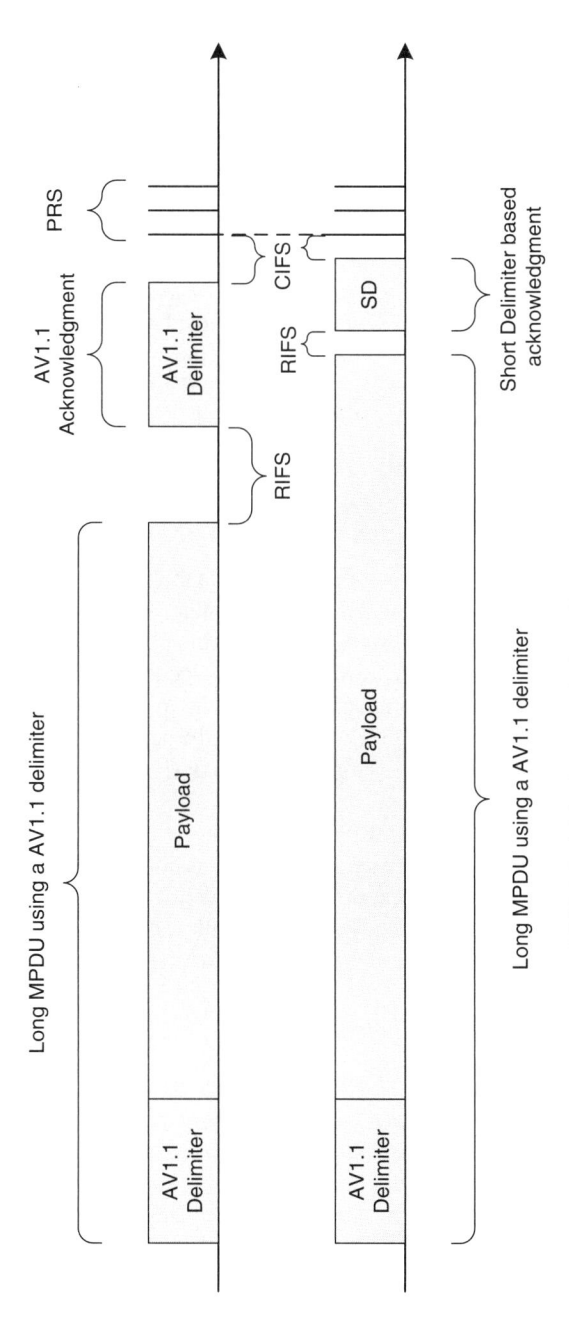

FIGURE 16.4 Short Delimiter efficiency improvement.

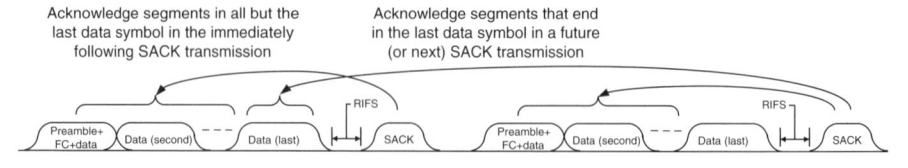

FIGURE 16.5 Delayed acknowledgment.

(CIFS_AV) can be reduced to 5 and 10 μs, respectively. Reduction of RIFS requires delayed acknowledgment, which is described in Section 16.5.2. Backwards compatibility when contending with legacy AV 1.1 devices is maintained by indicating the same length field for virtual carrier sense in both cases so that the position of the PRS contention remains the same. A field in the Frame Control of the Long MPDU indicates the Short Delimiter format to an AV2.0 device so that it can correctly determine the length of the payload.

16.5.2 Delayed Acknowledgment

The processing time to decode the last OFDM symbol and encode the acknowledgment can be quite high, thus requiring a rather large the Response Inter-Frame Space (RIFS). In AV1.1, since the Preamble is a fixed signal, the preamble portions of the acknowledgment can be transmitted while the receiver is still decoding the last OFDM symbol and encoding the payload for the acknowledgment. With the Short Delimiter, the preamble is encoded in the same OFDM symbol as the payload for the acknowledgment, so the RIFS would need to be larger than for AV1.1, eliminating much of the gain the Short Delimiter provides.

Delayed acknowledgment solves this problem by acknowledging the segments ending in the last OFDM symbol in the acknowledgment transmission of the next PPDU, as shown in Figure 16.5. This permits practical implementations with a very small RIFS, reducing the RIFS overhead close to zero. AV2 also allows the option of delaying acknowledgment for segments ending in the second to last OFDM symbol to provide flexibility for implementers.

16.5.3 TCP and UDP Efficiency Improvements

The combination of Short Delimiter and Delay Acknowledgment can provide a significant improvement in efficiency. Figures 16.6 and 16.7 show the improvement in throughput for the AV2 PPDU format with the Short Delimiter and Delay Acknowledgment compared to the AV 1.1 PPDU format for CSMA channel access and TDMA channel access, respectively. For example, the TCP throughput on a channel that supports 70 Mbps throughput with the AV 1.1 PPDU format will support 92 Mbps with the AV2 PPDU format in CSMA. This assumes the PHY rate for the payload portion of the PPDU is constant. A payload size of approximately 20,000 bytes was used for this analysis.

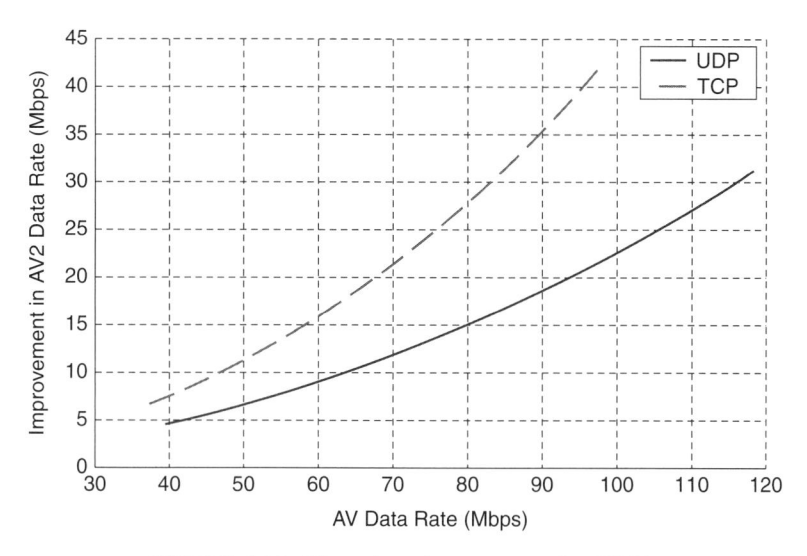

FIGURE 16.6 Throughput improvement for CSMA.

16.6 IMMEDIATE REPEATING

AV2 supports repeating and routing of traffic to not only handle hidden nodes but also to improve coverage (i.e., performance on the worst channels). The basic repeating and routing function used by AV2 is the same as in IEEE 1901 and Green PHY (refer to Section 14.2.6).

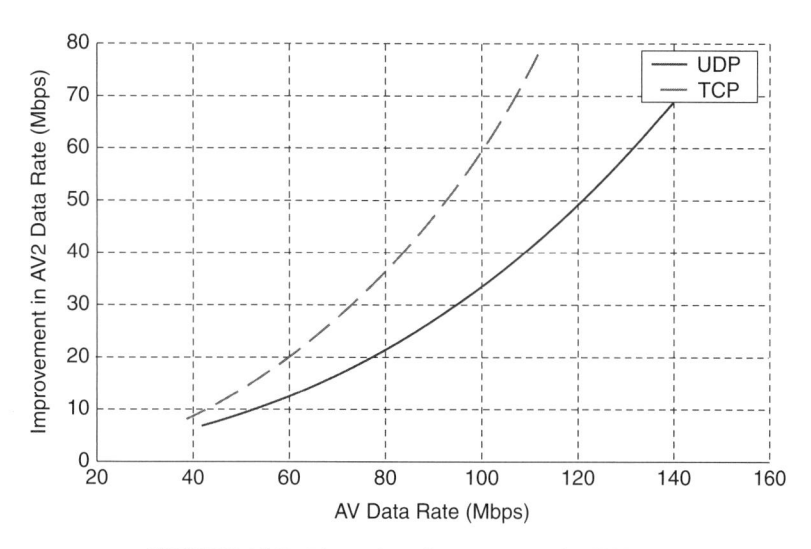

FIGURE 16.7 Throughput improvement for TDMA.

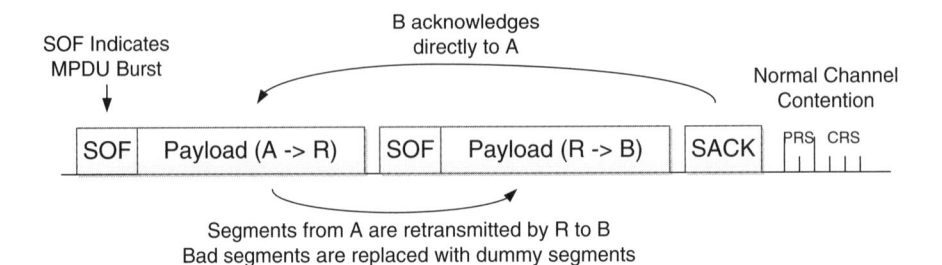

FIGURE 16.8 Immediate repeating channel access for CSMA.

With HomePlug AV2, hidden nodes are extremely rare. However, some links may not support the data rate required for some applications such as a 3D HD video stream. In a network where there are multiple AV2 devices, the connection through a repeater typically provides a higher data rate than the direct path for the poorest 5% of channels.

Immediate repeating is a new feature in AV2 that enables high efficient repeating. Immediate repeating provides a mechanism to use a repeater with a single channel access, and the acknowledgment does not involve the repeater. As shown in Figure 16.8, Station A transmits to the repeater R. In the same channel access, repeater R transmits all payload received from station A to station B. B sends an acknowledgment directly to A. With this approach, latency is actually reduced with repeating, assuming the resulting data rate is higher, the obvious criteria for using repeating in the first place. Also, resources required by the repeater are minimized since the repeater uses and immediately frees memory it would require for receiving payload destined for it. Also, the receiver has no retransmission responsibility for failed segments.

Figure 16.9 shows the result of a coverage analysis based on data measured in 25 homes for the 2–86.13 MHz frequency band where the channel characteristics for all combinations of paths between five locations were measured in each home. The analysis shows the coverage for networks of 2, 3, 4, and 5 AV2 SISO nodes where all nodes are capable of repeating. For each connection in the network, the best throughput of the direct path or one of all available two hop repeating paths is selected. Note the 2 node case has the same results as the direct path for 3, 4, and 5 nodes network where repeating is not supported. Note that this analysis shows the coverage throughput for the 98%, 99%, and 100% coverage points improve by approximately 57%, 122%, and 409%, respectively, in the case of a five nodes network with repeating.

16.7 POWER SAVE

The power save protocol in the HomePlug AV 2.0 Specification is the same as in the HomePlug Green PHY Specification. Refer to Section 15.3.1 for more detail.

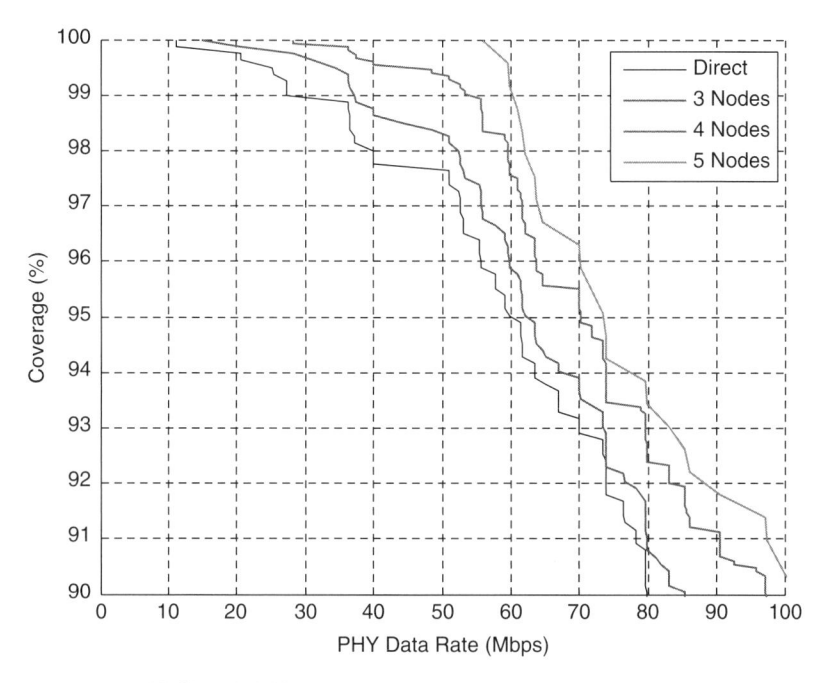

FIGURE 16.9 AV2.0 SISO PHY rate with repeating.

16.8 SUMMARY

This chapter provided an overview of the HomePlug AV 2.0 Specification which builds on the strengths of HomePlug AV but uses key extensions, such as MIMO, augmented bandwidth, increased modulation density and repeating, to achieve 1.5 Gbps throughput.

As illustrated especially in the last several chapters, HomePlug AV has become the foundation of several other derivative works such as the IEEE 1901 Standard as well as the HomePlug GreenPHY and AV2 Specifications.

It is the hope of the authors that this book, H*omePlug AV and IEEE 1901: A Handbook for PLC Designers and Users* will prove useful as a PLC reference and in giving an accessible exposition of the core technologies that have revolutionized high speed Powerline Communications over the past decades.

APPENDIX A

ACRONYMS

The following acronyms are used in this book:

Acronym	Meaning	Primary Reference
ACC	Access	14.4.2.1
ACK	ACKnowledgment	5.2.2
ACLSS	AC Line-Cycle Synchronization Status	5.4
ADOF	Advanced Diversity OFDM Frame	14.3.4
AFE	Analog Front End	4.1.1
AGC	Automatic Gain Control	4.1.1
ARQ	Automatic Repeat reQuest	6.13
ARRL	Amateur Radio Relay League	3.3
ATS	Arrival Time Stamp	6.2
AVLN	Home AV Logical Network	1.5
AWGN	Additive White Gaussian Noise	3.5
B2BIFS	Beacon To Beacon Interframe Spacing	
BBF	Bidirectional Burst Flag (BBF)	5.3.1
BBT	BeaconBackoffTime	
BDF	Beacon Detect Flag	5.3.1
BEHDR	Beacon Entry Header	5.2.2.8
BENTRIES	Beacon Entries	5.2
BENTRY	Beacon Entry	5.2.2.8
BIFS	Burst Interframe Space	6.11
BLE	Bit-Load Estimate	9.6.1
BMRAT	Broadcast and Multicast Repeating Assignment Table	14.2.6.2
BPCS	Beacon Payload Check Sequence	5.2.2.9

(*continued*)

HomePlug AV and IEEE 1901: A Handbook for PLC Designers and Users, First Edition.
Haniph A. Latchman, Srinivas Katar, Larry Yonge, and Sherman Gavette.
© 2013 by The Institute of Electrical and Electronics Engineers, Inc. Published 2013 by John Wiley & Sons, Inc.

(*Continued*)

Acronym	Meaning	Primary Reference
BPCnt	Beacon Period Count	15.3.1
BPL	Broadband over Power line	1.3.3.1
BPSK	Binary Phase Shift Keying	3.1
BPST	Beacon Period Start Time	8.3.1
BPSTO	Beacon Period Start Time Offset	5.2.2.8.7
BSF	Bridging Station Flag	11.3.1
BTO	Beacon Transmit Offset	12.3.9
BurstCnt	Burst Count	5.3.1
C&IWG	Compliance and Interoperability Working Group	1.3.2
CAPn (CAn)	Channel Access Priority n	15.3.2.2
CBC	Cipher Block Chaining	6.10
CCo	Central Coordinator	2.2
CDESC	Connection Descriptor	9.3
CENELEC	European Committee for Electro-technical Standardization	1.3.3.2
CEI	Channel Estimate Indication	11.2.1
CFP	Contention Free Period	9.5
CFPI	Contention Free Period Initiation	8.3.1
CFR	Code of Federal Regulations	1.3.3.1
CFS	Contention Free Session	5.3.1
CIFS	Contention Interframe Spacing	16.5.1
CIFS_AV	Contention Interframe Spacing AV	8.4.1.2
CINFO	Connection INFOrmation	9.3
CINR	Carrier power to Interference power plus Noise power Ratio	14.3.4.2
CISPR	Comité International Spécial des Perturbations Radioélectriques	1.3.3.2
CLS	Connectionless Service	9.5
CLST	Convergence layer SAP type	9.3
CM	Connection Manager	2.2
CN	Central Network	2.2
CNF	Confirm Message (suffix)	13.2.6
COMSOC	(IEEE) Communication Society	1,1
CP	Contention Period	5.3.1.9
CPF	CP Flag	
CPLT	Control PiLoT	15.3.3
CRC	Cyclic Redundancy Code	5.1.1
CSCD	Current Schedule Count Down	5.2.2.8.2
CSMA	Carrier Sense Multiple Access	1.1
CSPEC	Connection Specification	9.3
CW	Contention Window	8.4.2
D2PAM	Double 2PAM	14.3.5.8.1
DAK	Device Access Key	10.3.4
DBC	Distributed Bandwidth Control	15.3.2.2
dBm	DeciBel (milliwatt reference)	4.7.2

(*Continued*)

Acronym	Meaning	Primary Reference
DCPPCF	Different CP PHY Clock Flag	5.3.1.9
Dev	Device	6.12
dibit	Bit Pairs	4.5.2.4
DOF	Diversity-Orthogonal Frequency Division Multiplexing (OFDM) for Frame body	14.3.4
DPLL	Digital Phase Locked Loop	8.2.1
DPW	Device PassWord	10.4.1
DT	Delimiter Type	5.1.1
DT_AV	Delimiter Type AV	5.1.1
DTEI	Destination Terminal Equipment Identifier	5.3.1.1
DUR	Duration	5.5
Eb/N0	Energy per bit over Spectral Noise Density Ratio	3.5
ECC	Error Correction Code	14.3.5.5
ECSF	Extended Carrier Support Flag	14.2.1
EFRS	Extended FEC Rate Support	14.2.4
EIFS	Extended Inter Frame Spacing	8.4.1
EIFS_AV	Extended Inter Frame Spacing AV	8.4.1.2
EIFS_X	Extended Inter Fame Spacing_X (EIFS or EIFS_AV)	8.4.1.2
EKS	Encryption Key Select	5.2.2.8.8
ELGISF	Extended Larger Guard Interval Support Flag (ESGISF)	14.2.2
ESGISF	Extended Smaller Guard Interval Support Flag (ESGISF) 14.2.2	
ETSI	European Telecommunications Standards Institute	1.3.3.2
EVSE	Electric Vehicle Supply Equipment	15.3
FCCS	Frame Control Check Sequence	8.4.1.2
FCCS_AV	Frame Control Check Sequence AV	5.1.1
FDM	Frequency Division Multiplexing	1.1
FEC	Forward Error Correction	1.1
FDA	Final Destination Address	11.5.2
FFT	Fast Fourier Transform	1.1
FIR	Finite Impulse Response	16.4
FL	Frame Length	14.3.4
FL_AV	Frame Length AV	5.3.1.4
FMI	Fragment Management Information	13.2.7
FMSN	Fragment Management Serial Number	13.2.7
FTEI	Final destination Terminal Equipment Identifier	11.5.2
FSM	Finite State Machine	6.7.1
GF	Galois Field	14.3.5.4.1
GLID	Global Link Identifier	5.2.2.8.1
GPP	GreenPHY Preferred	9.4.1
HCD	Handover CountDown	5.2.2.8.9
HFID	Human Friendly IDentifier	10.3.1.2
HLE	Higher Level Entity	7.5.1.1

(*continued*)

(*Continued*)

Acronym	Meaning	Primary Reference
HMCCD	Hybrid Mode Change CountDown	5.2.2.8.13
HoIP	Handover In Progress	5.2.2
HPAV	HomePlug AV	2
HS-ROBO	High Speed ROBO mode	5.3.1.4
HSTA	Hidden Station	7.5.2.1.3
ICV	Integrity Check Value	6.2
IDWT	Inverse Discrete Wavelet Transform	14.3.3
IFFT	Inverse Fast Fourier Transform	4.1.1
IFS	Inter Frame Spacing	5.6.1.7
IFS	Interframe Spacing	5.3.1.5
IGF	Intermediate Grant Flag	5.5.1.5
IIR	Infinite Impulse Response	16.4
IND	Indication Message (suffix)	13.2.6
INL	Interfering Network List	12.3.3.1
KBC	Key Being Changed	5.2.2.8.8
KCCD	Key Change CountDown	5.2.2.8.8
LBDAT	Local Bridge Destination Address Table	11.3.1
LDPC-CC	Low Density Parity Check-Convolutional Code	14.3.1
LDPCCs	Low Density Parity Check Codes	
LID	Link IDentifier	5.4.1.2
LLID	Local Link IDentifier	9.4.1
LRT	Local Routing Table	14.2.6.1
MCF	Multicast Flag	5.3.1.1
MFBF	Mac Frame Boundary Flag	5.3.2
MFBO	Mac Frame Boundary Offset	5.3.2
MFL	Mac Frame Length	6.2
MFS	Mac Frame Stream	6.7.1
MFT	Mac Frame Type	6.2
MIB	Management Information Base	14.3.5.10
MIMO	Multiinput Multioutput	16.2
MINI-ROBO	Mini ROBO mode	5.3.1.4
MITM	Man In The Middle	11.5.7
MME	Management Message Entry	13.2.8
MMENTRY	Management Message Entry Data (a.k.a. MME)	13.2.8
MMQF	Management Message Queue Flag	5.3.2
MMTYPE	Management Message Type	13.2.6
MNBC	Multinetwork BroadCast	6.9.1
MNBF	Multinetwork Broadcast Flag	5.3.1
MoCA	Multimedia Over Coaxial	1.4
MPDU	MAC Protocol Data Unit	4.2
MPDUCnt	MAC Protocol Data Unit Count	5.3.1.3
MRD	Marketing Requirements Document	1.3.2
MRTFL	Max Reverse Transmission Frame Length	5.3.1.3
MSDU	MAC Service Data Unit	6.1
MTYPE	IEEE Assigned Ethertype value of 0x88e1	13.2.4

(*Continued*)

Acronym	Meaning	Primary Reference
NACK	Negative ACK	
NACK	Negative ACKnowledgment	6.7.1
NAVLN	Neighbor AVLN	12.5
NBE	Number of Beacon Entries	5.2.2.8
NCNR	Noncoordinating Network Reported	5.2.2.3
NCo	Neighbor CCo	8.1
NEK	Network Encryption Key	10.4.1.5
NID	Network Identifier	5.2.2.2.1
NMK	Network Membership Key	10.4.1.3
NMK-HS	NMK and Secure Security Level	10.4.2.1.2
NMK-SC	NMK and Simple Connect	10.4.2.1.2
NMK-SL	NMK and Security Level	10.3.1
NN	Neighbor Network	2.3
NOP	No operation	6.7
NPSM	Network Power Save Mode	5.2.2.1
NPW	Network Password	10.4.1
NSCCD	Number of Beacon Slots Change Count Down	5.2.2.8.1.2
NTB	Network Time Base	7.5.1
NTEI	Next TEI	14.2.6.1
ODA	Original Destination Address	6.1
OFDM	Orthogonal Frequency Division Multiplexing	1.1
OPSF	Oldest Pending Segment Flag	5.3.2
OSA	Original Source Address	6.1
OSI	Open System Interconnect	2.2
OTEI	Original TEI	11.5.2
OUI	Organizationally Unique Identifier	5.2.2.8.1.5
PAM	Pulse Amplitude Modulation	14.3.1
PB	PHY Block	5.3.2
PBB	PHY Block Body	5.3.1.5
PBH	PHY Block Header	
PBSz	PHY Block Size	5.3.1.4
PCMWG	HomePlug Policy Creation and Management Working Group	12.5
PCo	Proxy Coordinator	2.3
PEKS	Payload Encryption Key Select	5.2.2.8.8
PEV	Plug-in Electric Vehicle	15.3.3
PHY	Physical Layer	1.1
PhyClk	PHY Clock	
PID	Protocol ID	10.4.2.6
PLID	Priority Link Identifier	15.3.2.2
PMN	Protocol Message Number	10.4.2.6
PN	Proxy Network	2.3
PPB	Pending PHY Blocks	
PPDU	PHY Protocol Data Unit	
PRN	Protocol Run Number	10.4.2.4

(*continued*)

(*Continued*)

Acronym	Meaning	Primary Reference
PRS	Priority Resolution Slots	4.6
PSCD	Preview Schedule Countdown	5.2.2.8.2
PSD	Power Spectral Density	4.7.1
PSP	Power Save Period	15.3.1
PSS	Power Save Schedule	15.3.1
PSSI	Power Save State Identifier	15.3.1.1
PSTA	Proxy STA	11.5
PxN	Proxy Network	11.5
QAM	Quadrature Amplitude Modulation	4.5.5
QMP	QoS and MAC Parameters	9.3
QoS	Quality of Service	2.3
QPSK	Quadrature Pulse Shift Keying	3.5
RBAT	Remote Bridged Address Table	11.3.2
RCE	Request Channel Estimation	14.3.4.2
RCD	Relocation Countdown	5.2.2.8.10
RCG	RTS-to-CTS Gap	5.5.1.6
RDR	Route Data Rate	14.2.6.1
REQ	Request Message (suffix)	13.2.6
REQ_TM	Max Tone Maps Requested	11.2.2
RIFS	Response Inter-Frame Space	16.5.1
RIFS_AV	Response Inter-Frame Space AV	5.3.1.4
RLO	Relocation Offset	5.2.2.8.10
RNH	Route Number of Hops	5.2.2.8.10
ROBO	Robust Modulation	4.5.4
RRTF	Request Reverse Transmission Flag	5.4
RRTL	Request Reverse Transmission Length	5.4
RSC	Recursive Systematic Convolutional	4.5.2
RSF	Reuse SNID Flag	14.2.5
RSOF	Reverse Start-of-Frame	5.1
RSOF_FL_AV	Reverse SOF Frame Length	5.7.1
RSP	Response	13.2.6
RSP	Response Message (suffix)	13.2.6
RSR	Request SACK Retransmission	5.3.1.6
RT	Region Type	5.2.2.8.3
RTS	Request to Send	5.1
RTSBF	RTS broadcast flag	7.5.1.2
RTSF	RTS Flag	5.5
RWG	Regulatory Working Group	1.3.3
RxWSz	Receive Window Size	5.4.1.2
SACKD	Sack Data	5.4
SACKI	Sack Information	
SACK	Selective Acknowledgment	5.4
SACKT	SACK Type	6.13
SAE	Society of Automotive Engineers	15.3.3
SAF	Sound ACK Flag	5.6.1

(*Continued*)

Acronym	Meaning	Primary Reference
SAI	Session Allocation Information	5.2.2.8.1
SC	Simple Connect	10.4.2.5
SCF	Sound Complete Flag	5.6.1
SCL	Sound Control	11.2.3
SISO	Single-Input Single-Output	16.2
SLAC	Signal Level Attenuation Characterization	15.3.3.1
SlotID	Beacon Slot ID	5.2.2
SlotUsage	Beacon Slot Usage	5.2.2
SNID	Short Network Identifier	1.2
SNR	Signal-to-Noise Ratio	3.5
SOF	Start-of-Frame	5.3
SPCS	Sound Payload Check Sequence	5.6.2
SRC	Sound Reason Code	5.6.1
SSN	Segment Sequence Number	5.3.2
STA	Station	9.4
STA_Clk	Station Clock	7.5.1
STD-ROBO	Standard ROBO mode	5.3.1.4
STEI	Source Terminal Equipment Identifier	5.3.1.1
STPF	Start Time Present Flag	5.2.2.8.1
SVN	Sack Version Number	5.7.1
SYNCM	Minus Synchronization	4.3
SYNCP	Plus Synchronization	4.3
TCC	Turbo Convolution Codes	1.5
TDM	Time Division Multiplexing	14.4.3.2
TDMA	Time Division Multiple Access	2.3
TDMS	Time Division MultiplexingSlots	14.4.3.2
TDMU	TDM Unit	14.4.3.2
TEG	Technology Evaluation Group	1.3.2
TEI	Terminal Equipment Identifier	10.3.2
TEK	Temporary Encryption Key	10.4.1.6
TMD	Tone Map Data	5.6.1.9
TMI	Tone Map Information	14.3.4
TMI_AV	HomePlug AV Tone Map Index	5.3.1
TWG	HomePlug AV Technical Working Group	1.3.2
TXOP	Transmission Oppurtunity	9.3
UDTEI	Ultimate Destination TEI	14.2.6.1
UIS	User Interface Station	10.4.8
UKE	Unicast Key Exchange	10.4.1
USAI	Unassociated STA Advertisement Interval	APPENDIX B
UUID	Universally Unique Identifiers	10.4.2.6
Var	Variable	5.4
VF_AV	HomePlug AV variant fields	5.1.1
VPBF	Valid PHY Block Flag	5.3.2
XOR	eXclusive OR	14.3.4.1
ZPAD	Zero Pad	5.6.2

APPENDIX B

HomePlug AV PARAMETER SPECIFICATION

Table B-1 lists some of the main parameters in HomePlug AV specification.

TABLE B-1 HomePlug AV Parameter Specifications

Parameter	Value	Section Reference
Allocation Time Unit	10.24 μs	5.2.2.8.1
Beacon To Beacon Interframe Spacing (B2BIFS)	90 ± 0.5 μs	8.3
Burst Interframe Spacing (BIFS)	20 ± 0.5 μs	6.11
CIFS_AV	100 ± 0.5 μs	8.4
CTS-MPDU Gap (CMG)	120 ± 0.5 μs	5.5
EIFS_AV	2920.64 ± 5.0 μs	8.4
Extended Inter Frame Space (EIFS)	1695.0 ± 5.0 μs	8.4
MaxFL_AV	2501.12 μs ≤ MaxFL_AV ≤ 5241.6 μs	5.3.1.4
Maximum Beacon Scan Time (MaxScanTime)	4 s	10.2
Maximum CCo Beacon Scan Time (MaxCCoScanTime)	2 s	10.2
MAX_TONE_MAPS	7	11.2.2
MinCSMARegion	1500 μs	8.3.2, Chapter 12
Minimum Beacon Scan Time (MinScanTime)	2 s	10.2
Minimum CCo Beacon Scan time (MinCCoScanTime)	1 s	10.2

(continued)

HomePlug AV and IEEE 1901: A Handbook for PLC Designers and Users, First Edition.
Haniph A. Latchman, Srinivas Katar, Larry Yonge, and Sherman Gavette.
© 2013 by The Institute of Electrical and Electronics Engineers, Inc. Published 2013 by John Wiley & Sons, Inc.

TABLE B-1 (*Continued*)

Parameter	Value	Section Reference
Priority Resolution Slot (PRS)	35.84 ± 0.5 μs	8.4
RIFS_AV	30–160 μs	5.3.1.4
RIFS_AV_default	140 ± 0.5 μs	5.6.1.7
RTS/CTS Gap (RCG)	120 ± 0.5 μs	5.5
Slot Time	35.84 ± 0.5 μs	8.4
Unassociated STA Advertisement Interval (USAI)	1 s	10.2

REFERENCES

1. HomePlug AV 1.1 Specification.
2. IEEE^{TM} Std 1901^{TM}-2010: IEEE Standards for Broadband over Power Line Networks: Medium Access Control and Physical Layer, Dec. 2010.
3. Wi-Fi.
4. IEEE 802.11x.
5. IEEE Std 802.3^{TM}-2005, IEEE Standard for Information Technology-Telecommunications and Information Exchange between Systems—Local and Metropolitan Area Networks—Specific Requirements—Part 3: Carrier Sense Multiple Access with Collision Detection (CSMA/CD) Access Method and Physical Layer Specifications.
6. Multimedia Over Coaxial (MoCA) Website.
7. Code of Federal Regulations, Title 47, Part 15 (47 CFR 15).
8. ANSI C63.14-1992 Document.
9. Directive 2004/108/EC of the European Parliament and of the Council, December 15, 2004.
10. CISPR-22.
11. CENELEC prEN50561-1.
12. HomePNA.
13. ARRL Report: http://p1k.arrl.org/~ehare/bpl/HomePlug_ARRL.pdf.
14. Ministry of Internal Affairs and Communications Ministerial Ordinance No. 118, Official Gazette, October 4, 2006, (Extraordinary Issue No. 227).

HomePlug AV and IEEE 1901: A Handbook for PLC Designers and Users, First Edition.
Haniph A. Latchman, Srinivas Katar, Larry Yonge, and Sherman Gavette.
© 2013 by The Institute of Electrical and Electronics Engineers, Inc. Published 2013 by John Wiley & Sons, Inc.

15. Katar, S., Mashburn, B., Newman, R., and Latchman, H., Allocation requirements for supporting latency bound traffic in HomePlug AV Networks, *IEEE Global Telecommunications Conference (GLOBECOM)*, 2006.
16. Katar, S., Krishnam, M., Mashburn, B., Newman, R., and Latchman, H., Beacon schedule persistence to mitigate Beacon loss in HomePlug AV Networks, *Proceedings of the IEEE International Symposium on Power Line Communications*, pp. 184–188, March, 2006.
17. Katar, S., Yonge, L., Newman, R., and Latchman, H., Efficient framing and ARQ for high-speed PLC systems, *International Symposium on Powerline Communications and its Applications*, 2005.
18. HomePlug 1.0.1 Specification.
19. Lee, M.K., Newman, R., Latchman, H.A., Katar, S., and Yonge, L., HomePlug 1.0 powerline communication LANs—protocol description and comparative performance results, *Special Issue of the International Journal on Communication Systems on Powerline Communications*, pp. 447–473, May, 2003.
20. IEEE 802.1D™.
21. FIPS 180-2, NIST, Secure Hash Standard, Aug. 26, 2002 (including the change notice dated February 25, 2004, concerning truncation).
22. ITU-T Rec. X.667 | ISO/IEC 9834-8, Information Technology—Open Systems Interconnection—Procedures for the Operation of OSI Registration Authorities: Generation and Registration of Universally Unique Identifiers (UUIDs) and their Use as ASN.1 Object Identifier Components, September 2004, http://www.itu.int/ITUT/studygroups/com17/oid/X.667-E.pdf.
23. Leach, P. and Salz, R. IETF RFC 4122, A Universally Unique IDentifier (UUID) URN Namespace, July 2005, http://www.ietf.org/rfc/rfc4122.txt.
24. HomePlug 1.1 Specification.
25. PKCS #5 v2.0 Standard, Password-Based Cryptography Standard.
26. Eastlake, D., 3rd, Schiller, J., Crocker, S., *Randomness Requirements for Security*, RFC 4086, June.
27. Schneier, B., *Applied Cryptography: Protocols, Algorithms, and Source Code in C*, 2nd edition, John Wiley & Sons, 1996.
28. Katar, S., Mashburn, B., Afkhamie, K., Latchman, H., and Newrnan, R., Channel adaptation based on cyclo-stationary noise characteristics in PLC systems, *International Symposium on Power Line Communications and its Applications*, 2006.
29. Mashburn, R., Latchman, H., VanderMey, T., Yonge, L., and Tripathi, K., Signal processing challenges in the design of the HomePlug AV powerline standard to ensure co-existence with HomePlug 1.0.1, *IEEE Workshop on Signal Processing Advances in Wireless Communications*, New York, NY, 2005.
30. Galli, S., Koga, H., and Kodama, N., Advanced signal processing for PLCs: Wavelet-OFDM, *IEEE International Symposium on Power Line Communications and its Applications (ISPLC)*, Jeju Island, Korea, April 2–4, 2008.
31. Zyren, J., The HomePlug Green PHY specification & the in-home Smart Grid, *Consumer Electronics (ICCE), 2011 IEEE International Conference*, pp. 241–242, January 9–12, 2011, doi: 10.1109/ICCE.2011.5722562URL: http://ieeexplore.ieee.org/stamp/stamp.jsp?tp=&arnumber=5722562&isnumber=5722481

INDEX

16/18 Code Rate, 271
30–50 MHz Frequency Band, 269
4096 QAM, 271
AC Line Cycle Synchronization for TDMA Allocations, 28
AC Line Cycle Synchronization in Coordinated Mode, 247
ACC, 296, 297, 300, 325
Access, 2, 3, 6, 11, 63, 268, 270, 274, 294, 295, 297, 298, 299, 301, 325
ACK, 91, 107, 108, 109, 110, 111, 192, 203, 325, *See also* Sound ACK
ACLSS, 66, 67, 247, 325
Acronyms, 325
Acting as an AV Bridge, 220
Additional Guard Intervals, 270
Addressing-Related Field(s) Reverse SOF, 95
Addressing-Related Field(s) RTS/CTS, 88
Addressing-Related Field(s) SACK, 85
Addressing-Related Field(s) Sound, 91
Addressing-Related Field(s) Start-of-Frame (SOF) Frame Control, 78

Admission Control and Scheduling (Persistent and Nonpersistent), 148
ADOF, 276, 282, 283, 285, 325
AES, 14, 82, 114, 133, 150, 193, 194, 195, 201
AES encryption algorithm and mode, 207
AES encryption key generation, 208
AFE, 33, 34, 275, 276, 313, 314, 325
AGC, 33, 34, 313, 314, 325
Amplitude Map, 22, 58
ARQ, 118, 325, 334
Arrival Time Stamp, 129
ARRL, 22, 325, 333
Assignment of LIDs, 157
Association, 179
Association of Hidden Station, 227
Association procedure, 231, 309, 310, 311
Association request, 176, 227, 228, 229
ATS, 101, 129, 325
Authentication, 14, 181
Authorization, 195
Auto Connect, 159, 161, 164
Automatic Generation of AES Keys, 208

HomePlug AV and IEEE 1901: A Handbook for PLC Designers and Users, First Edition.
Haniph A. Latchman, Srinivas Katar, Larry Yonge, and Sherman Gavette.
© 2013 by The Institute of Electrical and Electronics Engineers, Inc. Published 2013 by John Wiley & Sons, Inc.

Automatic Repeat Request (ARQ), see ARQ
Auto-Selection of CCo, 122
AV Bridge, 220, 221, 222
AVLN, 178
AVLN, Behavior as a CCo in an, 177
AVLN, Behavior as a STA in an, 176
AWGN, 325

B2BIFS, 140, 332
backoff: channel access, 146
backoff: power, 316
Backup CCo, 125
bandpass, 274
Bandpass PHY, 274
Bandwidth allocation, 168
Bandwidth Management, 168
Bandwidth Scheduling Rules, 246
Bandwidth Sharing, 307
Baseband PHY, 274
BBF, 79, 325
BBT, 172
BDF, 82, 325
Beacon, 64
Beacon Frame Control, 65
Beacon MAC Protocol Date Unit payload format, 64
Beacon Management Information (BMI), 68
Beacon Payload, 65
Beacon Payload Check Sequence (BPCS), 77
Beacon Payload: Addressing, 66
Beacon Period, 135
Beacon Period Configuration, 171
Beacon Period Structure in Coordinated Mode, 142
Beacon Period Structure in CSMA-Only Mode, 139
Beacon Period Structure in Uncoordinated Mode, 141
Beacon Period Synchronized to AC Line Cycle, 27, see also AC Line Cycle Syncronization
Beacon Time Stamp (BTS), 65
Beacon Transmission offset (BTO), 65
Beacon Type, 65
BeaconBackoffTime, 172
beamforming, 313, 315

Behavior as a CCo in an AVLN, 177
Behavior as an STA in an AVLN, 176
BEHDR, 69, 325
BENTRIES, 64, 325
BENTRY, 68, 325
Bidirectional Bursting, 115
Bidirectional Bursting During CSMA, 116
BIFS, 80, 90, 98, 115, 325
BLE, 81, 325
BMI, 68
BMRAT, 273, 325
BoD, 5
BPCnt, 304, 326
BPCS, 64, 77
BPL, 6, 326
BPSK, 19, 326
BPST, 138, 140, 326
BPSTO, 74, 326
Bridging, see AV Bridge
Bridging, 16
Bridging, 219
Bridging: communicating through an AV bridge, 221
Broadcast/Multicast and Partial Acknowledgment, 119
BSF, 221, 326
BTO, 65 326
Buffer management, 106
BurstCnt, 78, 81, 326
Bursting, 92, see also Sound
Bursting, bidirectional, 115
Bursting-Related Field(s) Reverse SOF, 96
Bursting-Related Field(s) SACK, 86
Bursting-Related Field(s) Start-of-Frame (SOF) Frame Control, 79

C&IWG, 5, 326
CAn, 148, 326
Capability of CCo, 16
CAPn, 308, 326
Carrier Sense Mechanism, 144
CC_ messages, 254
CCo:, 254
Cco: Auto-selection of, 122, 124, 125
CCo: Backup, 125
CCo: Capability, 68, 123, see also Beacon Payload
CCo: Failure Recovery, 125
CCo: Handover, 68, see also Beacon Payload

INDEX   **339**

CCo: Network Management Functions, 127
CCo: Power Save, 307
CCo: Selection, 122
CCoCap, 68
CDESC, 154, 155, 256
CEI, 212, 214, 215, 216, 217, 218, 219
CENELEC, 7, 72, 326
Central Coordinator, 121, 260
CFP, 136, 326
CFPI, 326
CFR, 6, 21, 333
CFS, 81, 326
Changing the NMK, 203
Changing the Number of Beacon Slots, 242
Channel Access, 16, 133
Channel access: CSMA/CA, 143
Channel access: TDMA, 148
Channel Adaptation, 25
Channel Adaptation: Bit-loading, 27
Channel Characteristics, 19
Channel Estimation, 211
Channel Estimation with Respect to the AC Line Cycle, 219
Channel Interleaver, 44
CIFS, 319
CIFS_AV, 146, 320, 326, 332
CINFO, 152, 326
CINR, 326
CISPR, 7, 326
CLS, 157, 326
CLST, 78, 83, 153
CM, 16, 326
CM_ messages, 262
CN, 15, 326
Coexistence, 223, 294
Coexistence Resources, 298
Coexistence Signals, 294
Communicating through an AV Bridge, 221
Communicating with an Unknown DA, 222
Communication Between Associated But Unauthenticated STAs, 112
Communication between Neighboring CCos, 239
Communication Between STAs not Associated with the Same AVLN, 112
Communication with a Known DA, 222
Computing the INL Allocation, 238
COMSOC, 2, 326
Concatenated Encoder, 279

Connection Admission Control, 171
Connection Identifiers, 157
Connection Monitoring, 161
Connection Reconfiguration, 164
Connection Services, 157
Connection services: connection monitoring, 161
Connection services: connection reconfiguration, 164
Connection services: connection setup, 159
Connection services: connection teardown, 161
Connection services: Global Link reconfiguration triggered by CCo, 167
Connection Setup, 159
Connection Specification (CSPEC), 152
Connection Teardown, 161
Connectionless Service, 157
Connection-Oriented Service, 157
Connections and Links, 154
Connections and Links During Bidirectional Bursts, 118
Constituent Encoders, 41
Contention Procedure, 146
Convergence layer, 13
Convergence Layer SAP Fields, 83, *see also* Start-of-Frame (SOF) Frame Control
Convolutional Codes Defined by Low-Density Parity-Check Polynomials (Optional), 280
Coordinated Mode, 234, 236
CP, 136, 326
CP_ messages, 260
CPF, 326
CPLT, 309, 326
CRC, 279, 326
CRC Encoder for FL, 279
CRC-24, 64
CRC-32, 77, 85, 94, 102
CSCD, 71, 326
CSMA, 143, 326
CSMA Channel Access, 143
CSMA-Only Mode, 233
CSPEC, 152
CTS, 88
CW, 146, 326

D2PAM, 285, 326
DAK, 194, 326

DAK-encrypted NMK, 189
Data Encryption, 114
Data Plane: Two-Level Framing, Segmentation, and Reassembly, 30
DBC, 308, 326
DCPPCF, 68, 78, 82, 139
Delayed Acknowledgment, 320
De-squeeze, 167
Detection and Reporting of Active HomePlug 1.0.1, 224
Detection Status Field(s) Reverse SOF, 97
Detection Status Field(s) RTS/CTS, 89
Detection Status Field(s) SACK, 87
Detection Status Field(s) Sound, 93
Detection Status Field(s) Start-of-Frame (SOF) Frame Control, 82
Determining a Compatible Schedule, 237
Dev, 327
Device Access Key (DAK), 194
Device Password (DPW), 194
dibit, 327
Direct Entry of the NMK, 198
Disambiguated TEIs, 180
disassociation, 179, 192, 193
Discover Process, 130
Distributed Bandwidth Control, 308
Distribution of NMK Using DAK, 199
Distribution of NMK Using Other Key Management Protocols, 202
Distribution of NMK Using Unicast Key Exchange (UKE), 200
Distribution of Power Save State Information, 306
Diversity Copier, 33
DOF, 283, 285
DPLL, 135, 327
DPW, 194
DT, 63
DT_AV, 63, 327
DTEI, 78, 327
DUR, 90, 327
Dynamic Channel Adaptation, 214

ECC, 281, 327
ECSF, 269
Efficient Notching, 316
EFRS, 271
EIFS, 327
EIFS_AV, 145, 327, 332

EIFS_X, 145, 327
EKS, 82, 327
ELGISF, 271
Empty Tone Filling, 50
Encrypted Payload Message, 209
Encryption Key Uses, 204
Encryption Keys, Pass Phrases, Nonces, and Their Uses, 194
Encryption of RSOF Payload, 118
Encryption-Related Fields, 82, *see also* Start-of-Frame (SOF) Frame Control
ENDTIME, 238
ESGISF, 270
ETSI, 7, 327
EVSE, 304, 327
Exchange of MMEs Through a PCo, 230
Extended Frequency Band, 315

FCC, 6
FCCS, 327
FCCS_AV, 327
FDA, 227
FDM, 298
FEC, 12, 32, 281, 327
FEC Type Field, 281
FFT, 269
FL, 276, 277, 279, 280, 281, 282, 283, 284, 289, 290
FL_AV, 78, 80, 81, 92, 93, 96, 97, 98, 116, 117, 146, 327
Flow Control, 106
FMI, 205, 251, 253
FMSN, 253
FN_MI, 253
Format of Sound MPDU Payload, 94, *see also* Sound
Forming a New AVLN, 181
Fragment Management Information, 252
Frame Control, 38
Frequency Band, 21
Frequency Band: Amplitude Map, 22
Frequency Band: Tone Mask, 22
FSM, 110
FTEI, 227

General AV Frame Control, 63
Generation of AES Keys, 208
Generation of Nonces, 208
Generator for RCE Frame, 278

Get Full AVLN Information, 178
Get Full STA Information, 178
GF, 327
GLID, 156, 327
Global Link Reconfiguration Triggered by CCo, 167
Global Link(s), 156
GPP, 304, 327
Green PHY Preferred Allocation, 307
GreenPHY, 302
Group of Networks, 237
Guard Interval Length, 36, 211

H1, 13
HCD, 74, 327
HFID, 178, 327
Hidden stations, 227, 328
HLE, 13, 327
HMCCD, 76, 328
HoIP, 68, 328
HomePlug 1.0.1 Coexistence, 223
HomePlug 1.0.1/1.1 Coexistence Mode Changes, 224
HomePlug 1.0.1-Compatible Frame Lengths, 225
HomePlug AV, 12
HomePlug AV and IEEE 1901, 2
HomePlug AV Coexistence Modes, 223
HomePlug AV: Network Architecture, 14
HomePlug AV: Protocol Layers, 12
HomePlug AV2, 312
HomePlug Green PHY, 302
HomePlug Powerline Alliance, 4
HomePlug Regulatory Working Group, 6
HomePlug Specifications, 4
HPAV, 16
HPAV_InHomeNNW, 236
HSTA, 130, 328
Human-Friendly Station and AVLN Names, 178

ICV, 101, 328
Identification of Hidden Stations, 227
IDWT, 275, 328
IEEE 1901, 268
IFFT, 34, 328
IFS, 328
IGF, 90, 328

Immediate Grant-Related Fields, 90, *see also* RTS/CTS
Immediate Repeating, 321
Initial Channel Estimation, 213
Initialization Vector, 114
INL, 237, 328
Instantiation of Proxy Network, 229
Interfering Network List, 237
Interleaver, 281
ISP Fields, 296
ISP Parameters, 301
ISP Resource Allocation, 299
ISP Signaling Scheme, 295

Joining an Existing AVLN, 188

KBC, 74, 328
KCCD, 74, 328

Last Symbol Padding, 50
LBDAT, 220, 328
LDPC, 274, 328
LDPC-CC, 274, 328
Leaving an AVLN, 192
LGF, 75
LID(s), 79, 156
Line Cycle Synchronization, 135, *see also* AC line cycle synchronization
Link Identifiers, 156
LinkID, *see* Link Identifiers
Links, 154
Links: connectionless, 156
Links: global, 81, 156
Links: local, 159
LLID, 156, 328
Local links, 156
Long MPDU Generation, 104
LRT, 272, 328

M1, 13
Mac Data Plane, 99
Mac Frame Generation, 101
MAC Frame Streams, 102
Mac Layer, 303
Mac Protocol Data Unit (MPDU) Format, 61
MAC-Level Acknowledgments, 144
Maintenance of Tone Maps, 217

Management Message Entry Data (MME), 254
Management Message Format, 250
Management Message Type (MMTYPE), 251
Management Message Version (MMV), 251
Management Message(s), 250, 301
Management Message(s) CC_, 254
Management Message(s) CM_, 262
Management Message(s) CP_, 260
Management Message(s) manufacturer-specific, 266
Management Message(s) NN_, 260
Management Message(s) vendor-specific, 267
Manufacturer-Specific Messages, 266
Mapping, 49
Mapping for BPSK, QPSK, 8 QAM, 16 QAM, 64 QAM, 256 QAM, 1024 QAM, 51
Mapping for ROBO-AV, 53
Mapping Reference, 51
Matching NIDs, 189
Max Tone Maps, 94, *see also* Sound
MaxCCoScanTime, 172, 332
MaxRxSSN, 109
MaxScanTime, 172, 332
MaxTxSSN, 107
MCF, 79, 328
Method for Authentication, 181
Methods for Authorization (NMK Provisioning), 195
MFBF, 84, 328
MFBO, 84, 328
MFL, 101, 328
MFS, 328
MFSCmd, 107
MFSCmdData, 79
MFSCmdMgmt, 79
MFSRsp, 107
MFSRspData, 79
MFSRspMgmt, 80
MFT, 101, 328
MIB, 293, 328
MIMO, 312
MinCCoScanTime, 172, 332
MinCSMARegion, 141, 332
MinRxSSN, 109

MinScanTime, 172, 332
MinTxSSN, 107
MITM, 328
MME, 254
MMENTRY, 239, 251, 253
MMEPAD, 254
MMQF, 84
MMTYPE, 251
MMV, 251
MNBC, 113
MNBF, 79, 89, 328
MoCA, 9
Modulation normalization scales, 53
Modulation-Dependent Parameters, 278
MPDU, 61, 328
MPDU Bursting, 114
MPDU Format, 61
MPDUCnt, 78
MRD, 5
MRTFL, 79
MSDU, 5
MTYPE, 251
Multinetwork Broadcast (MNBC), 113

NACK, 107, 108, 109, 111
NAVLN, 249
NBE, 78
NCNR, 66, 67
NCo, 329
NCTEI, 74
Neighbor Network, 233
Neighbor Network Bandwidth Sharing Policy, 248
Neighbor Network Coordination, 67, *see also* Beacon Payload
Neighbor Network Instantiation, 240
NEK, 194
NEK Provisioning, 203
Network Encryption Key (NEK), 194
Network Identification, 178
network identifier of an AVLN, 178
Network Membership Key (NMK), 194
Network Operation Mode, 67, *see also* Beacon Payload
Network Password (NPW), 194
Network Status, 298
Network Time Base Synchronization, 127
NewEKS, *see* EKs
NewHM, 76

NewHSTA, 227
NewNumSlot, 76
NewNumSlots, *see* NumSlot
NewSNID, 76
NF_MI, 253
NID, 14
NMK, 194
NMK Provisioning, 195
NMK-HS, 197
NMK-SC, 197
NMK-SL, 178
NN_ messages, 260
Nonces, 195
NOP, 107
North American carrier and spectral masks, 58
Notch and Power Control, 293
NPSM, 66
NPW, 194
NSCCD, 76
NTB, 127
NTB_STA, 128
NTEI, 272, 273
NumDisNet, 74
NumDisSTA, 73
NumSlots, 67
NumSym, 78

ODA, 250
OFDM, 2
OPSF, 84, 110
Original Destination Address (ODA), 250
Original Source Address (OSA), 251
OSA, 251
OSI, 9, 13, 334
OTEI, 227, 228
OUI, 77
Overview of the PPDU Encoding/Decoding Process, 277

P1901
Packet Classification, 151
Participation in Multiple Networks Beacon Payload, 68
Participation in Multiple Networks Start-of-Frame (SOF) Frame Control, 82
Passive Coordination in CSMA-only Mode, 248
Payload, 39

Payload Demodulation Reverse SOF, 97
Payload Demodulation Sound, 92
Payload Demodulation Start-of-Frame (SOF) Frame Control, 80
Payload Symbols, 54
Payload-Level Encryption, 207
PB, 10, 30, 329
PB136, 40
PB520, 40
PBB, 83, 329
PBCS, 83
PBH, 83
PBSz, 92
PCMWG, 248
PCo, 260
PEKS, 209
PEV, 304
PEV-EVSE Association, 309
PHY, 32
PHY Block-Level Encryption, 207
PHY Clock Correction When Participating in More Than One Network, 129
PHY Clock Synchronization, 30
PHY Encoder, 278
PHY Encoder Major Specifications, 292
PHY: Channel Interleaver, 44
PHY: clock and Network Time Base synchronization, 65
PHY: clock frequency tolerance, 294
PHY: empty tone filling, 50
PHY: last symbol padding, 50
PHY: mapping function, 49
PHY: PPDU formats, 35
PHY: PPDU structure, 36
PHY: puncturing, 42
PHY: ROBO Interleaver, 46
PHY: ROBO modes, 46
PHY: scrambler, 278
PHY: symbol timing, 36
PHY: tone mask, 22
PHY: Turbo Convolutional Code Encoder, 39
PHY: Turbo Convolutional Encoder, 41
PHY: turbo interleaving, 42
PhyClk, 30, *see also* PHY clock
Physical Layer, 32, 302
PID, 205
PLC, Role of, 8
PLID, 103

344 INDEX

Plugfests, 5
PMN, 202
PN, 40
PN15, 278
PN7, 278
postamble, 291
Power Backoff, 316
Power Save, 304, 322
Powerline Channels : Design Approach, 18
Power-on Network Discovery Procedure, 172
PPB, 81
PPB_Threshold, 155
PPDU, 34
PPDU Format(s), 35, 276
PPDU: Structure, 36
PPDU: Symbol Timing, 36
Preamble, 37
precoder, 313
Preloaded NMK, 198
Primitives. APCM, APCP, ETH, MD
Priority Contention, 148
Priority of Channel Estimation Response, 219
Priority of Management Streams, 103
Priority Resolution Symbol, 56
Procedure to Establish a New Network in Coordinated Mode, 240
Procedure to Share Bandwidth in Coordinated Mode, 244
Procedure to Shut Down an AVLN, 246
Protocol Failures, 204
Protocol ID, 202
Protocol layer diagram, 12
Protocol Message Number, 202, 329
Protocol Run Number, 202, 329
Provisioning NEK for Hidden Stations (Authenticating the HSTA), 230
Provisioning NEK for New STA, 203
Provisioning NEK for Part or All of the AVLN, 203
Provisioning the NMK to Hidden Stations, 229
Proxy Beacons, 229
Proxy Coordinator (PCo) Messages, 260
Proxy Network Shutdown, 232
Proxy Networking, 225
PRS0, 148
PRS1, 148

PSCD, 71
PSD, 315
PSP, 304
PSSI, 306
PSTA, 255
Puncturing, 42
PxN, 226, 330

QAM, 271
QMP, 152
QoS, 16
QPSK, 24
Queue-Related Field(s) Reverse SOF, 95
Queue-Related Field(s) RTS/CTS, 89
Queue-Related Field(s) SACK, 86
Queue-Related Field(s) Sound, 92
Queue-Related Field(s) Start-of-Frame (SOF) Frame Control, 79

RBAT, 221
RCD, 75
RCE, 277
RCG, 90
RDR, 272
Reassembly, 106
Receive Buffer Management, 109
Recovering from the Loss of a PCo, 232
Regulation, Europe, 7
Regulation, the rest of the world, 8
Regulation, the United States, 6
Relative Power Levels, 56
removing a station from an AVLN, 193
Removing a Station from an AVLN, 193
Repeating and Routing, 272
Repeating and Routing of Broadcast/ Multicast MSDUs, 273
Repeating and Routing of Unicast MSDUs, 272
REQ_TM, 213
ReqType, 230
Request SACK Retransmission, 119
Request to Send (RTS) clear to Send (CTS), 88
Reverse SOF (RSOF) Frame Control, 95
Reverse Start-of-Frame (RSOF), 95
RF (Radio Frequency), 55, 309
RF (Random Filter), 205, 209
RIFS, 318
RIFS_AV, 81

RIFS_AV_default, 93
RNH, 272
ROBO, 11
ROBO HS-ROBO, 46
ROBO Interleaver, 46
ROBO MINI-ROBO, 46
ROBO Modes, 46
ROBO STD-ROBO, 46
ROBO_AV, 46
RRTF, 86
RRTL, 86
RSC, 41
RSF, 272
RSOF, 95
RSOF_FL_AV, 95
RSP, 125
RSR, 78
RT, 72
RTS, 88
RTS Flag, 91
RTS/CTS, 88
RTS/CTS, 146
RTSBF, 129
RTSF, 91
RWG, 6
RX_IN_SYNC, 111
RX_RE_SYNC, 111
RX_WAIT_SYNC, 110
RX_WINDOW_CHG, 111
RxWSz, 109

SACK, 85
SACK Data, 87
SACK Retransmission-Related Fields, 81, *see also* Start-of-Frame (SOF) Frame Control
SACKD, 86
SACKI, 87
SACKT, 87
SAE, 309
SAF, 213
SAI, 72
SC_Join, 186
SCCD, 76
SCF, 92
Scheduler and Bandwidth Allocation, 168
SC-Join and SC-Add, 190
SCL, 215, 216
Scrambler, 40, 278

Security, 193
Security and Network Formation, 172
Security Level, 196
Segmentation, 104
selecting a new AVLN, 181
Selective Acknowledgement-Related Field, 97, *see also* Reverse SOF
Selective Acknowledgment (SACK), 85
Setting of Virtual Carrier Sense (VCS) Timer, 145
Setting the Value of SlotUsage Field, 244
SHA, 66, 193, 208
Short Delimiter, 317
Shutting down an AVLN, 246
SISO, 313
SLAC, 310
SlotID, 66
SlotUsage, 67
Smoothing, 129
SNID, 63
SNID Reuse, 271
SNR, 271
SOF, 77
SOF Payload, 83
Sound, 91
Sound ACK, 93, 330, *see also* Sound
Sound Complete Flag, 93, *see also* Sound
Sound Frame Control, 91
Sound Reason Code, 93, *see also* Sound
SPCS, 94
Squeeze, 167
SRC, 216
SSN, 84
STA, 64
STA_Clk, 127
Start-of-Frame (SOF), 77
Station Roles, 16
Stayout region, 72
Std-ROBO, 46
STEI, 66
StepSize, 44
STPF, 70
SVN, 86
Symbol mapping: 8-QAM, 53
Symbol mapping: except 8-QAM, 52
Symbol Shaping, 55
Symbol timing, 36
SYNCM, 37
SYNCP, 37

System block diagram, 13
System Clock Frequency Tolerance, 294

TCC, 33
TCP Efficiency Improvements, 320
TDM, 298
TDM_SLOT_LEN, 301
TDM_UNIT_LEN, 301
TDMA, 14, 331
TDMA Allocations Reverse SOF, 97
TDMA Allocations RTS/CTS, 89
TDMA Allocations SACK, 87
TDMA Allocations Sound, 93
TDMA Allocations Start-of-Frame (SOF) Frame Control, 81
TDMA Channel Access, 148
TDMA with Persistent and Nonpersistent Schedules, 29
TDMS, 298
TDMU, 298
TEG, 5
TEI Assignment and Renewal, 179
TEI: disambiguated, 180
TEI: leases and renewals, 181
TEI_MAP, 272
TEK, 195
Temporary Encryption Key, *See* TEK
Temporary Encryption Key (TEK), 195
Termination, 41
TMD, 97
TMI, 39
TMI_AV, 78
Tone Map, 49
Tone Map Intervals, 218
Tone Mask, 57
Transceiver Block Diagram, FFT, 33
Transceiver Block Diagram, Wavelet, 275
Transfer/Handover of CCo Functions, 125
Transition Between Neighbor Network Operating Modes, 234
Transitioning from Being a STA to Being an HSTA, 231
Transitioning from Being an HSTA to Being a STA, 231
Transmit Buffer Management, 107
Transmit Power, 56
Turbo Convolutional Code, 24
Turbo Convolutional Encoder, 41
Turbo Interleaving, 42
TWG, 5, 331
TX_IN_SYNC, 108
TX_INIT_MFS, 108
TX_RE_SYNC, 109
TXOP, 153

UDP Efficiency Improvements, 320
UDTEI, 273
UIS, 210
UKE, 194
Unassociated CCo Behavior, 175
Unassociated STA Behavior, 174
Uncoordinated Mode, 234
Universally Unique Identifier, 331
USAI, 173, 174, 332
User Interface Station (UIS), 210
User-Appointed CCo, 124
UUID, 331

VCS Timer, 145
VCS-Related Fields, 90, *see also* RTS/CTS
Vendor-Specific Messages, 267
Version-Related Fields Reverse SOF, 97
Version-Related Fields SACK, 87
VF_AV, 63, 331
Virtual Carrier Sense (VCS) Timer, 145
VLAN Tag, 251
VPBF, 84

Wavelet, 274
Wavelet Process, 282
Windowed OFDM, 23, 55

ZPAD, 94,